The Component

RIVER PUBLISHERS SERIES IN SOCIAL, URBAN, ECONOMIC AND ENVIRONMENTAL SUSTAINABILITY

Series Editors

MEDANI P. BHANDARI
Akamai University, USA; Summy State University,
Ukraine and Atlantic State Legal Foundation,
NY, USA

DURGA D. POUDEL, *PhD*
University of Louisiana at Lafayette,
Louisiana, USA

SCOTT GARNER, *JD, MTAX, MBA, CPA*
Asia Environmental Holdings Group (Asia ENV Group),
Asia Environmental Daily, Beijing / Hong Kong,
People's Republic of China

JACEK BINDA, *PhD*
Rector of the International Affairs,
Bielsko-Biala School of Finance and Law,
Poland

HANNA SHVINDINA
Sumy State University,
Ukraine

The River Series on Social, Urban, Economic and Environmental Sustainability is a series of comprehensive academic and professional books which focus on the societal side of sustainability. The series focuses on topics ranging from theory to policy and real-life case studies and applications.

Books published in the series include research monographs, edited volumes, handbooks and textbooks. The books provide professionals, researchers, educators, and advanced students in the field with an invaluable insight into the latest research and developments.

Topics included in this series are as follows:-

- Climate Change Mitigation
- Renewable Energy Policy
- Urban sustainability
- Strategic environmental planning
- Environmental Systems Monitoring and Analysis
- Greening the World Economy
- Sustainable Development

For a list of other books in this series, visit www.riverpublishers.com

The Component

A Personal Odyssey towards Another Normal

Kas Oosterhuis

River Publishers

Routledge
Taylor & Francis Group

NEW YORK AND LONDON

Published 2024 by River Publishers

River Publishers

Alsbjergvej 10, 9260 Gistrup, Denmark

www.riverpublishers.com

Distributed exclusively by Routledge

605 Third Avenue, New York, NY 10017, USA

4 Park Square, Milton Park, Abingdon, Oxon OX14 4RN

*The Component: A Personal Odyssey towards Another Normal /
Kas Oosterhuis.*

Routledge is an imprint of the Taylor & Francis Group, an informa business

ISBN 978-87-7004-015-0 (hardback)
ISBN 978-87-7004-062-4 (paperback)
ISBN 978-10-0381-210-4 (online)
ISBN 978-1-032-63174-5 (ebook master)

While every effort is made to provide dmependable information, the publisher, authors, and editors cannot be held responsible for any errors or omissions.

to Ilona Lénárd, my partner in life and business

Contents

Contents

Preface

After 16 years of leading the Hyperbody Research Group for complex geometry and interactive architecture as a professor from practice at the Faculty of Architecture at the TU Delft (2000–2016), and after two exciting years of living in Qatar and working at the Department of Architecture and Urban Planning at Qatar University as a full professor (2017–2019), I temporarily moved with Ilona to our 1930 "peasant baroque" country house in Nagymaros in Hungary, along the banks of the Danube. Nagymaros is a 5000-inhabitant small town at the Danube Bend amid two major natural reserves, the Börzsöny mountains in the North, and the Pilis mountains south of Nagymaros. During Corona, we had ample time to enjoy the outdoors, hiking the numerous hiking paths, often accompanied by friends. Corona gave me space and time to write this book, to spend 1200 hours drafting and editing *The Component*. Writing *The Component* was a long overdue job, more than one decade after finishing my last book *Towards a New Kind of Building* (2010), and two decades after *Hyperbodies, Towards an E-motive Architecture* (2003), and *Architecture Goes Wild* (2002). The subtitle for *The Component: A Personal Odyssey Towards Another Normal* indicates that this book is a personal account of four decades of thinking, designing, building, and theorizing. I have left out too personal anecdotes and refrained from criticizing colleagues in person but did not hesitate to position my work in relation to that of my contemporaries.

The Component can be read non-linear. One may jump in at any chapter, even at any paragraph with a heading. I wrote the book as an orchestrated yet non-linear series of subjects that are connected but are understood as concise statements in themselves. Similar to my often-repeated observation, parametric design is building relations between components that have their own identity; each paragraph contains a single identifiable idea, project, or statement and is related to the immediate neighboring paragraphs. The reader may start at Chapter 10, which holds a dozen provocative simple rules of law for a more fair and just distribution of rights, responsibilities, land, and money. The reader may want to read backward or jump at will from chapter to chapter. Some may be triggered by the title of a chapter or a paragraph and want to start there.

I was lucky to have my visual artist spouse Ilona and my architect friends, professors Nino Saggio, Philippe Morel, and Dinos Spiridonidis, to review my draft text and make their valuable comments. They helped me stay close to the subject, avoid walking side paths, and refrain from unverifiable assumptions, while appreciating my efforts to write this one million characters thick-as-a-brick book.

For the e-book version, I linked 800+ names, projects, and products to relevant web pages, YouTube, or Vimeo videos, and my photo and video repository. For the printed version, I replaced the links with 1.2 × 1.2 cm QR codes that the reader can scan with their mobile phones or tablets when navigating the book, which gives the augmented reading experience. Reading books today is a combined analog and digital experience, similar to my take on science, art, and architecture. The analog and the digital are merging into a new Nature 2.0, the inevitable extension of Nature 1.0. The real and the hyper-real are merging, as in our work art and architecture have merged, as the different scientific disciplines are merging, thanks to the emergence of the digital. Merging does not mean retroactive AI-driven alchemy, as in MidJourney, Dall-e, or Stable Diffusion, but merging into an until yet unseen synthetic architecture, forwards to unknown territories, into the parallel universe of Another Normal.

 www.oosterhuis.nl

 www.hyperbody.nl

 www.lenard.nl

 www.riverpublishers.com

List of Abbreviations

1B1D	One building one detail
AA	Architectural Association
BIM	Building information model
CCA	Canadian Center for Architecture
CENG	College of Engineering
CIG	Central Industries Group
CIPEA	China International Practical Exhibition of Architecture
CNC	Computer numerical control
CVS	Concurrent versioning system
DAUP	Department of Architecture and Urban Planning
DBFO	Design, build, finance, and operate
DECC	Doha Exhibition and Conference Center
DRL	Design Research Laboratory
ETFE	Ethylene tetrafluoroethylene
F2F	File-to-factory
FEM	Finite element method
FEW	Food, energy, and water
FLW	Frank Lloyd Wright
FSI	Floor space index
GC	Generative Components
GFA	Gross floor area
GSM II	Game Set and Match II Conference
GSM	Game Set and Match
GSM4Q	Game Set and Match IV Qatar
IAL	Interactive Architecture Lab
ISS	International Space Station
IW	Interactive Wall
M2M	Machine-to-machine
MANIC	Multimodal Accommodations for the Nomadic International Citizen

MBS	Mohamed bin Salman
ME	Mechanical Engineering
NAAB	National Architectural Accrediting Board
NICs	Nomadic international citizens
NPRP	National Priorities Research Program
NSA	Nonstandard Architectures
NURBS	Non-uniform rational basis spline
ONL	Oosterhuis_Lénárd
P2P	Peer-to-peer
PDRP	Parametric design to robotic production
PLA	Polylactic acid
POC	Production on command
PP	Polypropylene
QF	Qatar Foundation
QNCC	Qatar National Convention Center
QNCC	Qatar National Convention Center
QU	Qatar University
RK	Rem Koolhaas
RKS	Rotterdamse Kunst Stichting
TGB	Très Grande Bibliothèque
UAT	Ubiquitous autonomous transportation
UBA	Ubiquitous booking app
UBI	Ubiquitous basic income
UEP	Ubiquitous energy production
UFP	Ubiquitous food production
UHD	Ubiquitous home delivery
UI	User interface
UIB	Ubiquitous immediate banking
ULR	Ubiquitous land right
URP	Ubiquitous robotic production
USR	Ubiquitous shared responsibility
UTC	Universal Time Coordination
UWP	Ubiquitous water production
UWT	Ubiquitous waste treatment
VINEX	Vierde Nota Ruimtelijke Ordening Extra

VOR	Virtual Operation Room
Web of N-H	Web of North-Holland
WHO	World Health Organization
ZHA	Zaha Hadid Architects
ZKF	Zero Knowledge Proof

1

Here and Now

The opening statement from my most recent book <u>Towards a New Kind of Building</u> is: "The world is changing. So is architecture, the art of building, primarily due to evolving communication and manufacturing methods that have changed drastically and with increasing speed." Drafting this new book during the COVID-19 pandemic, during the continuing denial of the ever-increasing climate crisis, during the continuing blatant racism, during an ever-increasing schism between the poor and the super-rich, and during the war that Russia has imposed on Ukraine, I need to add "and due to evolving social dynamics." In response to the pandemic, the climate crisis, blazing war rhetoric, the alternative truths of fake news spreaders, poverty and injustice, and an increasing number of warmongers, scientists, artists, and architects are imagining new rules for the new economy, eventually crystallizing into new tangible projects. New cultural rules based on new forms of social distancing, climate justice, and war-mongering are likely to become more permanent in the years to come. People will change their spatial behavior, how and where they meet, how and where they communicate, how they move from one place to another, and how they shop, relax, and work, and, as a direct result of their adjusted social behavior, we rebound to see substantial changes in the way materials are mined, how products are designed and fabricated, and how urban environments are reimagined and built.

I started writing at the beginning of the COVID-19 pandemic, throughout the years of relative abstinence of travel and consumption, to the public denial, especially from the side of freedom of speech propagandists, of the permanence of the virus. The virus is here to stay, and we will need to protect ourselves structurally by repeated vaccinations and by respecting the rights of those with underlying medical conditions. While writing, it became clear to me that I need to question how the new societal rules might relate to my design strategies as developed over the last four decades. Is there a relationship at all, and if yes, what exactly is its nature? My initial hunch was that there is indeed a strong correlation between rule-based design strategies like the conceptual and parametric design to robotic production and assembly methods, both in the realms of art and architecture, and the new rules for the new society, whose correlation inspired me to author this book. Most significantly, the correlations are to be found in my proposals for cities and event structures, ranging from my provocative graduation project Strook door Nederland from 1978 to our latest event sculpture proposal, the Seven Daughters. As the main focus of this book is on the leading role of the component in science, culture, art, and architecture, I will look at the COVID-19 virus as an invasive actor with a far-reaching influence on everything it touches. When I started writing in January 2020, I did not intend to consider the effects of viral outbreaks concerning my design strategies, but now, almost two years later, I have no choice. The resonance between the swarm behavior of viral infections – and that of numerous other natural phenomena – and the communication between the interacting components of the new kind of building is simply too strong to ignore.

At the peak of the COVID-19 crisis, concerned designers argued for a permanent reduction of travel, production volume, and global trade, arguing for a return to more essential values of daily life, work, and leisure activities, for a re-evaluation of the meaning of work and home. Yet, I have a different take on it, since what one is arguing for is to do less of the same, to return to a simpler, yet, in principle, the same, way of living one's life. What kind of economy would that entail? What financial system would support such a downsizing of the economy? Can the worldwide negative economic growth support an increasingly large world population? Arguments like these seem nostalgic to me, and white supremacist, while the world is much bigger than the USA and Europe! Currently, the whole continent of Africa produces just 5% of the global residential and industrial carbon-dioxide emissions, while they are most vulnerable to the climate changes caused by global emissions. Should their economy shrink also? Asia and Africa are emerging superpowers, rightfully requesting a level playing field. While global warming is already out of control, when Asia and Africa produce their fair share of greenhouse gases at the same rate as the Western world, Earth will unbearably heat up. Therefore, even a drastic reduction in our current level of carbon-dioxide emissions would not make a substantial difference. This is what I feel, and what causes mankind's indifference to measures that it needs to take? When a problem is too big, one gets paralyzed, and the only option seems then to dance on the edge of the volcano.

As opposed to doing less of the same or nothing at all, there is an open window to move toward a highly capital-intensive techno-social economy instead, which levels out the downsides of mass production and neo-capitalist inequality, while securing a good life for all 7, 8 to eventually ten billion inhabitants of Spaceship Earth. The growth of the techno-social economy is in the level of capital invested and not in the number of products it produces. To get to a new normal, well-balanced situation, one needs to thoroughly reconsider the methods of agricultural and industrial production. Ubiquitous robotic production, production on command, scalable food, energy, and water production, universal basic income and ownership rights, automated transport systems, ubiquitous booking of homes and workplaces, and peer-to-peer decentralized transactions will form the basis for an economy that I call "another normal." Networked, scalable, and distributed production and consumption patterns will lead to more intimate ties between consumers and producers. Production of the constituent components of the techno-social global culture will be produced only where, when, and as needed. We must replace centralized mass production methods with decentralized mass customization of almost everything, of the unique components that make up the products. The paradigm of customization will invade all levels of society, as it is scalable from the very small to the very big, eventually leading to an intensified, information-rich, and capital-intensive decentralized economy, an economy of the swarm. This book is about the components that make up the swarm.

Components are, in principle, acting protocells in real time. Like monads, components are the smallest constituent parts that interact to form the bigger whole. The word component has been already prominent in many of my earlier writings, notably in *Towards a New Kind Of Building* (TaNKoB). More recently, in 2019, I organized the fourth Game Set Match Conference GSM4Q at Qatar University in Doha, Qatar; the motto of the conference was not coincidentally "Nomads and Monads." The notion of the component offers a focused viewing angle, a vantage point for the analysis of the relations between components at varying scales in different disciplines, a framework for a critical self-analysis of my design intentions, and a referential framework to scrutinize early design concepts in the actuality of today's pandemic-struck society. While typically each design project has layers of meaning, the design, production, and assembly of unique interacting components represent just one of those layers, albeit an elementary one. I deliberately do not assign a central position for the component, while there is no such order between the layers of meaning that would justify such a central position. As in the hive, in the flock, and in the swarm, there is nothing and no one central.

The system of ordering among the layers of meaning does not paint a static picture but is intrinsically a complex adaptive system. This book is 1) a selective flashback of four decades of my work in practice and academia (What have we done?), 2) a critical account of where I stand right now in 2020 (Where are we now?), and 3) an anticipatory look into the coming decade (Where do we go?). My last book *Towards a New Kind of Building* dates back one decade ago. The pandemic gave me the time and space to draft this book. I typically wrote the previous books in a shorter period. In just one summer, I wrote Hyperbodies, Towards an E-motive Architecture for Antonino Saggio's cute series IT Revolution in Architecture. TaNKoB took me half a year, but for this book, I have been working during the two pandemic years, on average two hours per day. A typical attention span is typically not longer than two hours, for anything from writing to watching movies or participating in a Zoom session. In the years after I wrote TaNKoB, two of my featured projects, the cultural center <u>Bálna Budapest</u> (2013) and the <u>Liwa Tower in Abu Dhabi</u> (2014), were completed, representing the uncompromising applications of what I discussed in the TaNKoB book, executed in our compact and efficient design practice ONL (Oosterhuis_ Lénárd) BV. We managed to design and execute larger   projects with a small staff. When discussing the production of drawings for the Bálna Budapest, one architect advisor to the client insisted that I should have at least 30–40 people working on the project, while I had a permanent staff of three young architects in the ONL Hungary office in Budapest. The outside world had the impression that we run a much bigger office than we actually had; in reality, the biggest size was around 12 people in our Rotterdam office – six staff plus six trainees. We had a similar team size at <u>Hyperbody TU Delft</u>. Starting as a professor from practice with one assistant, we managed to grow into a team of 12 people, assistant professors, Ph.D. candidates, and teaching assistants, at its peak; after the financial crisis of 2008, step by step, we had to reduce to only 6 people by 2016, the year of my retirement due to TU Delft age policies.

I have dedicated the "What Have I Done?" chapter to highlights of one decade as a freelancer, three decades of ONL in Rotterdam, and almost two decades of Hyperbody at TU Delft, where I fulfilled a professorship at the Digital Architecture chair between 2000 and 2016. Later on in this book, where I discuss the components of interactive architecture, I have included links to the editorials I wrote for the <u>iA Series of Interactive Architecture</u>, a series of five issues, in a cute small format, especially for use by students.

A swarm behavior of interacting components and a component-based design to production form the solid basis for ONL and Hyperbody projects. Having made radical customization systems work, through in-house programming and file-to-factory production methods in the past four decades, it seems fair to say that in hindsight, my work anticipated the societal changes that the current COVID-19 pandemic promises to bring about. Making things work has from the start been my leading design principle. Making things work means designing a system, which does what it is meant to do, like the scripts for the Saltwater Pavilion, the Web of North-Holland, the A2 Cockpit in the Acoustic Barrier, and the interactive installations of ONL and Hyperbody. Scripts for data-driven production by CNC (computer numerical control) machines and the components for interactive installations either work or do not work at all. One wrong number or text in the script, one line of code missing, and it does not work at all; in scripts, the smallest details have to be right. That is what makes the scripts verifiable, and that is why scripts potentially form the basis for an open transparent society. The digital is superior to the analog; in the digital, there are

no metaphorical stories, there are no lies or excuses, and it simply has to work. Making things work is why my keen interest as a designer goes into a clear digital definition of the basic component that assembles the constructs.

Component-based design works as a principle at any scale, from furniture design (the Body Chair), art projects in public space (Ilona Lénárd's TT Monument), pavilions (the iWEB),

buildings (the A2 Cockpit), master planning (Manhal Oasis), to interactive installations (Festo Interactive Wall). I will

demonstrate that procedural customization and component-based thinking work well for concrete art as well, as in the autonomous paintings of Ilona Lénárd, as described in my

Ubiquitous Symmetry paper. As a matter of principle, ONL and Hyperbody projects are based on hands-on, in-house developed design techniques, and they pass the tests of proof of concept. None of the designs and installations is based

on assumptions or hypothetical solutions that only exist in theory. "You can do anything as long as it works," I typically tell my TU Delft students, which is the modus operandi of the autonomous artist. The artist cannot hide behind anyone else; the artist is the sole responsible person for the tangible result, which is the ultimate vulnerable position to take. There is no one else to blame when things go wrong or otherwise are not much of a success. There is no one to rescue the artist from a self-chosen road to nowhere, from lines of thoughts that are not picked up by others, for which the artists might have hoped. When it does not work, so be it, try harder or leave it, and concentrate on the next challenge. And nothing else pleases the artist more than when others appreciate their unsolicited work, and, eventually, it is picked up by clients.

At ONL, we always solved the equations for the parametric design to robotic production ourselves. I did not want to throw problems over the fence, relying on the likes of Arup and Happold to solve the geometry and the preparation for production. Yet, that is still a widespread practice in architectural design offices, while they are chained in the linear building chain from design ideas to the engineers, from the building permit plans to the tender procedure, from the tender to the contractors, and finally from the contractor to the manufacturers, who actually make the components. At ONL, we worked with the highest precision in parametric modeling, and we wrote the scripts to produce from our data directly and took responsibility for the correctness of the data. We managed to unchain the traditional building chain by teaming up with the manufacturers directly, in a non-linear network rather than in a linear chain, both for the smaller works of art in public spaces and for the larger buildings. In that sense, I have always been more of an artist than an architect. Not surprisingly, from the start of ONL (Oosterhuis_Lénárd) as a joint office between an architect and an artist, our motto has been "The fusion of art and architecture on a digital platform." Just like any artist, we work primarily to educate ourselves, and only secondarily to satisfy a client. We do not "serve" the client, as says the professional code of conduct for architects. Artists are entrepreneurs who work differently from commercial architects and project developers, who in the first place are in it for the money and use architecture as the vehicle to accumulate their wealth. We go for the design challenge of uncharted territories in the first place and use the remuneration to support our efforts. We could only afford this attitude since we have been a small enterprise through the years. I realize in hindsight that my design philosophy and our working methods, effectuated in such compact design teams as at ONL and Hyperbody, seriously have limited the number of clients we were able to attract.

The "What Have We Done?" chapter includes my view on Patrik Schumacher's Parametricism. Since Zaha Hadid's untimely death in 2016, Patrik Schumacher has effectively taken over the ZHA office and has emerged as their theoretical leader. He successfully merged Hadid's flamboyance with parametric design and libertarian political ideas. However, none of the three components has much to do with the other. Hadid's style has no congruence with parametric design, while parametric design has not much, if anything at all, to do with neo-libertarian thinking. The combination feels a bit treacherous to Hadid's original style, parametric design, and techno-capitalism alike. Hadid is famous for

 her calligraphic sweeps, first angular and later in smoother strokes, but in no way parametric. I have written more about Zaha Hadid's calligraphic sweeps in my blog on my website. It seems to me that in the libertarian mantra of unfettered free competition, the "me, myself and I," productivity growth, and the money, but not the content, is the main subject of interest. On the other end, parametric design is instrumental to come to an inclusive participatory form of architecture, which is in its essence inclusive and collaborative, not competitive. Parametric design in itself is toneless, if not linked to robotic production methods. In the following pages, I argue for an architectural practice that is based upon equally emotive as systemic methods of thinking, designing, and producing, making things work for a fair share of the global population.

In general, I prefer the term associative modeling over parametric
 modeling. In the commercial white paper Associative
 Information Modelling (2015), I explain why clients should choose ONL for their design challenges. Associative modeling
 is a very accurate term, coined by the British architectural computer programmer Robert Aish, who developed the software Generative Components for Bentley Systems.
 During 2014–2015, ONL was briefly operating under the name VAA.ONL, a fusion between Van Aken Architecten in Eindhoven and ONL. Unfortunately, there had been little response to the white paper, not in the least because the merger with Van Aken Architecten did not work out well. In 2015, Van Aken went bankrupt, and ONL was almost dragged along into the abyss. Fortunately, the new owner – who had been one of the previous

directors – saved our fate and we got our portfolio back into our own hands.

In the "Another Normal" chapter, I write about the synthesis of components into a coherent building body. My view on the synthetic contrasts with the bio-digital mimesis approach, which claims to design after and with nature, to "mothering nature" as Neri Oxman puts it. Oxman wishes to work with natural agents as active co-builders, thereby maneuvering herself in a fringe position; her empathic delicate bio-inspired approach is not fit to be scaled up to the robust scale of the global building industry. I am critical of growing building materials using mushrooms or growing structures using living silkworms (MIT Silk Pavilion), despite their intrinsic beauty. I do not wish to jump into disproportional illusions to clean polluted air (Daan Roosegaarde's Smog Free Tower), nor do I embrace literal forms of biomimicry while I am more interested in the inner drive of a system than in observable similarities. We might try to imitate superficial features of natural systems, like the wings of birds, but that does not respect the intrinsic logic of natural systems or the logic of synthetic systems. The fixed wings of airplanes have a design that is distinct from the flapping wings of birds. Biomimetic designs are typically too literally depending on, or working with, existing nature, while I am aiming at constructing a hitherto unseen, unknown, and alien form of synthesized and digitally informed nature. The total mass of technical extensions to our bodies, constituting our exo-brains, exo-hands, and exo-eyes, has reached the point that it exceeds the total biomass on earth. A blanket of cities, infrastructure, vehicles, furniture, household objects, computers, and cellphones covers the surface of Earth; satellites populate the skies, and artifacts are making up everything that we see and appreciate around us. We live inside our bodily extensions, inside this physical augmented reality. All of these extensions belong to the synthetic world of design, engineering, production, and building. Design to build is synthetic by nature; there is no need to imitate biological systems. Because of the difference in scale, the speed of evolution, and the choice of materials, synthesized structures have a logic of their own. Synthetic architecture applies an as-of-now simpler logic of synthesizing the building body from its constituting components than how DNA and cell structures construct living nature. When I look at the workings of a single biological cell and its immediate environment, I may compare its complexity with a complete city and its connections to neighboring cities. A cell is like a city, complete with its factories, energy centers, storages, universities, libraries, and homes, while the inhabitants primarily act as messengers and information carriers. As in the city, vehicles, data carriers, tolls, borders, filters, and semi-permeable membranes follow the internal logic and rules of the city game. In the synthetic world, the essential building components are the monadic cells that evolve in a complex adaptive process, following a logic of their own, in close contact with their immediate neighbors. They are what they are, and they do what they do – they are the agents of the swarm. We have, by far, not arrived at the same level of complexity in our human-assisted synthetic world as in the pre-industrial natural world; yet, we are developing fast. Since we have started embedding real-time interaction, neural networks, and artificial intelligence into the building components of our world-making, we have already come a long way toward a natural complexity that we might label as *nature*, the sequel. I do not want to speculate on future developments, nor do I want to fall into the trap

 of a Trumpist superlative turbo-language, i.e., not use words like amazing, incredible, or never seen before. My down-to-earth approach follows the adage "What you see is what you see," coined by the American artist Frank Stella when being asked for an explanation of his work, as to avoid making up stories that are bigger than the work itself. Similarly, ONL's and Hyperbody's building components are what they are, and they do what they do.

Also in the "Another Normal" chapter, I discuss how science inspires me in the first place, especially the science of natural physics exploring the quantum nature of the universe and of the smallest subatomic vibrations. I like to think that there is a relationship between the elementary processes in the universe and data-driven component-based design, not only on a theoretical level but also hands-on establishing the connections between the components. Interactions and connections between components are even more relevant to understand a system than the physical components themselves. Art and science inform synthetic architecture, synthesizing from the bottom-up to the larger whole, by the design of the associated components, and their mutual relationships, as actors in the swarm. Of main interest are the rules of law, the rules of the design game, and in terms of its visual appearance, the shape of the law as I demonstrated in the Strook door Nederland. Learning from science and art, I have adopted a working method that celebrates the swarm. Multiple components that are similar but not the same are flocking together in the design constellations, looking at each other and building relationships with each other. As early as the early nineties, I realized that associative design is the natural instrument to deal with rules and variables. The parametric design systemic is open not only to self-chosen variables but even more relevant, as we will see toward the end of this book, for the principles of inclusive design, whereby each player plays the rules of the interactive design game at a level playing field, experts, and laypeople alike.

Measuring, verifying, and evaluating data are keys to the development of any of the designs and, in a wider context, applicable to the new rules of the economy of "Another Normal," discussed in the concluding "The Chinese Patient" chapter. Social distancing and ubiquitous booking are based on the engineering principles of "measuring is knowing." What can be more pleasing than to live in a world without waiting in lines at the airports, in shops, and heavy traffic? What can be more attractive than paying on the fly instead of waiting in line at the cashier's desk? Processing of data in real time is key to making the new economy work. The underlying technologies of real-time data processing build the foundations for the new nomadic citizen. At Qatar University, I developed the concept of the Ubiquitous Booking App, for the MANIC research project. This application allows you to book anything, anywhere for any time, merging lease and ownership in one overarching global system. Universal basic income, ubiquitous land ownership, ubiquitous food, energy, and water production, ubiquitous AI-driven transportation, ubiquitous robotic production, and ubiquitous home delivery are the supporting pillars of the concept of ubiquitously available *Multimodal Accommodations for the Nomadic International Citizen.*

11

In the "Ubiquitous Components" chapter, I dive deeper into the role of the component in other fields of knowledge, in science, language, music, and art. Not being an expert in those fields, I take the risk to be under-informed and jump to conclusions that are not well-funded. Yet, it is instrumental to at least try to figure out to what extent my approach to the component in architecture resonates with the role of the component in other fields of expertise. What is common in my explorations into other fields is that in my view, all man-made artifacts are a natural extension of nature, not opposed to what we romantically perceive as nature. Components are the physical form of an executable, components act, do things, perform, they make things work. An executable file in the world of computation is a component that runs the program inside one computer, whereas the computer is a higher-level component designed to communicate with other computers. As in the nanoworld, as in the universe, in our organic tangible world, all components are actors, executing a program. Every identifiable component, whether big or small, I consider to be an information-processing vehicle. I consider a house an information processing unit. I consider a factory an information processing unit. The input, the process, and the output vary, but all are information-processing vehicles like we are ourselves. As we live in a dynamic world, every single component is an actor that establishes bidirectional relations between each other. The key message of this book is that components can no longer be dealt with as static elements, as dead bodies. Components are actors that are very much alive, whether living very slowly in the form of visually immobile objects or at a high speed like packages of information traveling at the speed of light. Components, whether slow or fast, must in principle be kept alive and kicking, to be actors interacting with neighboring actors. To design means to design the interactions between the actors.

The "Components versus Elements" chapter is a mirror image of Rem Koolhaas' Venice Architecture Biennale of 2014. I discuss his 15 "fundamentals," one by one; while he registers elements, I am looking for components. While his elements are typically mass-produced products without a predefined context, my components are unique customized parts that only fit in a particular place, in a bi-directional relationship to their immediate neighboring components. My components are live actors, establishing connections to their immediate neighbors, like the parts of a three-dimensional puzzle. Koolhaas' elements are typically cold clashing with each other, in a deconstructivist style, whereby the elements do not know each other and are not even interested in each other, while my components are well aware of each other and acknowledge that they need to inform each other with respect to their dimension, shape, and performance to connect. For me, architecture is a synthesis of precog-style components, components that know where they are going to fit in.

The "Component" chapter hosts a detailed description of a series of ONL projects seen from the perspective of their constituent, passively acting components. It expands on the development from mass production to mass-customized design-to-production methods. Every new project means another step in the direction of the "one building one detail" paradigm that characterizes the most recent ONL projects. The series of subsequent projects shows the self-initiated inevitability of radical customization. This inevitability is a self-imposed regime that rules my thinking whenever I design furniture, or buildings, or participate in joint art projects. This is what I do, this is what I am, and this will be my legacy.

Nonstandard architecture is an inclusive approach since it allows for the exact description of both traditional rectangular and smooth fluid architecture. Interactive architecture adds another level of inclusiveness, while it allows for the programming of both static and dynamic structures. Component-based design is the ultimate form of open-ended inclusiveness, describing the structures at their most generic level. Component-based design is not an ultimate form of specialization, but, in fact, it is the most generic form of architecture. Components are the acting parts of the real world, not an abstraction from reality as elements are.

In the "Where Are We Now?" chapter, I discuss projects that I designed and initiated in the years 2017–2019, in my role as an educator and a researcher at the Department of Architecture and Urban Planning at Qatar University, as a partner in art projects with visual artist Ilona Lénárd, and in my continuous capacity as ONL's lead architect. I chose

"Where Are We Now?" as the title of my lecture at the international GSM4Q conference at QU, which I organized in February 2019. GSM4Q is the sequel of the previous GSM I (2002), GSM II (2006), and GSM III (2016) conferences that I initiated at the TU Delft. The GSM4Q lecture is an account of the works developed and executed in Qatar, design studios and workshops with my students at Qatar University, competition projects developed with a team of Qatar-based architects, art projects by and with Ilona Lénárd, research projects like MANIC funded by Qatar University, and ONL projects done in collaboration with a Dutch team of young architects. Some of these yet to be

published projects like the Maidan Monument for Hundred Heavenly Heroes in Kyiv can be found on my blog site www. oosterhuis.nl. The "Where Are We Now" chapter deals with the actual situation that we find ourselves in. In this chapter, I summarize my GSM4Q lecture at Qatar University. I played the somewhat ominous David Bowie video of his surprise swan song "Where Are We Now" as the introduction to my lecture, as a farewell gesture to Qatar University, while my contract was not renewed for as of yet unknown reasons.

The "Where Are We Now?" chapter contains a series of case studies, showing how we have combined art, design, and architecture strategies that we developed over the years into coherent real-world proposals. In the self-analysis of the projects, the notion of the component frames the train of thought. One of our ongoing projects is the further development of the urban design instrument Participator, in its basis an application that allows the exploration of urban design ideas by a multidisciplinary team, laypeople, and experts alike. The design of open participatory design systems is a logical next step after parametric design for unique individual buildings. Open design systems allow for quantifiable and qualifiable variations of a structured idea, following the same logic of the file-to-factory process. Participatory design systems already exist on the Internet for customizable products like rings, glasses, furniture, and wiki-houses, usually with a limited scope of interaction and a limited influence on the outcome of the participatory design process. Variations are more of the same, featuring only minor differences. Participator offers a design instrument for deeply different outcomes; besides the regular numeric sliders and randomizers, we have embedded a tool to sketch 3D trajectories that attract functional units.

In the "Proactive Components" chapter, besides interactive ONL projects, some key Hyperbody projects (Muscle Tower, Muscle Body, Festo Interactive Wall, POD[64], and Pop-Up Apartment) were published on the Hyperbody Wiki pages and the Hyperbody YouTube channel. The Hyperbody website is not developing further since I formally closed Hyperbody in 2016. Since my retirement as a professor from practice at the TUD, the information-rich website is functioning as the Hyperbody Archive site. There is valuable information, notably on student projects under Education in the top menu bar. At Hyperbody, we stimulated the students to work with graphic programming interfaces like Nemo/Virtools, Max MSP, and Grasshopper. The notion of the component is a key unit in Grasshopper, the graphic interface to Rhinoceros 3D. David Rutten of TU Delft developed Grasshopper, released in 2007, five years after we at Hyperbody started working with the graphic interface Nemo – later renamed Virtools – for the design of interactive environments. David Rutten was aware of Hyperbody's interactive virtual reality designs executed with Nemo/Virtools when Robert Mc Neel hired him to write a graphic interface for Rhino. In Grasshopper, as in ONL and Hyperbody projects, the components are actors acting upon the geometry, not unlike the sliders in the Attractor Game that I developed with my ONL team in the late nineties using Visual Basic software. For the interactive Hyperbody installations, we used Max MSP to capture and manipulate real-time streaming data that inform the interaction.

The "Where Do We Go?" chapter describes the future outlook, extrapolating the state of the art of today, unveiling design strategies for the immediate future of buildings and urban environments, toward a more information-dense, capital-intensive, socially inclusive, and abstract-art-driven world. Where do we go? What practical strategies can lead us into Another Normal? How will we work and live? Are we becoming international nomadic citizens? Will we live on Mars? In this chapter, I question the role of interacting components in a variety of social, technical, biological, global, virtual, artistic, and architectural systems, as well as in design software. I remain focused on how component-based systems receive or retrieve information, how they process information, and how they send information. More than anything else, I look into the forces that are driving the components. The concept of the interacting component turns out to be scalable and ubiquitously applicable.

In the concluding chapter "The Chinese Patient," I sketch out scenarios for a reimagined post-capitalist pandemic-proof society. I describe new rules for the new economy that might come into effect sooner than later, that is, if the unprecedented global response to the 2020 pandemic equally triggers the will to fight the climate crisis. As we speak, it looks like the CEOs of the international companies are not ready for the transition; they want to get as much profit out of the fossil fuel economy as possible and use the war that Russia has invoked in Ukraine as an excuse. For the time being, the CEOs and the privileged upper class prefer to dance on the edge of the volcano; they do not take substantial action, despite continuous warnings from the United Nations and actions from the younger generations. Only a techno-socialist revolution can turn the tide, not a techno-libertarian revolution. Parametric design to robotic production is one of a dozen strategies that must converge to make the world a better place. In this last chapter, I have listed a dozen other radical rule-based strategies one needs to consider to turn the tide. Earth itself will survive easily, but improving the outlook for humans needs immediate action.

First, I discuss the necessity for a universal basic income (UBI) policy. Wealth-sharing policies are mandatory to reduce the ever-increasing inequality between the rich and the poor. I fail to see how (neo)libertarian policies could lead to a reduction of the inconvenient gap between the super-rich and the starving poor. In the USA, during the peak of the pandemic, 15% of the American population has been depending on food banks, depending on charity instead of having a right to food. I do not see the logic of the CEOs of big companies to filter billions of dollars out of our pockets first and then arbitrarily redistribute a small part of their wealth to charity. When every citizen would have a basic income, there would be at least a beginning of a level playing field. Additional measures are necessary to construct a fairer and more just world. Second, people will need simple, straightforward land property rights, coined as ubiquitous land rights (ULR). In the Middle Eastern countries, the state grants birthright to native citizens to own a piece of land of approximately 1000 m². On that land, they can build their own house, grow their food, and have their own business. Subsequently, it is important and inevitable to introduce simple rules for ubiquitous food production, energy production, and water production (UFP, UEP, and UWP). These three strategies interweave with each other into the FEW Nexus. In the FEW Nexus, the proper treatment of waste becomes an essential source of renewable energy. In the footsteps of the "Chinese Patient," who was locked in a Chinese psychiatric hospital for requesting better food, clothes of wool, transportation by limousine, and to transform China into one big villa park, I discuss the need for ubiquitous autonomous transportation (UAT). As we speak, leading companies evaluate automated vehicles. Hundreds of thousands of electric vehicles that are on the roads today are already fit to operate completely driverless, on the road and off the road. Cityscapes will change drastically when all vehicles are AI-driven, and the lanes reserved for cars can become fewer and narrower, leaving ample space for trees giving shade, enlarged pavements for (electric) bikes, and a vibrant public life. Crucial is the ubiquitous robotic production (URP) rule, creating the framework for the democratization of design and production, making the production of almost anything affordable for everyone, including but not limited to food, energy, water, and household products. Production on command (POC) means producing only when, where, and as needed. We should refrain from producing for the "market"; instead, we should produce for the demand. New products are simulated and prototyped before they are produced on a larger scale. Only when there is a demand, regardless of how small the demand is, the product is produced. Crowdfunding shows a feasible way forward. Last but not least, simple rules are needed for fine-mazed decentralized ubiquitous home delivery (UHD) systems to optimize the way we do shopping. It is more efficient for the number of kilometers

17

driven and hence in the amount of carbon dioxide produced, to have almost everything delivered to your home address than to go shopping in your vehicle. Home delivery, by small self-driving robotic vehicles (drones, <u>Gita</u>, vans), is today's digital reincarnation of the traditional milkman, baker, greengrocer, butcher, fishmonger, and scissor sharpener, who used to deliver to your (grand)parent's door until the sixties of the last century. Small-scale home delivery blends effortlessly with the principles of the walkable city. Fun shopping for social reasons, in historic city centers, would remain a leisure activity. Add to the above strategies a universal workplace strategy supported by the ubiquitous booking app (UBA), and we go full cycle. The new nomadic international citizen feels everywhere at home; wherever they are, they customize their environments to make them their own. They work from anywhere in the world, working for the kind of company that is operating more like a distributed network rather than bound to a fixed place. The ubiquitous booking app gives the new citizens an instrument to organize their private and professional lives and their social interactions.

The above rules of play are based on a digitally well-equipped society, where everyone and everything is an actor in the dynamic adaptive system, and where everyone and everything interacts with their immediate neighbors, like the acting members of an interacting swarm. Being digitally well-informed and digitally well-connected is a key factor to build a successful society. Not only will the people need to be well informed but also the building components that interact with their informed environments must be equally well-informed. We will live our future lives as actors on the Internet of Things and People, seamlessly merging the physical and the virtual. Being well-informed is a prerequisite for things and people to play, to live inside the evolution of the game of life, and to shape "Another Normal."

2

What Have We Done?

The Formula

Self-instigated design proposals are a free offer to society. It is up to others to pick up on unsolicited ideas or ignore them. Especially in graduation projects, students can grasp the opportunity to develop such a potential contribution to the advancement of society. An unconventional graduation project can have a similar impact as a disruptive work of art, provided that the student follows his/her intuition and does not follow old recipes from teachers or copy examples from practicing professionals. Some graduation projects die in beauty, and others stir up the collective conscience of the profession. According to the Hungarian mathematician Albert-László Barabási, who wrote the book The Formula, the Universal Laws of Success, there are five statistical laws responsible for coming across:

1) performance drives success, but when performance cannot be measured, networks drive success,
2) performance is bounded, but success is unbounded,
3) previous success × fitness = future success,
4) while team success requires diversity and balance, a single individual will receive credit for the group's achievements, and
5) with persistence, success can come at any time.

My graduation project titled "The De Strook door Nederland" was pretty much one of those inconveniently disruptive proposals, which created a disturbance and sowed division among my tutors. At first, I had chosen Izak Salomons as my thesis mentor. I chose him while he had once proposed a sensitive and at the same time rational master plan for the new satellite town Almere near Amsterdam, which in my view made much more sense than what has been built. What has been realized is the so-called "cauliflower" urban planning, with endless branching eventually ending in dead-end streets, the exact opposite of the open grid – very seventies indeed. But Salomons was so shocked by some of my statistical analyses, notably that of the expected number of suicides in the new town I was planning, that he refused to guide me any longer and advised me to turn to Rem Koolhaas instead. A small group of students, among others Kees Christiaanse (later KCAP) and Herman de Kovel (later DKV), circled Rem Koolhaas (from here on referred to as RK) at the end of the seventies. RK pressured us to be more ambitious and to work hard, following the libertarian mindset that he had taken with him from the Architectural Association and his long stay in New York. Superstudio and Archizoom were my

 heroes at the time for their radical conceptual designs. Brian Eno just released his Music for Airports (1978), where he introduced a superpositioning of slowly shifting, endlessly looping motifs, each with its own intervals, thereby creating a musical score that never repeats itself. Eno invented a graphical notation for the score that was tempting to apply almost one-to-one

 to the spatial design graphics of urban planning, especially suited for linear cities. It was also the time of Bob Wilson's Einstein on the Beach (1976), an opera featuring a series of four acts (including different scenes) and five "knee plays" or entr'actes, whereas both the music of Philip Glass as well as the choreography of Lucinda Childs are highly repetitive and mesmerizing. We attended the live performance at the old Schouwburg Theater in Rotterdam. The opera Einstein on The Beach had a huge impact on our

 thinking and, in retrospect, has influenced my graduation project a lot. Only Barabási's fifth law seems to apply to my career, which got a boost almost 20 years after finishing my graduation project. The international breakthrough of the waste transfer station Elhorst/Vloedbelt (1994) and the

 Saltwater Pavilion (1997) meant a temporary success but did not go unbounded and did not reach a critical mass of

fitness. The Saltwater Pavilion was especially radical for both its exterior appearance and interior experience. The building body appears more as a sculpture without any reference to what is commonly perceived as a building, while the interactive interior was experienced as disturbing and incomprehensible by many; only a few art-loving contemporaries could appreciate it fully. Somehow, our studio remained a fringe actor, and we got only a few commissions, especially when compared to those who followed the footsteps of RK and eventually became one of the many successful OMA spin-offs. When Frits van Dongen and I visited RK's office at the Boompjes in Rotterdam to find out whether it was a good idea to work with RK – it must have been around 1982 – he showed us a small model of his Boompjes tower slabs, which I did not appreciate for its frivolous postmodern toppings. Not overly subtle, I characterized it as "Amsterdam School" when he asked for my opinion. Rotterdam-born RK was not exactly happy to hear that; he silenced, then sniffed, and we instantly knew that we were not going to work for or with RK/OMA, not ever. Not following Rem Koolhaas eventually meant choosing the hard way in the Netherlands. Riding dominant architectural waves never was my stronghold, nor did I – I guess due to my somewhat introverted personality – build a public image of myself, which made me less visible in the Dutch architectural scene. Our work was featured in Bart Lootsma's bestselling book SuperDutch, but it was the studios of MVRDV, UNStudio, and West8 that skyrocketed after the turn of the centuries, while my work was, in the words of MVRDV's Winy Maas, often found too specific. I must strongly refute the specificity of my projects, while my approach toward the constituting components of architecture, toward the digital file-to-factory process, and embedded interactivity have proven to be ultimately generic and generally applicable. There still is the possibility of a late success, but, as Barabási's fifth law states, only if I persist. I will.

At the end of the seventies, I was ready for something radically different, liberating myself from the humanitarian stranglehold of the anthropocentric ideas of Aldo van Eijck and Herman Hertzberger, editors of the Forum magazine. Both of them dominated the architectural discourse until the end of the seventies at the TU Delft.

RK, who ridiculed Herman Hertzberger as a wooly social worker lamenting while bumping around in his Citroen 2CV, advised us to read the novel *The Fountainhead* by the libertarian propagandist <u>Ayn Rand</u>, thereby implicating that we take an example from Howard Roark and adopt a hedonistic, egocentric working ethic. RK reportedly was impressed by the success of young – that is younger than him – American entrepreneur-architects. The discourse paved the way for more audacious conceptual large-scale thinking. My resistance against Hertzberger's claim of the human scale was that he had mistaken the scale of a human for the human scale. A walk on the beach proves my point; there is no human-sized scale to measure your dimension against, while the vast endless sea and sand fuse the very small and the very big. There are the sand particles and there is the ocean, both part of the same coherent global system, and well appreciated by humans. From then on, I realized that the human scale must be seen as inclusive of all scales, not confined to the absolute size of the human. The human scale is not anthropocentric, the human scale is relative and scalable in space and time.

My design thinking gradually evolved into rule-based thinking, into spatial programming, defining new components to constitute a building, a city, or a society. From rule-based design to component-based design meant just a small and logical step. Rather than superficially surfing the waves of popular ideas, we went on to instigate new waves. Instead of surfing the waves, we chose the path of the fusion of art and architecture on a digital platform. I must confess that I simply disliked OMA's projects as published in the addendum to Delirious New York. Especially Hotel Sphinx, with that postmodern crown, that traditional facade treatment, and that obvious hedonist scenario, it went against everything that I was interested in. For me, programming was meant to lead to a digital, verifiable universe, not to metaphors, stories, and scenarios. Yet, I got energized by RK's appeal to be more ambitious but did not want to follow his footsteps. My international breakthrough came with the waste transfer station Elhorst/Vloedbelt in Zenderen when I was awarded the national OCÉ/BNA prize (1994) and the International Architectural Record/Business Week Award (1994). The

Saltwater Pavilion was nominated for the Mies van der Rohe prize in 1998 but did not get through to win the prize. The invitation to occupy the newly established chair of Digital Architecture at the Department of Architecture at the TU Delft came as a direct spin-off from the Saltwater Pavilion. We had numerous international publications on our work, but in my home country, I found little resonance. In hindsight, we simply were running too far ahead of the troops. A similar fate befell us later with the <u>A2 Cockpit in the Acoustic Barrier</u> at Leidsche Rijn near Utrecht (2005). After the completion of the A2 Cockpit, we have not received any new building commissions in the Netherlands. Obviously, it is too radical for its unique parametric design to CNC production systemic thinking and too disturbing for the Dutch building market. Partly because of the global financial crisis, many design offices went through an all-time low from 2008 onwards, and for us, in particular, we lost momentum. The universal laws of success applied to our formula did not work for us.

The Shape of the Rule

The Component is a treatise written as a personal odyssey; it is neither a catalog nor a monography. Yet, the best way to follow the evolution of our design strategies for the notion of the component is through several featured projects, realized ONL projects, Hyperbody research projects, and art projects by Ilona Lénárd, specifying their constituent components. The first project, setting the tone, is my TU Delft graduation project from 1979, completed after nine years of study, the average duration of an architecture study at the time. RK was absolutely right to stir us up. The notion of the strip as the abstract organizing component for urban planning was a hot topic in the years of my graduation project. The <u>Continuous Monument</u> project by Superstudio, An Architectural Model for Total Urbanization (1969), was a prime example of how to radicalize architectural thinking. Earlier, the Italian group Archizoom proposed the "<u>No-Stop City</u>" (1969), an endless habitable single-story space, referring to big factories or American-style shopping malls. They described the city as an assembly line of social issues, representing the ideology and theory of the metropolis. Some years later, the two couples Rem Koolhaas/Madelon Vriesendorp and Elia/Zoe Zenghelis took the idea of such a seductive metropolitan strip for their project <u>Exodus, or the Voluntary Prisoners of Architecture: The Strip</u> (1972), adding a macabre and subversively attractive element to the formal esthetic. They were foreseeing that architecture would disappear within the metropolis. Although the projects by Archizoom, Superstudio, and the future OMA were more of a social-cultural critique than realistic urban planning proposals, I thought of it differently and proposed the strip as a comprehensive and feasible strategy to restructure the Netherlands. In my graduation project, the <u>Strook door Nederland</u> (1979), the main components are the legislative rules of play. Rules of play represent the immaterial substance that rules future material crystallization into built structures, similar to how quantum vibrations at the sub-atomic scale set the rules for the materialized world as we as humans see and feel. Rules of play are the driving forces that make molecules and buildings alike emerge from being invisible. National rules of play typically have the shape of the country where the legislation takes place. In my project, I established a link between a national rule of law and a specific shape, appearing as the shape of the law. That shape of the national rule, which I imagined taking effect overnight in a late Parliament session, is not following the contours of the Netherlands but comes with its own shape, a strip of 5 × 200 km, its bounding box, outside of which the rules do not apply. The idea is that a national government could think of a law as a design instrument, not just following natural borders, but willfully imposing a shape on the law. I considered my Strook as a beneficial Sword of Damocles hovering over Holland, casting its shadow upfront. The rules of law only apply to life inside the Strip. The Strook is the carrier of the rule, while a set of internal legislative rules represent the agents, the acting components, and the actors acting upon the playing field. Its simple formal shape induces a high degree of local serendipity when executed, as do city grids. The Strook is projected onto the map of the Netherlands, perpendicular to the major communication arteries like the main East-West highways and the West-European rivers that end up in the Dutch delta and the rivers Rhine, Maas, and Waal. The Strook would be subdivided into one million individual plots of 1000 m² each. The Strook includes both land and water, meaning that one could get a plot of water for further development. The Strook Door Nederland is an abstract yet concrete design game, perpendicular to mainstream developments that go with the flow; I tagged the Strook a national thwart. I proposed soft and hard options for the rules of law; the soft version is the easiest to implement and the hard version would change the face of the Netherlands completely. The soft version would have a visual effect like crossing the border to Belgium. Crossing the 5-km-wide Strook would give the experience of crossing a virtual border. All of a sudden, the houses would look different, basically because they are subject to a different rule of law. My kind soft rule was to free all building activities from esthetic supervising committees. In Holland, we have a committee that is called the Schoonheidscommissie, which translates as the Beauty Committee. Its name was later changed to Commissie van Welstand, "welstand," literally meaning that something stands well, bearing the double meaning of both propriety and prosperity. The Schoonheidscommissie was originally meant as free but compulsory advice to non-professional builders on how to make a proper design, but, today, it has been watered down to a mechanism to discuss the quality of a design among architect peers. Debunking the committee would be a nice first step; the world inside the Strook sure would look different than in the rest of Holland. The hard rule is a much more drastic measure, I proposed to concentrate all new building structures inside the Strip. This rule is intended to increase the contrast between a high-density metropolis and the natural swampiness of the Dutch river delta; there would be a hard border between the Strook and the surrounding areas. The Strook would become a sort of long-stretched Manhattan, in between unspoiled nature at either sides, not unlike Leonidov's Magnitogorsk linear city. The hard rule is a critique of endless extensions strangling the old city centers. Instead, I was in favor – and I still am – of an intensified contrast between a highly cultured city and an uncultured landscape, a hard-edged metropolis in the wilderness. As in virtually all my later design projects, some simple calculations form the basis for the dimensions and the shape. Within the high density, there would be a broad spectrum from detached villas to super high-rise towers. The face of the Netherlands would change drastically by just one single rule of play. It shows the potential power of a simple rule that creates complexity.

Needless to say, my graduation project was not exactly lauded by the traditional urban planner who was assigned to be my urban planning supervisor, but my main architecture tutor Rein Saariste and guest critic Rem Koolhaas were supportive. In his appreciation, Koolhaas described my plan as a rare combination of abstraction and concreteness. What I wanted to get at with the Strook was to find a balance between simple top-down rules and emergent bottom-up freedom, unleashed by those simple rules. The idea of simple rules inciting diversity has become a leading component in our work ever since. But not all simple rules lead to diversity – rules have to be radical and kind at the same time. There are liberating rules and restrictive rules. Restrictive measures are mostly exceptions to a rule, whereas in liberating measures, the exception *is* the rule.

In January 2021, the concept for a new city in the desert named NEOM was launched by Crown Prince Mohamed bin Salman (MBS) of Saudi Arabia. Saudi Arabia vows to invest 500 billion USD in the project. NEOM is planned to be a free trade zone and is bound to attract investors worldwide. The backbone of NEOM is THE LINE, a 170-km-long linear city. In some ways, THE LINE has some striking similarities with the Strook Door Nederland. A strong simple set of rules is set to generate a rich outcome. One straight underground communication line that crosses valleys and mountains, seas, and lakes and connects numerous green neighborhoods. NEOM is technique oriented, featuring an underground linear rapid transit system and underground services to support a series of car-free park cities. NEOM is presented as a solution for Saudi Arabia's post-oil era. Saudi Arabia realizes, as does the UAE and Qatar, that in 25 years, the oil business will be history. NEOM is promoted as a vehicle to invest in clean technology, sustainable tourism, and a healthy lifestyle, in the emancipation of Arab women. The fascination for the strip and the line as an urban planning instrument continues. In the last chapter "The Chinese Patient," I coin some ideas to construct a more fair and just society. As a thought experiment, a compilation of these ideas may be applied to the NEOM strip to check its viability.

XYZ

The first built project under my name is the Villa XYZ in LEGO, exhibited in the exhibition "L'Architecture Est Un Jeu Magnifique" in the Centre Pompidou in 1985. I was invited to participate by Reyn van der Lugt, the Director of the Rotterdamse Kunst Stichting (RKS) at the time. In a 7-m-long drawing, the only Rotring drawing that I have saved for a possible future display, I precisely quantified and qualified each block as a constituent component. I ordered exactly 50.000 building blocks, the maximum I could get from the sponsoring LEGO company in Denmark. Since I have an aversion to pedestals – not an uncommon aversion among sculptors – I chose to turn the pedestal into an integral part of the design concept. Most of the bricks I used for the base of the 1:25 scale model represent a $25 \times 25 \times 25$ m^3 cut-out of the crust of the Earth. The components on top of the surface of the base are a conceptual representation of the constituent parts. The LEGO villa fits in my XYZ series of designs as published in the Wiederhall XYZ issue (1986). Along the horizontal X-axis, I arranged the functional components, while the Y-axis represents a path to its immediate neighbors. The Z-axis represents communication with the world, a slim towering antenna. All components are set against a black background, representing a hole in the sky, thus linking the house to the universe. The LEGO villa is placed in the force field between its immediate neighbors, the Earth, and the Universe.

I listed the exact types and numbers of LEGO blocks and their conceptual meaning in a database. Identifying, quantifying, and qualifying the constituent components, means being faithful to the principles of falsification and verification as put forward by the Austrian-born philosopher Karl Popper. I could not know back then in 1985 that such a verifiable dataset was exactly what would form the basis of our future parametric design to robotic production working method. I wanted the conceptual design decisions and the definition of the physical components as transparent and reproducible as possible, thereby demystifying the traditional metaphorical vagueness of design concepts. Not many visitors got the point of my meta-villa – they expected a 1:25 scale model of a villa like most other exhibitors submitted, among others: Jacques Herzog, Jean Nouvel, Jo Crépain, Mecanoo, and William Alsop. A conceptual design based on simple rules typically undermines the spectator's comfort zone, as concepts have the power to unsettle. As in the Strook, I aimed for maximum effect with minimal means. Or, as I wrote in the exhibition catalog: The most simple and audacious scheme will have the most powerful impact. And that scheme is verifiable, based on quantifiable and qualifiable components.

Λn Endpoint and a New Beginning

The <u>BRN Catering</u> building in Capelle a/d IJssel (1987) is the first

building I realized under my name, co-designed with my mentor architect and then business partner <u>Peter Gerssen</u>

(1932–2017), who was the son-in-law of <u>Huig Maaskant</u> – he married Maaskant's daughter Carla – who was Rotterdam's

most prominent post-World War II reconstruction architect, responsible for numerous landmark buildings in Rotterdam and the Netherlands, today recognized as the godfather of Rotterdam-based architects Rem Koolhaas/OMA and MVRDV. Gerssen, one of Maaskant's project architects, designed several uncompromising works, notably the

<u>Adriaan Volkerhuis</u>. The design of the Adriaan Volkerhuis means a radical departure from the until-then conventional base-shaft-capital concept of high-rise buildings. The Adriaan Volkerhuis rises cold-blooded out of the ground and just stops there at the top, without any form of a decorative crown. The Adriaan Volkerhuis *is* the shaft, and the whole building is subject to one strong conceptual design theme that governs all further design decisions. A second radical design concept of the Adriaan Volkerhuis is the choice

for the basic spatial component of 7.2 m × 7.2 m, in line with the then-popular structuralist approach. The third uncompromising feature is the choice for an extreme form of prefabrication, with as many of the same building components as possible, a most effective use of mass production methods. I appreciate the Adriaan Volkerhuis building for its audacity, straightforwardness, and chilling beauty. The design concept for BRN Catering is directly inspired by the design principles of the Adriaan Volkerhuis, and maybe even more so by the revolutionary Fläkt building that Gerssen designed after he had started his own office. Fläkt is composed of three components: a central column, a wall panel, and a pie-shaped floor component. The design of BRN Catering combines a tall storage space on the ground floor and two office floors on top of it. Like Fläkt, the design is based on a minimum number of as large as possible prefabricated components. BRN is a minimalist, expensive-looking, yet particularly cost-effective cute little building. I modeled the building in 3D using the then-popular Dutch Arkey software. Using the software taught me to be painfully exact. Besides maximum prefabrication, besides free-span interior spaces, which allow for total flexibility in the layout, BRN Catering features the integration of smart climatic control systems that are well-integrated into the structure. At that time in 1987, it was known as the most energy-efficient building in the Netherlands.

BRN Catering is a design exercise in stretching the extremes, making look bigger, what is already big, and making look smaller, what is already small. The panels of the glass facade are horizontally stretched to emphasize the length of the building, while the connection details are minimized. BRN Catering is the first mullion-less facade in the Netherlands. All mass-produced components are custom-designed, and nothing is taken from a catalog. A component exists by its connections to neighboring components. The actual component includes the connection detail. The main structural components are the 3.3 (height) × 3.6 (width) m^2 load-bearing concrete panels with integrated glazed wooden window frames and integrated electric heating. The main cladding components are the darkened glass panels that cover the building as a whole, including the storage spaces at the ground floor level. BRN is a minimalist building with maximum performance.

From that time on, my design efforts are aimed at defining the main components including their mutual connections as precisely as possible. At the same time, I realized that there would not be an easy way out of extreme minimalism. I felt that I could easily become trapped in a minimalist dead-end street. Learning from the arts, and specifically from contemporary artists like Frank Stella, who went from minimal esthetics into a more maximalist yet equally concept-driven direction, I decided to take a different, more emotive, and intuitive road. BRN Catering marked my supreme point of minimalism; I could not have reduced the figure of a building more. BRN Catering has simple rules leading to simple outcomes. BRN Catering marked in many ways an endpoint and a beginning. I decided to end the partnership with the 17 years older Peter Gerssen and started a joint venture with visual artist Ilona Lénárd, my partner in life since 1977, and from then on in business as well. I started looking for a conceptual way out of minimalism. Simple rules that lead to complex outcomes are what we were looking for to find our way out of minimalism. Complexity is based on simple rules, as opposed to complicatedness, which is the inevitable outcome of fuzzy rules that lead to many costly exceptions.

The method of procedural intuitive sketching using 3D graphic software showed us the way forward. Intuitive digital sketching meant clicking with the mouse from point to point in a 3D environment, freely swapping from front to back and from left to right, and rotating up and down. Not consciously placed points, but more like automatic drawing. Well before I finished BRN Catering, Ilona Lénárd made a series of wall objects titled the Metawell series, based on a selection of numerous intuitive digital sketches.

The Metawell series is a form of generative art, driven by analog algorithms. Together, we decided to apply the idea of intuitive sketching to conceptual spatial design, intending to consider the 3D sketch models for larger-scale buildings. Although not consciously generated, digital sketching digitally implies that the design parameters are registered and fit to inform the physical construction of the models. Intuitive digital sketching using variables to generate a diversity of outcomes meant for us the decisive step to parametric design. While intuitive sketching provided us with the method to go uncontrolled toward a rule-based complexity, parametric design gave us the instrument to quantify and qualify diversity and complexity and link it to the production of a series of unique constituent components.

Parametric design as we practiced from the early nineties catapulted us one level up from mass production of elements in a series of the same to mass customization of a series of unique components. Culturally speaking, it meant a jump from the late fifties and early sixties into the late eighties and early nineties. Building upon the great achievements of serial prefabrication, parametric design to robotic production is the natural successor of mass customization, opening the world of serial production of unique components. Both mass production and mass customization find their basis in producing a series of clearly defined components. For mass production, it is crucial to produce a series of components that are as much as possible the same, and for mass customization, the new paradigm is to produce a series of unique components that are not the same, albeit *similar*, unique in shape, dimensions, and performance – similar but not the same. The parameters may change but not the procedure. Both mass production and mass customization are highly systemic. Mass customization is inclusive of endless variations by nature, while mass production is based on the rigidity of one component with a fixed size and performance, and hence very inefficient for exceptions to the basic component. Hence, our favorite slogan, in its most compact form, promotes the superiority of mass customization: "the exception *is* the rule."

In general, the majority of today's designs are still based on mass production, thirty years after we picked parametric design to robotic production and put our bet on customization. Architects and contractors typically are slow to pick up new developments, even when the innovations in design and production would demonstrably work to their advantage. The building industry is, and has always been, a conservative factor in the global economy. The building industry is a slow and viscous amalgam of thousands of small and medium-sized and big businesses that cannot change direction easily. This is also the reason why the CO_2 emissions that are caused by the building industry are so hard to control. One needs to change thousands of businesses to substantially reduce the CO_2 emission of one single building. If only governments would have the guts to adopt a simple rule of play that demands a systemic transition toward parametric design to mass customization of all production processes in the building industry, that rule of play would imply that also the architects would need to drastically change their design methods and switch to component-based parametric design, such that the connections to the robotic CNC production methods can be established. Component-based design is the prerequisite for the circular economy that is based on the dry assembly of components, whereas no traditional wet technologies are allowed. In the last decades, architects made the change toward computer-aided design before; parametric design to robotic production is the next transition architects will have to go through.

From the Saltwater Pavilion onwards, the buildings that I have realized are based on the design principles of mass customization. They could not have been built without the then-available data-driven design and production technologies, that is, not for the same tight budgets. As a direct result of the chosen path, the emotional feeling of what is beautiful has drastically changed since we adopted digital customization techniques.

Maximum minimalism is traded in for minimal maximalism. Simple rules creating simplicity are traded in for simple rules creating complexity. One building with one parametric detail is the ultimate simplexity design method we invented for the Web of North-Holland (2002). The buildings we have realized since represent that new kind of beauty. "We are changing your view on what is beautiful or not" is the title of my 2011 TEDx Delft lecture. As we speak, my current work is culminating in visually complex, super-controlled parametric designs like the Dutch Pavilion for the Dubai World Expo 2020 and the Seven Daughters project for Qatar, which projects will be discussed in greater detail in the "Where Are We Now?" chapter. The Expo 2020 building is based on one basic component that adjusts its shape, dimensions, performance, finishing, and appearance to its unique position in the spatial lay-out, without any exceptions to the rule. Just that, just there, just then. When, where, and as needed. Based on carefully chosen simple rules, a sheer limitless variety can be unleashed. We could not have come to this radical statement for mass customization without having laid the basis for it through our early conceptual designs and earlier realized mass production designs.

Studio Theo van Doesburg

New developments in art and science usually have my keen attention, typically more than what is new and trendy in architecture. Developments in other fields matter to architecture and building technology. Especially after coming to live together with Hungarian-born Ilona Lénárd who studied Art at the Willem de Kooning Academy in Rotterdam, I became deeply interested in art, especially in the process of making art. I found out that the way artists think matches perfectly with how I think in terms of architecture. After 10 years of supporting and following each other's work, we decided to find ways to work more closely together. Out of a pool of 30 applications, we were selected to live and work for one year in the Van Doesburg artist house, Rue Charles Infroit 29, Meudon Val-Fleury, Paris. Our state-funded sabbatical year in Paris has been fundamental in building a vision of the intimate relation between art and architecture. During our stay in the Van Doesburg studio, we decided to go for a maximum fusion of art and architecture. At the same time, I explored design software to support the fusion. Working with our personal computer, the legendary Atari ST turned out to be instrumental in finding common ground. We found that we could bridge the gap between the disciplines of art and architecture by working on the same digital platform.

The sabbatical year allowed us to get to know French architects, artists, and philosophers personally. Notably, we talked to the philosopher Paul Virilio and architect Claude Parent, the founders of the group Architecture Principe (1963–1966), inventors of the architecture of the oblique, publicizing the sloping surface to enhance the fluidity of movement of people, to stimulate deeper human relationships. We met with the then-young Dominique Perrault, the architect of the Bibliothèque Nationale de France, alias the Très Grande Bibliothèque (TGB), François Deslaugiers, the architect who worked

with Von Spreckelsen, Paul Andreu, and Peter Rice on La Grande Arche,
 Nicolas Schöffer the Hungarian-born spatio-dynamic artist,
author of La Ville Cybernétique (1969), and Paul Andreu,
the lead architect of Aéroports de Paris, designer of the
Charles de Gaulle Airport. Each of the encounters with these
 geniuses in their field has influenced my thinking in one
important way or another, especially the impact of Parent
and Schöffer can be directly traced back to our work. Parent
 for the notion of sloping surfaces as behavior structuring
components in architecture, and Schöffer for his interactive
cybernetic installations, notably the Tour Cybernétique
 (1954) in Liège. The imposing presence of La Grande Arche in
La Défense made a major impression on me. I was impressed
by its conceptual boldness, its physical HyperCUBE-ness,
and especially by the way it is placed as an autonomous object,
constructed as one big urban component, literally a building block of
the city. Hercules could have picked up the HyperCUBE and placed
it elsewhere; that strength is the post-stressed structural coherence
of the HyperCUBE. The Grande Arche sits on just a few giant pillars,
carefully positioned to fit precisely between the metro tracks under
the sloping platform of La Défense.

Similar to the almost empty office of Claude Parent that we once
visited, the vast office space of François Deslaugiers once was occupied
by a large team of collaborators to design and engineer the airy open
elevator shaft and the tensile structure in the void of the Grande Arche
was now sadly empty. One could feel the looming economic crisis.
Listening to a lecture by Paul Virilio in the Maraqais, I was fascinated by
his up-tempo way of nose-picking ideas, and finger-shooting them into
the audience. Virilio was a considerate man – he sent us New Year's
cards after we exchanged business cards. My meeting with Dominique
Perrault was a remarkable event. I proposed to him to join forces on
the design for the TGB, which at the time was just launched as an
open international competition. Although he showed much interest
in the radicality of my BRN Catering design, he decidedly rejected
that idea, while leaning proudly backward, puffing a fat cigar. Perrault
gloriously won the international competition and built his masterpiece

some years later along the banks of the Seine. Perrault was tried and tested in engaging with local politicians, which apparently is the way to secure commissions. Many friends came to visit us in Meudon. We were especially happy to meet art historian Wies van Moorsel and her husband the late architect and art historian Jean Leering. Wies van Moorsel is the niece of Nelly van Doesburg, as her only remaining family member she inherited the studio-house and donated it to the State of the Netherlands.

In the year before our stay in Meudon, I was invited to teach a master unit at the AA in London. I went city-hopping from Rotterdam to London every week. I still feel sort of angst for the often-bumpy rides over the Channel. It made me very aware of the atomic nature of materials when traveling at higher speeds. Airplanes can be hit hard by otherwise ephemeral clouds. In retrospect, it is evident that the year in Meudon has built the foundations of my vision of architecture. And equally evident, my weekly trips to London inspired me to ponder about materiality, clouds, swarms, speed, and friction. I became deeply interested in the concept of the swarm and its constituent components. I started to think about a combination of extreme prefabrication of the building components as developed for the BRN Catering building with a possible unique shape for each of the constituent components, as the birds in the swarm are a family of components, similar yet with a unique individuality. More and more, I wanted to design buildings that are a swarm of building components, flocking together to shape and materialize the building body.

Script for a Movie

During our stay in Meudon, I typed down my thoughts on my beloved Atari 1040ST computer and printed them out on my dot matrix printer, which I used in combination with a then-expensive compact telephone annex fax machine. The importance of the new invention of the personal fax machine should not be underestimated. Just think of it: cramming a sheet of paper in it on one side, and watching it creep out of another machine on the other side, thousands of kilometers away – a real teleportation machine, a "Beam me up" time machine. I used that machine extensively to communicate with the commission to design the Van Doesburg exhibition in Museum Boymans van Beuningen in Rotterdam, invited by curator Evert van Straaten, the later long time Director of the Kröller Müller Museum. Not much later, after we established our office in the Groothandelsgebouw in Rotterdam in 1989, that precious 3.000 Dutch florins fax machine was suddenly gone mysteriously, while no one else had the key to the office but me.

Some fascinations that kept me busy at the time must be described to understand my later work. The script for a movie I wrote back then on my Atari in 1988 in Meudon was originally meant as a scenario for a documentary film in collaboration with Willem Kars from the Rotterdam-based graphic design studio Hard Werken, mediated by my long-time architect friend Ton van Hoorn, who is known for his pioneering groundwork for computer-aided design and elementary design methods. Unfortunately, we did not manage to get the funds to start making the movie. An important part of the script dealt with the notion of the universal container and the matter that is contained. There is a clear link between my fascination for the container back then and my actual obsession with the component. Not just the components themselves are of interest, but the connections between the components. A sea container in itself as an autonomous object is an element, while a container that is connected functions as a component, an acting part of a larger whole. The distinction between element and component is a key ingredient to understanding our future ventures. I have dedicated chapter 5, titled "Components Versus Elements," to elaborate on this topic.

The movie script dealt with a wide variety of subjects, as can be read from the titles of the episodes:

1)   Material Density,
2)   Superclusters,
3)   Unification Theory,
4)   Transparent Steel,
5)   Tempo Alienation,
6)   Implosive Cybernetics,
7)   Artificial Ecology,

8) Flatland,
9) Tension between the Dimensions,
10) Fly-overs,
11) Road Graphics,
12) Free Flow,
13) Automotive Styling,
14) The Flying Container,
15) Dramatized Meteorology,
16) Acupuncture Urbaine,
17) Cultural Entropy,
18) Monocultural Exaggerations,
19) China One Big Villa Park,
20) Stretching to the Extremes,
21) Centripetal and Centrifugal Forces,
22) Space Colonies,
23) Digital Space,
24) Intuition and Logic,
25) Data-explosion,
26) Delayed Conflict,
27) Hidden Seducers,
28) The Black Monolith,
29) Artificial Intelligence,
30) The Spiral of Progress,
31) Invisible Technology,
32) The Absent Detail,
33) Hidden Delights,
34) Smarties,
35) The Shaped Container,
36) Working Space,
37) Statistics,
38) Aliens,
39) Subdermal Architecture, and
40) The Open Volume.

Quite a rich and diverse list of challenging topics, I definitely must publish it once, unabridged and unedited after these past 30 odd years. Many of the same subjects can be found in this book *The Component*; it seems that my early fascinations have stood the test of time. The wording and choice of subjects, varying from observations from daily life to musings on art, science fiction, and natural physics, are surprisingly consistent with the subjects that still fascinate me today. It features this particular blend of practice and theory that has always had my keen interest. It is, indeed, a rare combination of abstraction and concreteness.

The Universal Container

From another as of yet unpublished text of mine titled *The Universal Container*, also written in Meudon in 1988, translated from Dutch without further editing: "The container fascinates as a universal standard box. The standard measurements make it into the first internationally applicable building component. Typically, 8 ft wide and 8.5 ft high, with varying lengths from 10 ft, 20 ft, and 40 ft, the shipping container is a self-supporting structural element, stackable up to 10 containers. The corners are lashed together whereby a coherent built construct can be achieved. A containership today carries in one shipment approximately 10 x 10 x 10 = 1000 containers packed in the streamlined steel skin of the hull, in its body. In structural respect, the lashed containers constitute a spatial system of Vierendeel trusses, while the walls of the individual containers take care of the stability. The transport of massive amounts of containers is a daily scene in the Netherlands: transport by water on huge containerships, by road on the back of a truck, and by air in the belly of a cargo plane. The container is the ultimate component of worldwide trade. The container does not betray its content. In the container plates of glass, food products, electronics, or even illegal slaves are transported. Every box has its secrets, its own story, every box is the container of hidden delights." That idea of a systemic set of simple components that are the building blocks of a worldwide ecosystem excites me as much now as it did back then.

Packages of Information

The global ecology of sea containers seriously intrigues me. It may not be that surprising, since we have lived in Rotterdam, the world's biggest sea container harbor, for the bigger part of our productive life. The sea container is the main factor constituting the economy of Rotterdam. The sea container links Rotterdam to the other main ports on the globe, seaports and airports being the physical transport hubs of the global economy. Today, my fascination for ports is by and large replaced by the magic of portals. Established only in the last decades, multinationals like Facebook, Twitter, Instagram, Amazon, eBay, and Alibaba have acquired their status as digital portals on the global internet. These portals are now among the biggest companies ever; their success is facilitated by systematically identifying every traded component with a unique number. While tagged sea containers are the components of the global trade in physical goods, small identifiable packages of information are the components building the internet, trading intangible information. It is intriguing how these packages of information find their way to their destination, not necessarily taking the shortest route but navigating in batches via numerous network hubs. Data travels across the internet in packages. Each data package, the small container of information, can carry a maximum of 1500 bytes. The package has a wrapper, a header, and a footer. Any message breaks up into packages that travel across the network, while packages from the same message follow their unique path around the globe. Upon arrival, the computer receiving the data assembles the packages like a puzzle, recreating the message. The individual packages, the smallest components, contain the information of their immediate neighbors, information that is needed to pack the parts together. Not the components as isolated elements in themselves, but the connections between the components are the crucial part. The constituting components of the information puzzle reassemble into a coherent whole. Whether looking at the global ecology of the physical shipping container or the ecology of the virtual global internet, their building blocks are the irreducible acting components. Sea containers and data packages, created and recreated at any time at any place in the world, are transported from anywhere to any other place on the globe. Learning from global trade and the internet, it is mandatory for any (digital) design project to define the building components and their mutual

39

connections to set up a coherent design to production to assembly system. In the fifties and sixties of the last century, the systemic approach has been the prefabrication of mass-produced elements of the same. Decades ago, we entered the digital age; today, the systemic approach implies mass customization of a series of unique components.

It is not so much the sea container as a self-contained element in itself, but rather the container in relation to its closest neighbors that is key to understanding relational design, i.e., associative design, i.e., parametric design. Each container has a unique code, identifying the owner, the country code, the size, type, and equipment category as well as any operational marks, to be able to be tracked and traced. Similarly, each component of a construct needs to be tagged from the very start of the design to the production process and must remain traceable and verifiable during the multidisciplinary design-to-production-to-assembly process. According to the construction wiki Designing Buildings, a building component is defined as "a constituent part of a building (or other built asset) which is manufactured as an independent unit, subsystem or subassembly that can be joined or blended with other elements to form a more complex item. Generally, components are self-contained and sourced from a single supplier, typically the complete unit is provided by that supplier rather than its constituent parts. A combination of components may be described as an assembly."

A sea container comes in a number of different types, featuring 16 modular variations in its appearance and performance. The eight main types are, basically variations on the same:

1) dry storage container,
2) flat-rack container,
3) open top container,
4) tunnel container,
5) open side storage container,
6) double doors container,
7) refrigerated ISO container, and
8) tank container.

The different types facilitate the variations in content like transporting bulk materials, dry goods, liquids, or food stuff. The different types of containers can be stacked in any possible combination, thanks to the universal lashing detail. Containers are lashed together with a single hand movement. The method is as follows: first, the lowest container is anchored to the deck of the ship. Subsequently, other containers are stacked on top of it and to the side. From the underlying container, the stacker/twist lock reaches into the precast corner of the container above, and then the handle is turned 90º to fix the joint. An additional connector is the bridge clamp that connects from the outside of the corners, mostly used for connections that are stabilizing the 3D stack of containers sideways. Together, the containers form a solid block that acts as a whole. The constituent components live by their connections to their nearest neighbors. Sea containers are designed to be structurally strong, self-supporting modules that connect to their immediate neighbors, i.e., any other container of any of the eight different modules. When filled, connected, and transported, the containers become active members of the dynamic transport community. When disconnected, they are mere elements on standby, like cars in a parking lot, waiting to become activated to act as a player in the global transport game. The structure of the container is self-supporting, supports its load, and is designed to support another nine fully loaded containers on top of it. The modular container is strong like a cell in a biological body, like a building block to constitute a larger building body. Shipping containers are the robust building blocks of our global transport system, whereas local transport systems use smaller building blocks like euro pallets, cardboard boxes, and envelopes. What they all have in common is that they transport materialized forms of information. Like most other types of components, shipping containers are composed of smaller subcomponents: the cast iron corners with openings to the outside to host the connectors, welded to the steel beams and to the corrugated sheet steel surfaces at all sides to enclose the box, and the mobile connecting components, the twist locks, and the bridge clamps.

Ultimately, the shipping container is a carrier designed to exchange information – information in any possible material form. In general, all objects and materials that are transported are in their essence a form of information, to be processed or consumed in a particular ecosystem. Food is eaten or, in other words, processed by the body, furniture is arranged inside a building, and thus processed by that building, and oil is burned to feed vehicles and buildings. The shipping container system facilitates information exchange on the scale of our globe and is therefore a main driver of global societies. The global exchange of goods and information has a huge cultural component, thereby defying anti-globalist tendencies. No country comes first in a dynamic global network. Constituent components are connected and interlaced in the global network of information exchange. Some connections may be stronger and some may be weaker, but the very nature of the system is that they are connected in one single global system, and they are players on a global level playing field. The global transportation network is for analog information exchange what the internet is for digital information exchange.

Low-resolution Architecture

The invention by the American engineer Keith Tantlinger of the twist lock system for connecting shipping containers dates back to the early fifties. His technical invention must have, directly or indirectly, inspired structuralist architects. According to philosopher Simon Blackburn, the definition of structuralism is the belief that phenomena of human life are not intelligible except through their inter-relations. This definition nicely relates structuralism to the assembly of components into a consistent swarm of similar components. In the last decades, we have witnessed numerous architectural designs that reference the boxiness of shipping containers literally. The rise of structuralism in the sixties of the previous century certainly gave it a boost. The container is a most basic platonic volume, and as such celebrated by generations of architects, assumedly because of its pureness and stereometric simplicity. Buildings that look like stacked containers, such as today's so-called pixelated architecture, are not minimalist, while in minimalist architecture, all boxiness of the constituent components is avoided, in favor of one strong overall form, as I have demonstrated with the BRN Catering building. Despite the obvious relational qualities of groundbreaking low-resolution structuralist projects like the Burgerweeshuis by Aldo van Eijck (1959), Habitat 67 by Moshe Safdie (1967), and Centraal Beheer by Herman Hertzberger (1972), I have chosen the path from minimalist architecture to a high-resolution nonstandard architecture, establishing relations between series of unique complex components, which I considered the logical consequence of computer-controlled technology of mass customization.

I first noticed the literal use of sea containers to constitute a home at the experimental neighborhood De Fantasie in Almere, the Netherlands, originally meant for temporary structures only. De Fantasie is a site for tiny houses avant la lettre. Not back then, and certainly not

now, I have been interested in low-resolution solutions for buildings or in stacking container-like shapes to constitute buildings as a form of three-dimensional copy and paste. As of today though, the 1:1 assembly of customized shipping containers to form homes has become more and more popular and has eventually even led to a mainstream movement in architecture that is known as pixelated architecture. The label pixelated is not accurate though; it should have been named voxelated architecture instead since pixels are the smallest indivisible components for two-dimensional graphics, while it is the voxel that is the smallest indivisible component of spatial structures.

Low-resolution (low-res) voxel architecture went viral in the Netherlands from the moment OMA won the competition for the Timmerhuis town hall of Rotterdam in 2008. OMA's Timmerhuis was visually preceded by the Computational Chair (2004) using genetic algorithms by the then-young architects' collective EZCT. Although OMA's Timmerhuis is not computationally generated, it is traditionally modeled as a voxel cloud, and the reference with EZCT is inevitable. The 3D structure of the Timmerhuis intends to metaphorically look like a cloud, an assembly of $7.2 \times 7.2 \times 3.6$ m^3 cubicles stacked in the form of an extremely low-res reduction of the shape of the cloud. Disappointedly, a cloud that does not float does not even seem to float but stands firmly on the ground. Although OMA denies any reference to and inspiration from Hertzberger's Centraal Beheer, it is at least visually referential. RK/OMA appropriates the esthetics of a former fabricated enemy. After OMA was selected for the job, the low-resolution trend, in absolute contrast to my evolving take on high-resolution design, really took off. Now, in the year 2022, an alarming multitude of design proposals worldwide are flirting with low-res voxel esthetics. At the time, I did not see this evolutionary success for low-res architecture coming. Barabási's rule of law for success was working well for low-res architecture. For me, the fascination with the shipping container was its flexible content in a component-based global communication network, certainly not as an inspiration for literally voxelated designs. Instead, I was interested in the connective principles of container systems as a means to enhance complexity in architecture. I imagined architecture to evolve toward geometrically complex shaped containers, which we later in our Sculpture City project coined as sculpture buildings.

Voxelated projects by OMA's spin-offs, notably by MVRDV and Ole Scheeren, followed quickly after the Timmerhuis. Knowing designers of MVRDV from the beginning of their careers – Jacob van Rijs and Nathalie de Vries invited me once to be their design tutor for a TU Delft design project in the late eighties – I take their design intentions seriously. When working at OMA, it is hard not to become infected by RK's hyperbolic turbo language and his somewhat cynical chuckling approach to architecture, as if in each design, there should be an element of irony, an element of mockery, of appropriation, and of making fun of something. The Ravel Plaza design by MVRDV stands for something worthwhile reflecting upon, however not for the reasons Mark Minkjan of Failed Architecture was awarded his prize for. When looking at the design, I am not so much shocked by the abundance of greenery on the balconies as Mark Minkjan is, nor by an assumed absence of balustrades in the renderings, but I am straightforwardly disagreeing with their design intention that destructs, collapses, erodes, and explodes. I simply cannot be empathetic with the concept of eating away bits and pieces from a shape that once had been in good order. They chose to erode something they criticize to define their design goal. Thus, the design becomes a built form of critique on something they do not like. But, if one does not want it, why take that as a starting point at all? To tell a story? What story? Looking at the design, I see a striking resemblance with the collapse of the World Trade Towers. Why take a disaster, whether unconsciously or consciously made as a design decision, as the inspiration for the visual effect? Having watched many sci-fi movies and having seen the twin towers coming down as a real form of science fiction, I realize that the visual impact of the image of destruction can be so strong that it somehow is burned into the designer's mind. Something so strong that there is no escape from it, at least not when it takes a straightforward tower design as the starting point for the storytelling. My critique of the Revel Plaza design is that the irony and the inherent cynicism confirm destructive Hollywood- driven prophecies rather than offering a constructive alternative to them.

Erosion and Explosion

As to indicate that the Ravel Plaza design concept is not an isolated incident, I refer to other recent designs featuring similar gnawing characteristics. The OMA's Asian spin-off Ole Scheeren designed the MahaNakhon tower in Bangkok, Thailand, evokes the erosion of a once-straightforward tower design. It is claimed to stimulate social interaction and offer shared space. The upward-spiraling erosion evokes more visual emotion than a simple rectangular tower would have been able to produce, but that is not my point. The point is what sort of emotion Scheeren – using the power of the proposal – has chosen to trigger. He puts the object of criticism and the critique in one design scheme, thus being able to tell a postmodern story of erosion and destruction. A more extreme version of violent storytelling, of explosion rather than erosion, is the story line MVRDV has opted for when designing The Cloud tower in Seoul. The exploded mid-section, throwing out low-resolution voxels into all directions, is strongly reminiscent of the mediated impact of the 9/11 attack on the World Trade Towers. Winy

45

Maas denies that there is such a direct relation, and on their website, they apologized for the unintended reference, but perhaps the image of violent destruction had somehow hijacked the designer's mind, unconsciously expressing the troubled sign of times. My designer's scheme of things tells me that the design of a skyscraper must come from an internal logic of the design-to-production process itself, rather than wrapping an otherwise straightforward tower with a local cosmetic intervention based on lighthearted storytelling. A design must be the expression of its character, instead of being an illustration of a story. What troubles me particularly is the poisonous combination of the erosive approach to design with the esthetics of voxelization. The designs of OMA and their spin-offs MVRDV, Ole Scheeren, and BIG, and many followers, share the fascination for low-res pixelation, i.e., voxelization. I consider voxelization for design concepts on the grand scale of architecture as a deceivingly reductive simplification of building technology, counterproductive to support advancements in architectural theory and practice. The combination of sympathy with destruction and erosion with populist voxelization is outright alarming for its uncanny parallels to populism in politics using hyperbolic forms of expression in combination with an over-simplified language. When voxelization is combined with the concept of internal porosity, at least there is an internal logic to support the design concept. Porosity in low-res architecture means the number of open voxels to substantiated voxels. In 2012, the Why Factory led by Winy Maas asked their students at TU Delft to build towers using white LEGO blocks. These porous constructs do not depend on that destructive fascination for erosion or explosion. Porosity is here an integral character trait of the substance of the tower, reminiscent of the algorithms of a Menger Sponge, which features approximately 25% porosity. The basic spatial arrangement of the Menger Sponge is $3 \times 3 \times 3 = 27$ cubes, whereas 7 cubes, i.e., the ones in the middle of the faces and the center of the basic unit of 27 cubes, are void.

We went in another direction to create diversity and complexity, diametral opposite to low-res architecture. There is one thing ONL projects have in common with MVRDV though, and that is the reduction of the number of different materials that constitute the built object. Their early projects like the blue rooftop house <u>Didden Village</u> in Rotterdam serve as proof of this unlikely correspondence. The exterior of the Saltwater Pavilion is one material, one texture, and one color, for the roof, the side, and the belly, basically wrapped all around the object. The same approach we applied to all of our later projects: one building, one material, and even one detail, which is the desire to emphasize the object as a whole, that is, not as a collage of different elements. My scheme of things is based on high-resolution complex geometry combined with a strong internal structural logic, based on simple rules creating the diverse and the complex. In the words of Buckminster Fuller: "You never change things by fighting the existing reality. To change something, build a new model that makes the existing model obsolete." In my world-making, architecture is not a comment on anything existing but an attempt to synthesize new worlds from customized building components. As a practicing member of the Nonstandard Architecture and the Transarchitectures movement, I contributed to defining high-resolution liquid architecture. Liquid Architecture naturally emerges from computation that effortlessly deals with complexity, compared to how our brains operate. Our brains have a limited capacity to imagine complex shapes and curved trajectories in space and understand multiple rotations, since our vestibular system, our interior gyroscope, is designed to keep us upright on our feet. People do not fly – people are constantly balancing on their two small feet to avoid falling. Working with computation as a design instrument opened for us a new world of imagination; in cyberspace, we could fly indeed, constructing designs in weightless space, freely rotating around the construct. The term Transarchitectures was coined at the end of the nineties by Marcos Novak, at the time, based at the University of Texas, USA. Novak wrote his essay <u>Liquid</u> <u>Architecture</u> in Cyberspace for the groundbreaking book Cyberspace, First Steps (1992), edited by Michael Benedikt. The term Nonstandard Architecture was introduced at the first <u>Archilab Conference</u> (Orléans, 1999), curated by Marie-Ange Brayer, Zeynep Mennan, and Frédéric

Migayrou. The subsequent series sequels of the Archilab conferences (1999–2013) are the nonstandard equivalent to the notorious CIAM conferences (1929–1959). Nonstandard architecture is relational, connective, performative, inclusive, associative, and transformative.

The transformative aspects of the nonstandard movement are adequately referred to by Marcos Novak as Transarchitectures, a term he coined around the millennium shift. He rightfully stated that Transarchitecture means "a transformation or transmutation of architecture intended to break down physical and virtual opposition, proposing a continuum ranging from physical architecture to architecture energized by technological augmentation to the architecture of cyberspace."

The Archilab Conferences and other fundamental developments in architecture were completely ignored by Rem Koolhaas, curating the 2014 Fundamentals Venice Biennale. Koolhaas' timeline of academic movements in architectural theory that was printed out on one of the long walls in the Italian Pavilion ended at the end of the eighties, that is, when he left the AA. Everything that happened after 1988 was not dealt with at all, as if his leave from the AA marked the end of the architectural discourse. On the contrary, it is precisely the time that a new approach to architecture started, informed by the rise of the digital era, which did not fit in Koolhaas' scheme of things, which makes his Biennale particularly retroactive, in extenso discussed in the "Components Versus Elements" chapter. RK intentionally missed the point of new computational developments in architecture by completely ignoring the richness and beauty of complexity, computationally generated from sets of fundamental simple rules. RK got stuck in the simple rules-create-simple-elements paradigm. The fields of research and practice that we explore today are dealing with fundamentally new paradigms in architectural theory, emerging from an understanding of the real existing state of digital technology. RK never went beyond the analog, even ridiculing the digital by labeling AMO as a so-called virtual counterpoint to the analog OMA, thereby reclaiming the word virtual back into the realm of the analog. AMO mistook virtuality for the unbuilt, while the virtual in cyberspace effectively means the hyper-real, currently referred to as the Metaverse. There is nothing more real than the virtual since we know every bit and byte of the virtual world.

Maison Particulière

The same year (1988) that we were selected to live and work at the Van Doesburg house, I was invited by curator Evert van Straaten of the Rijksdienst Beeldende Kunst to design the Theo van Doesburg exhibition at the Museum Boymans van Beuningen in Rotterdam. The exhibition contained original stained-glass windows, drawings, models, and an animation of the Maison Particulière. The main object is an XYZ construct. A heavy concrete block represents the X-axis, the Y-axis is represented by the stainless-steel frame with four original stained-glass windows, while the animation screens that are embedded in the concrete block offer a deep vertical view, representing the Z-axis. The models are packed in three volumes with viewing windows representing another set of XY and Z-axes, positioned at a 45º angle to the exhibition space as Doesburg often chose in his works, as in the dance hall in the Aubette and a series of paintings. As part of the exhibition in the museum Boymans van Beuningen, I wrote the script for a computer-generated video of Theo van Doesburg's groundbreaking design for the Maison Particulière, executed in collaboration with, at that time young architect, Cornelis van Eesteren. The Maison Particulière is a cold assembly of loosely connected architectural elements, vertical walls, and horizontal floor slabs, levitated in a weightless space, in a seemingly accidental configuration. Theo van Doesburg has intended to visualize a dynamic universe of components that move freely in space. The connections between the parts are non-existing, emphasizing their mobility and weightlessness. Van Doesburg's absence of the connection detail implies that the constituent components are there only temporarily and locally, ready to move on to form other configurations. The animation I realized in collaboration with the Rotterdam-based model builder Peter Snel of Kappers Trimensi, who was the proud owner of a then-advanced, hundreds of thousand guilders worth graphic computer system. We modeled and identified the constituting components of the Maison Particulière and allowed each component to travel in 3D space along the *X*-, *Y*-, and *Z*-axes, aligned with the direction of the component. We made the components implode in that one particular configuration, that is, the Maison Particulière, and explode again to become a free-floating mobile component somewhere in the endless universe of all possible components. In the animation, as in the associative design and as in the Internet of Things, the constituting rectangular components, the colorful slabs, are given a unique identity to be addressed individually by attaching direction and speed of movement to the components. We programmed the implosive and explosive movements of the components to start slowly, accelerate vigorously, and slow down again when having reached their temporary and local destination. With this animation-concept video, the dynamic nature of Van Doesburg's universe is revealed. Not unlike Piet Mondrian before him, who started to paint

instances of such abstract yet concrete dynamic universes some ten years before, and just prior to Gerrit Rietveld's Schröder House in Utrecht (1924), Van Doesburg must have almost physically felt such a dynamic universe expanding and contracting inside his head – a dynamic universe that is constantly evolving, never the same, always in motion, always forming new configurations. Mondrian's paintings and Van Doesburg's

architectural assemblies are local instances of a dynamic component-based world, temporarily frozen in time. One of the quotes of Van Doesburg, stated in 1915, is: "Art is not being but becoming." Almost a full century after the Maison Particulière, the time has come to physically realize such dynamic assemblies at the grand scale of architecture. The animated video of the Maison Particulière is a statement that precedes ONL's and Hyperbody's interactive installations, whereby the constituent components are moving in real time. The constituent components of interactive installations are active agents with the capacity to transform shape and content. The sliding walls of Rietveld's Schröder House showed us the way.

Architectural Association

I flew into the AA in 1987, at the exact time that the deconstructivist Bernard Tschumi, Rem Koolhaas, and Zaha Hadid were about to leave. I typed and printed with my Epson dot matrix printer an ambitious letter to the dean, Alvin Boyarsky, in which I stated that I am against all sorts of "isms". And I still am, I see nothing good in framing movements in terms like modernism, brutalism, postmodernism, historicism, deconstructivism, and parametricism. In my opinion, one must describe what it is, not what it looks like. The shared characteristics may be described as brutal, modern, historic, parametric, etc. Except for parametricism, other "isms" are not coined by the architects themselves. Attempts to build movements usually are distracting from what is *de facto* happening under the skin. I once had lunch with Zaha Hadid during jury breaks and confessed to her that I appreciate science fiction series like *Thunderbirds* and *Star Trek*. She was not much interested in Sci-Fi esthetics; in her work, she seemed to be more prone to sort of streamlined Flintstone esthetics. Her soft bench designs of that time, which were on display at the AA, are elegant, somewhat heavy monoliths, and structurally an abstraction from how things are made, without visible details. The absence of detail is similar to that of Van Doesburg and Rietveld, but the freeform formal language means a departure from De Stijl and her suprematist heroes. Hadid does not want to express how things are made, how they are structurally supported, or how they are assembled, in favor of that one single monolithic expression, as a sculptor. The wide-angle distorted components in her compositions are abstracted, and dematerialized, in an attempt to evoke the feeling of flying and floating. Interestingly but not surprisingly, googling Hadid's The Peak Leisure Club project, Google's "similar images" search algorithm shows mainly pre-war winged airplanes, which tells us as much about the floating nature of her flying snippets and their inherent retro-active character. Somehow, it feels as if she was, in her early designs, not interested in adopting new technologies at all. That came only decades later, probably under the influence of Brett Steele and Patrik Schumacher, co-founders of the Design Research Laboratory (DRL) at the AA by the end of the nineties. The DRL was established one decade after I introduced Intergraph workstations running Microstation software into the AA. Intergraph was, at the time, the biggest hardware–software company after IBM. I introduced our disruptive digital approach of intuitive sketching with the computer into what was back in 1988 a quite diverse yet still fully analog academic environment. The most popular among students at the time at the AA was Peter Wilson's Diploma School unit. He seduced his students to draw Dutch windmills in great detail, showing the complex wooden beam structures and the workings of the milling process, and the students loved it. Other than Hadid, Wilson was deeply interested in the working of the constituent components. Yet, Wilson was nostalgically looking in the rear-view mirror, instead of investigating how things are constructed today. I taught Unit 12 in AA's Intermediate School with Andrew Herron, one of the sons of Archigram founder Ron Herron. Andrew and I took a train ride to Milton Keynes to successfully apply for funding from the British Intergraph branch, introducing computation as a driver for design strategies.

During an excursion that summer to Brazil, to see the achievements of Roberto Burle Marx, Oscar Niemeyer, and Lucio Costa, meetings were arranged with Costa and Niemeyer in person. Seated as school boys in a classroom at the building site of the Latin America Memorial in Sao Paulo, Niemeyer gave us an instructive lecture sketching out his featured works with a red felt-tip pen on large pieces of transparent sketch paper, and donated three of these signed sketches, dedicated "para los amigos holandeses," for the Dutch "comrades." At the AA, I organized a slideshow for students with images of my trip to Brazil. At the AA bar, Rem told me that Niemeyer is the domain of Zaha, implying that I should respect that and stay away from Niemeyer. I shared my Intermediate Unit with Dennis Crompton, aka "Dennis the Menace" as Zaha used to joke, as the technical tutor. Unfortunately, the delivery of the powerful Intergraph computers that were crucial to moving forward took more than half a year. As a result, I did not have sufficient time to unfold my educational plan to intuitively work with computers as our exo-brains. My idea to use computers as a design instrument to enhance the student's creative potential did not get off the ground.

In the pre-AA years, Ilona and I extensively experimented in our Rotterdam studio in an abandoned school building with intuitive digitally sketched components to form the genetic material for buildings that could be, at the same time, a piece of art and a piece of architecture. We use the computer as an instrument, an acting component in the design process, and a co-designer. A computer acts since it actively runs a program, which makes it a different kind of instrument than a piano that sits still and just responds to hammering the keys. Like pianos, computers have a keyboard, but the effect of hitting the keys of the computer sets in a digital calculation process is much more complex than the vibration of a string. While both the computer and the piano are extensions of the human body, the added value of computers is that they have a life of themselves, when turned on and connected to the Internet. Computers are the acting components in a global network of communication between things and people. Out of the potential of running programs inside the computer naturally emerges a new notion of what architecture is. Nonstandard architecture has emerged from computation.

Working intuitively by tweaking the parameters of simple 3D software, we anticipated future developments concerning parametric, robotic, interactive, proactive, and participatory architecture. The explicit purpose of the intuitive sketch is to surprise oneself, to free oneself from a linear chain of thoughts, from trodden paths. Intuitive sketching is empathetic with an open mindset, with the artistic procedures of hands-on thinking—working with base materials. When artists start working on a painting, especially an abstract painting, they do not know what the final result will be. They have chosen a direction but not a predefined endpoint. Abstract paintings are never really finished; the artist just stops painting when it says enough. Abstract paintings taught me that we need to operate the computer in a combined rational and intuitive way, using the computer as a designer rather than as a draftsman. We had to find a way out of working with computers as intended by the software engineers. The Autocad software is developed by and for draftsmen and engineers, not for designers. That explains why we are still stuck with lousy interfaces like the mouse, the keyboard, and the screen between the designer and the computer. And that also explains why so many talented designers who are intrigued by the limitless potential of computer logic find it hard to be a good draftsperson and a good designer at the same time. I have seen many talented students lose themselves in the technical aspects of the software and become less interested in the design itself. It also explains why in the long run many talented digital architects end up in other businesses than the building industry. Being aware of the trap of becoming absorbed by technical issues, we knew that we had to find ways to connect intuition to logic, to connect the file to the factory, the production to the assembly, and the design concepts to the computer scripts that eventually drive the robotic production of the constituent components. We had to discover the conceptual and procedural components that work for us.

Metawell Series

Working with visual artist Ilona Lénárd uncovered a way out of the otherwise beautiful cul-de-sac of minimalist architecture. By experimenting in the mid-eighties with our Atari 1040 ST computer, she uncovered a new freedom of expression, using a simple 3D program, Stad3D. The procedure for her Metawell series was as follows:

1) make spontaneous gestures with the mouse of the computer,
2) play with different parameters for the resolution,
3) materialize the sketches into something meaningful,
4) choose the outer contour of the work,
5) cut that contour out of a piece of Metawell corrugated aluminum,
6) make incisions to fold the plate into the third dimension,
7) choose colors and materials to paint the surfaces alternately with abstract gestural patterns and smooth surfaces,
8) decide what is up and what is down,
9) drill holes in the back, and then
10) finally decide how to position it on the wall.

Each choice in the process represents a strategic component in the decision network. Physical components (paint, plaster, and aluminum) and immaterial components (ideas, concepts, and choices) are interwoven to give shape to the procedure. A stack of decisions is made as in any script and in any procedure, whereas the initial concept is the open framework that allows one to fill in the details during the execution of the work.

Lénárd's "idiot savant" style of working, as she describes in her book Powerlines, is based on sketching without deliberate thinking, which is equivalent to dadaist automatic drawings and Coop Himmelblau's psychograms and somewhat comparable to the surrealist paranoiac method but not expressed as a surrealistic pseudo-realistic dreamscape. Intuitive sketching with the computer exists in an abstract environment that is, at the same time, concrete and verifiable. Lénárd refers to the idiot savant while she has direct access to her internal database of movements without consciously thinking about which gesture to choose, as has the idiot savant who can memorize a complete telephone book without further understanding. The computer sketch functions as an object trouvé, of which all spatial coordinates are precisely known and subsequently can be used for the execution of the work.

In a way that is similar to the boundless universes of Mondrian and Van Doesburg, the notion of a dynamic universe plays an important role in the recent painting works of Lénárd, notably the Tangle series and the Q series that are made in Qatar. Rules of play are the immaterial components that build up the universe of her procedural paintings, manifested by material components in the form of fast and spontaneous strokes with acrylic markers on large canvases. Brush strokes that are characteristic of a specific painting are repeated over and over again, each time slightly different, layer by layer uncovering a deep, personal universe. The viewer is invited to immerse oneself in the tangled web of abstract calligraphic components. A procedure is a series of actions that are sequentially related to each other. It is similar to scripting in digital architecture, writing the code line by line. A procedure sequentially strings together several components, addressing each of them individually, each of which is different in the spontaneous execution of the premeditated gesture. ONL's commissioned architectural work and Lénárd's autonomous artwork are interwoven through the notion of hybrid components interacting in sequential scripts, open frameworks that intentionally reserve space for surprise and serendipity, in a mix of intuitive and logical acts.

Based on the paradigm of the intuitive computer sketch, we organized a series of memorable events in the years after we were honored to live and work a full year in the Van Doesburg studio. We invented titles like Artificial Intuition (Gallery Aedes, Berlin and TU Delft, 1990), The Synthetic Dimension/ The Global Satellite (Museum De Zonnehof, Amersfoort 1991), Sculpture City (RAM Gallery, Rotterdam, 1994), The Genes of Architecture (Academies in Rotterdam, Berlin, Budapest, 1996), and ParaSITE (R96 Festivals Rotterdam, 1996). In the Artificial Intuition Workshop, we invited the participants, architects, artists, and landscape designers alike, to sketch freely in three dimensions using the bulky workstations generously provided by Intergraph, running Mechanical Engineering software. Back then, Galerie Aedes was located in Unter den Bogen, under the elevated railway tracks. Each time a train passed, the magnetic fields distorted the graphics on the Intergraph screens, fortunately swiftly recovering when they were gone. At the time in 1990, there was no architectural software available that could even come close to managing complex shapes. Intergraph's Mechanical Engineering software is developed to model complex three-dimensional trajectories of piping in oil refineries. Having seen this at the Intergraph sales office in Hoofddorp in the Netherlands, I immediately realized that intuitive 3D modeling would revolutionize the conceptual part of the design process. The program allowed us to sketch in 3D, to sketch intuitively, not as the result of a rational process, but spontaneously, energetically, and gestural. This was one of those breakthrough moments for conceptual digital design. A conceptual idea of the work procedure on the computer is an essential component in the design process. Components are not just pieces of hardware, while a component can perfectly well be a soft procedure.

The design process we were experimenting with resonated lightly with Salvatore Dali's <u>paranoiac-critical method,</u> in an abstract sense. Dali defined the paranoiac-critical method as irrational knowledge based on a delirium of interpretation. Different from Dali, however, and other than Koolhaas' fascination for the paranoiac-critical method as in his book Delirious New York, we reject a (sur)realistic depiction of the subconscious, and we opt for a non-referential abstraction of stereometric constructs. Whereas Dali depends on distorted images from a possible (in)organic reality, we use the intuitive computer sketch to trigger unprecedented imagination, to explore what could be a possible three-dimensional construct. In terms of verifiability, every construct in the virtual reality of the design software is equally realistic. We stated that once having modeled such seemingly impossible geometry in digital space, the data points can serve as the spatial coordinates for a real project with real dimensions and tangible material properties. Around the same time, Wolf Prix of Coop Himmelblau made their blindfolded <u>psychogram</u> <u>sketches</u>, serving a purpose similar to our intuitive computer sketches. The crucial difference from the Himmelblau method, however, is that we use the digital data directly to construct our designs, which we materialized in the real world, notably in the <u>Hydra</u> of the Saltwater Pavilion. The software allowed us to immerse ourselves into the sketched   components, navigate them, and embark on a journey inside these mesmerizing three-dimensional worlds. The advantage of a digital sketch over an analog sketch is that the digital geometry can be tweaked endlessly while maintaining its integrity and that the data can be used directly to build the constructs. We used to freely play with the variables to suit speculative interpretations, anticipating what later became known as associative design and parametric design.

High-resolution Complexity

In contrast to architects embracing low-res esthetics, my fascination for the container is not the visual resemblance. My explicit aim is to design buildings as a connected flock of interacting individual components.

My fascination is to design complex adaptive design systems, made possible by the digital technologies that became available as early as the eighties of the 20th century. In the beginning, we puddled through primitive 3D software and liked the parametric aspects, whereas, basically, all 3D software is parametric in itself as one can manipulate the geometry by changing values/parameters. Subsequently, our usage of the software evolved from Arkey for BRN Catering and City Fruitful, DesignCAD for art projects, Autocad for the Waste Transfer Station and housing projects, 3D Studio for Sculpture City, ParaSITE and Saltwater Pavilion, Revit for the Variomatic, Maya/3D Max for Web of North-Holland and the TT Monument, ProEngineer for the A2 Cockpit, Rhino in combination with Grasshopper for the Climbing Wall, and the parametric ceiling for Ecophon. Revit, ProEngineer, and Grasshopper are mature pieces of parametric software, which allow for complex relationships between different types of components. From the programming and scripting, we used Visual Basic for the Attractor Game/Reitdiep master plan, MAXscript, MAYAscript for the Venice Biennale 2000, and HTML for the interactive web application for the Variomatic house. AutoLISP routines are used to establish the connection from the file-to-factory process for the Saltwater Pavilion, the Web of North-Holland, and the A2 Cockpit. Grasshopper for Rhino meant a real game changer, because of the affordable price tag and the easy-to-use graphic interface. Grasshopper allowed us to develop complex participatory design systems, for the Climbing Wall, the Body Chair, and the Participator application.

Another not-so-user-friendly software of the eighties, the Master Architect/Microstation software of Intergraph eventually came with Robert Aish's brilliant, yet equally challenging-to-use associative software Generative Components (2003), marking the start of a new era where advanced software became the enabler of nonstandard architecture for established offices. At Hyperbody, we were among the first to recognize the importance of the Generative Components software. Hans Hubers of Hyperbody organized in 2005 a smart geometry workshop led by Robert Aish using generative components. Besides ONL's architects Sander Boer and Gijs Joosen,

Neri Oxman, the later professor at MIT's Media Lab, and Jeroen Coenders of White Lioness/Packhunt.io took part in the workshop. Around 1998, Gehry started to work with the top-of-market CATIA software by Dassault Systems, originally developed for aviation design and used it to control the complexity of the Guggenheim Museum Bilbao (1999) and the Walt Disney Concert Hall (2003) in Los Angeles. Gehry Technologies, led by Jim Glymph, designed their own software Digital Project on top of CATIA.

The same Jim Glymph who developed the Digital Project told me, after having seen my presentation of the Saltwater Pavilion in 1998 at the Transarchitectures conference organized by Odile Fillion in Monaco, that we Europeans are always one step ahead of Americans. Associative and parametric design software started to become a serious component in architectural thinking, enabling unheard forms of complexity. Gehry's complexity, however, is not the kind of complexity ONL and Hyperbody are interested in. At Gehry's, they start with making large paper models, then they digitize the models, and then use the digital data to post-model them in Autocad and parts of it in Digital Project. Gehry's complexity is a form of complicatedness; it is complex rules leading to complex outcomes. But the effect of Gehry using CATIA to solve his self-imposed geometrical complications gave an enormous boost to the profession. It needed big projects like the Walt Disney Concert Hall and the Guggenheim Museum Bilbao to reach the broader public. Suddenly everything seemed to be possible. Architecture Goes Wild (2002) is the telling title of the compilation of my theoretical essays from the previous century. Complicatedness can be technically solved by the likes of Arup and Buro Happold, but it is not the same as building smart. Building smart requires an integer process from scratch, from the conceptual idea to the definition of the high-resolution complexity of the components, from the file to the factory, establishing a direct and hence cost-effective link between the data of the designer and the capacities of the production machines to produce the constituent components. In much of the work of Prix, Gehry, and Hadid, the contractor relies on traditional labor-intensive work, especially when it comes to double-curved roof structures and interior finishing. The sweeping roof of the Heydar Aliyev Center in Baku by ZHA is built up from several layers on top of the main structure, and these layers have a complicated relationship with each other. On top of the main load-bearing MERO space frame structure, which is in itself a simple system and well fit for doubly curved shapes, is a secondary structure of different nature to hold the cladding panels; two completely different systems, one for the structure and one for the skin, whose doubling introduces unnecessary extra costs in the design phase, the engineering phase, and the execution phase. In the chapter "From Art to Architecture," I demonstrate how ONL's featured projects have synchronized the structure and the interior and the exterior skin into one coherent system by completely avoiding autonomous secondary structures. Hadid and Gehry often choose traditional methods for the interior finishing, mostly multilayered plastered gypsum board, out of sync with the main structure. In several Gehry or Hadid projects that I have visited, thick layers of plasterwork were needed to get to the desired smooth finishing for the curved interior walls covering the main structure. The real potential of complex geometry must be based on simple rules that create complexity, i.e., simple rules that include the connections between the components of the structure and the skin. While OMA's simple rules generate simple geometries, Gehry's complex rules generate complicatedness, ONL's simple rules generate an affordable form of complexity, whereby structure, interior, and exterior skin form one coherent relational system.

The unfortunate asynchronization between structure and skin is, to some extent, due to the traditional building chain, where contractors are made responsible for the production details. The traditional linear building chain, from the design to building approval and from the tender documentation to the selection of the contractor, is counterproductive since there is little or no opportunity for design feedback from the execution back to the design. Such asynchronous relationships and the lack of design feedback are two of the major reasons for unnecessary high construction costs. The alternative is to form a design & build consortium from scratch, whereby the designer and the manufacturer, not the contractor who is mostly a managing intermediary, are in a constant dialog. ONL's Web of North-Holland pavilion and the A2 Cockpit are designed and executed in a design & build consortium with the steel manufacturer Meijers Staalbouw, who formally acted as the main contractor. Only by teaming up with the main manufacturer could we benefit from the principles of parametric design to robotic production.

Saltwater Pavilion

The Waterpavilion project was the international breakthrough of both NOX architects (Lars Spuybroek, Maurice Nio) and ONL (Oosterhuis_Lénárd). Ashok Bhalotra (1935–2022) of Kuiper Compagnons in Rotterdam invited two offices to design one Waterpavilion. We were challenged to find out in the preliminary design process how to work together. It turned out to be an exhausting process with many twists and turns. After some exchange of ideas, we agreed on the concept to have a freshwater and a saltwater part, thus representing the water cycle from clouds to rain to river to sea to evaporation into freshwater clouds and to rain again, in an endless cycle. Based on my early digital sketches of a smooth rounded body with themed, immersive interior experiences, I was logically assigned the saltwater part, the part that is facing the inland sea. NOX's early colored pencil drawings showed building fragments loosely embedded in the dunes, circling a dark deep well filled with water. The proximity to the sea offered us the perfect opportunity to design a sculpture that could function as a building, as we envisaged in the Sculpture City project that we developed just some months before – a sculpture that was seaborne and stranded on the rocky shores of the inland bay area on the artificial island of Neeltje Jans.

The collaboration with NOX was a tough competitive one. Besides two designers, there was a two-headed client: Rijkswaterstaat with project leader Frans Walraven on the one hand, responsible to manage the interior storyline, and Frans van Spaandonk, director of the Neeltje Jans Expo, the manager for the building, on the other hand. From the start, the design process was one of general excitement, of big expectations, stuffed with ominous tensions. The solution that I at

61

a given moment, suggested was to connect the two themed parts, freshwater and saltwater, by one single elliptical section, by the most minimal connection possible – a simple rule that allowed for complexity, like the Apollo-Soyuz connection. In many ways, we worked together as the Americans and the Russians, speaking a different (architectural) language, cultured by different obsessions, yet sharing the fascination to create something unique. This practical solution of the minimal docking connection gave both of us the freedom to efficiently do our thing and yet make one building. We found a common "adversary" in the Rijkswaterstaat selected interior designer René van Raalte of the company Eden from Amsterdam. While we were fully dedicated to making a statement of fluidity, the interior designer proposed unfitting modernist orthogonal display cases of glass and steel, not even trying to match our fluid designs. At a certain moment, the overheated discussions became so intense that I developed acute appendicitis and had to be hospitalized the same day. The client decided to appoint another more soft-spoken designer from the same company, who proposed zoomorphic objects that none of us appreciated, and he missed the mark. Half a year before the scheduled opening, there still wasn't an acceptable vision of the interior experience. The client was desperate. To get out of the impasse, I proposed that we as the architects would design a basic, interactive atmospheric layer of the interior experience ourselves, which is meant as a background for more explicit information on the water cycle to be added later. The idea was reluctantly agreed upon. The realization of the Waterpavilion and especially of the interactive programming for the abstract water experience was a hard-fought achievement, which eventually catapulted us onto the international podium.

While we, from the beginning, modeled and visualized the concept design in 3D Studio, NOX relied heavily on color pencil sketches. Their fascination for digging into the sand, for the grace of water-sculpted dark caves and windows to the underworld, contrasted starkly with my fascination for vectorial bodies, weightless space, and spaceships to eventually land on the artificial island of Neeltje Jans. NOX, meaning night in Latin, emerges from the dark caves, while ONL lands down from space. We build spaceships that seek communication with the local environment, to eventually make a precision landing. Only later in the process, NOX hired a 3D modeler architect, a good friend of ONL architect Menno Rubbens (the present director of the Cepezed Projects). It gave their design a radical twist, and their models began to look more like our own initial 3D sketches that featured a smooth twisted volume. Halfway through the project, Spuybroek and Nio decided to split ways, while Spuybroek emerged as the sole owner of NOX. After the opening of the project in 1997, Spuybroek avoided mentioning in the numerous international publications that two parts together form one whole Waterpavilion. The book on the design process of the Waterpavilion as a whole that ACTAR wanted to publish never happened. The intention of Bhalotra to invite two architects to design one building was meant to be a sort of two-component glue. Bringing the two components together, a strong chemical reaction takes place. Navigating the interior from the undulating NOX part into the lemniscate loop inside the ONL part, many people, except probably architects and architecture critics, experienced it as one continuous flow.

In every possible way, we ended up designing opposites. While NOX designed a roof, ONL designed a body. While ONL chose black industrial polyurethane coating, NOX took silver-coiled bitumen. While ONL wanted the texture of the rough black skin like a moonscape, NOX had a shiny skin glittering between the artificial dunes. While NOX opted for a curved biomorphic shape, ONL developed a stretched vectorial body with rotational symmetry, more like a synthetic vehicle than an organic form. While ONL has the feature lines along the full length of the volume, NOX emphasizes the segments, like the scales of crustaceans. While ONL has an internal endless loop in the form of a 3D lemniscate, NOX wanted an unfathomable deep water well. While ONL integrated the climatic installation in the twisted volume that separates the Wetlab and the Sensorium, NOX placed the control room and the installation for the air conditioning outside their part of the building. While NOX chose monochromatic color schemes and projections, ONL opted for full-color glass fibers along the 3D curves defining the contours of the body, and full-color virtual reality projections onto the semi-transparent interior skin.

The textured black skin that wraps the Saltwater Pavilion surprisingly shares features with an early small project by MVRDV, notably the

63

Hoenderloo lodge (1996). MVRDV wrapped the small ticket box in one material, rusted steel, thus emphasizing the object as a whole, at the expense of expressing the traditional elements like roofs, walls, doors, canopies, gutters, etc., separately, but otherwise shaped as a traditional skewed shed. The Saltwater Pavilion takes the body-ness of the building further, as a streamlined black sculpture with rough skin. Despite all the magnified differences, architects act and live in the same era, and in hindsight, it will show that they probably have more in common than they like to acknowledge.

Both parts of the Waterpavilion are built by the same contractor and by the same steel construction company. The practical experience and insights of Henk Meijers of Meijers Staalbouw, the son of a local blacksmith, were indispensable for the cost-effective solution to be found for the execution of our designs. I was intrigued by the capacities of the CNC-driven cutting machines that Meijers had in his factory. It was decided to make maximum use of the machines. If it were not for Meijers, we would have not discovered the uncompromising beauty of the digital file-to-factory process as early as the mid-nineties. Meijers was happy to finally find an architect who was interested in linking the data of the designer to the machines of the manufacturer. He told us that he had bought the cutting machines capable of being programmed by routines already 15 years before but had not found an architect who was interested in synchronizing the design process with the capacities of the machines. Architects kept designing the traditional way, ignorant of a possible direct connection between the digital data of the design and the digital data driving the execution. Even now, a quarter of a century years after completing the Waterpavilion, most architects persist in traditional methods of design.

ONL developed in-house AutoLISP routines to share the data directly with the algorithms driving the machines, and we shared that knowledge with our NOX designers of the Freshwater Pavilion. The redefinition of what are the constituent components originated from the design-to-build process of the Saltwater Pavilion. Although not formalized as such, the Saltwater Pavilion marked the start of a *de facto* design & build relationship with the steel construction company, Meijers Staalbouw. To connect the design data to the machines, the architect has to be strict in the definition of the components, be painfully precise in the exact dimensions and mutual angles between the components, and anticipate how the components are connected. We learned that in the design-to-production process, the architect must take full responsibility for the exactness of the data. Only by scripting the 3D curvature of the structural Hydra loop could be defined as a manageable series of automated bending and cutting operations on tubular steels. By another script procedure, ONL's Menno Rubbens defined the wooden sandwich components describing the smooth fading in and out of the exterior fins. Only by a scripting-to-execution procedure could the mutual angles and shifts in position for the details that change from section to section be developed. The relation between the main structure and the skin must be incorporated in one comprehensive parametric detail. The structure of the Saltwater Pavilion is still traditionally based on vertical sections at a regular distance, but during the process, we learned that for future projects, a new strategy had to be adopted. The volume of the complex geometry body should be wrapped in a three-dimensional mesh, and the components should be directly linked to the topology of the structure and the skin. The Saltwater Pavilion represents the transition phase between traditional building conventions of plans and sections and topological nonstandard architecture. In the projects after the Saltwater Pavilion, we used the newly acquired insights to start from scratch with an uncompromising parametric design to robotic production process.

From then on, I realized that there is no progress possible in architecture without a bi-directional partnership with the manufacturing parties. Architects must come out of their autonomous designer bubble and team up with other disciplines to form multidisciplinary teams, ultimately design, build, finance, and operate (DBFO) teams, sharing full responsibility for all aspects of the design & construct process, which is similar to what artists do on a smaller scale. The artist couple Christo and Jeanne-Claude has *de facto* managed DBFO on a larger scale. Their Running Fence (1976) project and other projects that I have seen in real life, such as the Pont Neuf (1985) and the Reichstag (1995), have been a major source of inspiration for my graduation project, the Strook door Nederland. The influence of the procedure of wrapping a building in one material on our architectural projects cannot be underestimated. I consider Christo's team, which initiates, manages, and executes the whole operation, a useful template for architectural design teams. The crux is to take full responsibility for one's participation in a project, from design to execution, and get rewarded with one's proportional share in the profits – or losses – as well. On a smaller scale, young architects who embrace parametric design to robotic production naturally work this way; they typically manage both the design and the in-house production. The much-needed scaling up to substantially larger projects has only just started.

One Building One Detail

Since my graduation project, I have been intrigued by small private houses in a lush green environment, like summer residences away from the city. For most of our life, Ilona and I have lived, and still live, in such pleasant green environments. At ONL I developed numerous housing schemes for compact detached homes, but we have not succeeded to realize any of them yet. In my early years as a freelance architect, I was in charge of making the production drawings of the so-called Stolpwoning, the bell jar house, for Peter Gerssen. The shape of the Stolpwoning is a cube cut in half over the diagonal and placed on the ground on the hexagonal floor plan. Drawing the plans and sections was an extremely complex task since all dimensions had to be calculated without computers, and using angular arithmetic, nothing was straightforward orthogonal. The tetrahedral pyramid roof of the Stolpwoning stands at an angle of 54.74º to the ground floor. All dimensions needed two digits after the comma to be sure not to accumulate deviations. Drawing the Stolpwoning on a drawing board proved to be a serious challenge, and drawing the constituent components even more so. The first batch of the Stolpwoning is an assembly of wooden sandwich panels. It was my task to draw each panel in its exact measurements, ready for production. It was by doing the complex calculations that I developed my taste for triangulated structures, my appetite to work with computation, and to rely on

66

computation. The concept of the Stolpwoning itself is as simple as a cube balancing on its tip and then cut in half, but building it vertically from the hexagonal ground floor up meant having to deal with all the angles. The Stolpwoning also was an extremely energy-efficient home, of which eventually some hundred were built, using varying building techniques. Originally designed in prefab wooden panels, the most interesting version is built up from white steel sandwich panels with an extra-strong XPS foam body, which is produced by the cooling box builder HEIWO from Wolvega in Friesland. XPS foam sandwich panels are normally used for cooling boxes on large trucks. From Gerssen, I learned to look at manufacturers from other industries than the traditional building industry. The details are extremely simple; floors, roof, and interior walls were all of the same thickness, i.e., 10-cm thick sandwich panel in the same color, riveted together with a folded 1-mm thick steel plate along the length of all connections. It was, at the time, working on Stolpwoning that the concept of the one detail for one building was born. At ONL, we radicalized the principle of one material that is both structure and skin into our slogan: "one building one detail." The paradigm of the single comprehensive detail developed into the parametric design to robotic production of a series of unique elements subsequently applied for the skin of the Saltwater Pavilion (1998), the structure and the skin of the Web of North-Holland (2002), the structure and the skin of the A2 Cockpit (2005), the facades of Fside Housing (2007), the structure and the skin of the Bálna Budapest (2013), and the structure and the facade of the Liwa Tower (2014).

In 2001, ONL was selected to design the pavilion for the 2002 Floriade international horticultural exhibition in the Haarlemmermeerpolder, invited by Titus Yocarini, the cultural ambassador of the Province of North-Holland. The design process of the North-Holland Pavilion went through several challenging stages; the continuation of the project was, at a certain moment, hanging by a thread, but eventually, it ended up as a most radical design. The first concept design, North-Holland Pavilion, is based on the proven shipbuilding technique of welding prefabricated plates of steel together to form one unified body. Version 1 was unfortunately rejected, while the external project manager from

DHV would not believe that it could be realized within the province's tight budget of 1 million EUR for the structure of the pavilion, and we could not get a reliable quote either on such short notice.

We turned to shipbuilding techniques before, since shipbuilding fits so well with the notion of a body, one material, and one technique that structures the whole body, from bottom to top, from tail to nose. Ships, airplanes, and cars have bodies; I want buildings to have bodies too. Five years before the Web of North-Holland pavilion, in one of the intermediate proposals for the Saltwater Pavilion, I proposed to turn to shipbuilders to build the body, which would have been a logical choice since the shape of the Saltwater Pavilion is a fluid form that narrows at the front and the back, pumping up the volume in the middle, like streamlined vessels. The cost estimates that we received from shipbuilders were not encouraging though; the price tag was one and

a half times the budget. A second attempt to find the most logical method of constructing the building body was to use prefabricated concrete elements to assemble the body, enthusiastically supported by the structural engineer Jan van der Windt van Zonneveld Ingenieurs, but this solution was estimated similarly high by prefab concrete building companies. Today, when we would ask for an offer from Central Industries Group (CIG), the shipbuilding components producer with whom we have worked some years later, the price would be right. ONL introduced CIG-architecture, at that time operating under the name of Centraalstaal, into the realm of architecture through our project Fside housing (2005) and has worked with them since, for example, for the BMW Ekris showroom (2006) in Utrecht and the Bálna Budapest (2012). Fside housing CIG has successfully entered the architectural market and completed complex double-curved skins for Asymptote's link bridge (2009) for the YAS hotel in Abu Dhabi, Isozaki's sidra trees (2009) for the QNCC in Doha, Henn's Porsche Pavilion (2012) in Wolfsburg, and ZHA's billboard (2018) in London, and many other complex geometry projects for artists, mainly Frank Stella, and other renowned architects. Shipbuilding technique in architecture was unfortunately not yet a feasible choice back in 2001 for integrated structural skin structures as the North-Holland Pavilion version 1.0, and I had to find a strategy to secure the project. To keep the process alive, I proposed a shiny flattened regular ellipsoid as the main volume, to contain an interactive arena with a 360º projection, as in our Trans_ Ports installation for the Biennale 2000. I reused the aluminum ellipsoid from a model of the earlier competition project for the European Patent Center (1989). That simplified proposal hit the right chord to get things going again, but I knew already that I would design something more exciting than just a regular ellipsoid. Being reassured that the project would become manageable, the client went along with us.

I kneaded the multiple-symmetric ellipsoid into an asymmetrically shaped elliptical solid with explicit dents and a fold line at the side of the entrance, and a boldly cantilevering end at the other end. First, I made a physical foam model, by intuitively cutting and sanding the recessions and sharpening the edges into folding lines until I was satisfied with the interesting shape. Then, we modeled the shape in Maya using the data points from the 3D digitizer, which was similar to how Gehry operates digitizing their distorted carton surface models. Finding the proper way to construct the resulting complex shape of the North-Holland Pavilion version 2.0, renamed the Web of North-Holland, meant another challenge. The resulting model features convex parts that are gradually transforming into concave parts and are even more complex by the folding lines that fade in and fade out. When we had the model ready, I asked my TU Delft colleague Mick Eekhout of Octatube to prepare an offer for the load-bearing structure and the skin. After some weeks, Eekhout, assisted by our long-time friend Karel Vollers, known for his twisted glass facades and his Ph.D. research Twist & Build (2001), presented their vision. Instead of having found a feasible solution, a second bump in the design process was created. Eekhout/Vollers stated that our design was geometrically not possible without the very expensive technique of explosion forming the skin. They presented seven alternatives, each of them missing the point of our design, basically promoting Octatube's proprietary space frame system instead. Their non-solutions infuriated me, and I had to kindly show them the door. Their proposals undermined the integrity of our design, effectively questioning our expertise in front of the client. I requested one week to prove that we were right and that the structure plus skin as we had designed could be constructed. I called in the help of Henk Meijers and made a full-scale one-to-one mock-up of a representative complex building component in our studio workshop at the Essenburgsingel 94c in Rotterdam. We modeled the structure and the skin in detail, based on the idea of the synchronization of structure and skin. For the skin, we found that the 2-mm composite Hylite composite aluminum is so flexible and impact-resistant that we could use it by gently cold forming into place, as a low-budget alternative to the Octatube proposed an expensive technique of the process of explosion forming, which they used for the rounded corners of Asymptote's Hydra Pier pavilion at the 2002 Floriade. Indeed, the geometry of the skin that we proposed may be theoretically not perfect, but the visual and spatial effect was excellent and beyond expectation. The cold forming has a charm of its own, something we would never have reached when following theoretical strictness. Meijers immediately understood the logic of our proposal and offered a price that fitted nicely within the tight budget, and the project was saved!

But the struggle to get things done was not over yet. The next challenge was to calculate the main load-bearing structure, the complex but not complicated steel structure, which is a flattened and rotated tweak of six frequently triangulated dodecahedrons. The iWEB structure was needed to be calculated by the finite element method (FEM) coming from a car design, while traditional static calculation methods for regular buildings are unfit for complex geometries. We invited engineer Daniel Bosia from Arup to have a look at it, but we had to abandon that option because of the tight budget. I then asked John Kraus of D3BN (now RoyalHaskoningDHV), with whom we worked for BRN Catering, to take on the challenge. He knew that he could not do it using his regular software for static calculations, but, apparently, they just acquired the TNO Delft-developed finite element analysis software, Diana. It took them three weeks of heavy computing usage to get the calculation done, and it appeared that they could only manage to come up with a reliable calculation when we would twist the dodecahedron in such a way that the main forces could contribute vertically positioned ribs, instead of following the internal logic of the 3D mesh. Against our idea of the optimum network structure, we did the twist because we had no choice – time was running out, and the production needed to start soon. After taking the three big hurdles, we came to the final design of what we now call the Web of Noord-Holland. The Web of North-Holland is the most radical design we had done until then, yet still a compromise with respect to the ideal triangulated network. The data from us and D3BN were imported by the digital modeler and production planner Aart Voskamp of Meijers Staalbouw, a down-to-earth genius who scripted the execution details of the steel structure, calculated the dimensions and positions of the connecting plates and bolts, and wrote the algorithm to drive the CNC machines. A second compromise, which I still regret, is that we had to abandon the concept of the triangulation wrapped all around the volume inclusive of the floor. The steel construction company aka the main contractor insisted that we execute the floor in concrete. Fair enough, I gave in, because if it were not for Meijers Staalbouw, the Web of North-Holland would not have been built in this form.

We had worked with Meijers Staalbouw before when designing and building the Saltwater Pavilion, and we concluded back in 1997 that the next step after 3D modeling should be an uncompromising digital file-to-factory procedure based on parametrically designed components. A good parametric file-to-factory process encompasses all possible positions on the topology of the continuous surface of the volume, with no exceptions to the rule. After having executed the Saltwater Pavilion in a *de facto* design & build team, we knew that new systemic design strategies must be all-inclusive strategies. What is needed is a generic project-specific approach toward the connections between the structure and the skin and between the design and the manufacturing. Each project demands a project-specific systemic parametric framework that is based on the components integrating the structure, the skin, and their mutual connections.

For previous projects like the Saltwater Pavilion, for this project the Web of Noord-Holland, and for future projects like the A2 Hessing Cockpit and the Bálna Budapest, I have always remained faithful to the budget in the design & build team, both for our internal costs and to the expenses of others. Staying within budget has inspired us to become smarter and more efficient. I strongly believe that this is exactly why we were able to successfully develop many project-specific instances of the parametric design to a robotic production method. It was the only way to build the nonstandard, whereas others were doomed to turn to traditional building methods or to stay behind in the unbuilt virtual reality fantasies or indoor installations. None of my peers from Nonstandard Architecture or Transarchitectures did realize a complete building following the rules of parametric design to robotic production. The financial and technical hurdles that we had to take have led to the invention of the design-to-production concept of a single complex component that comprises all possible parts of the whole building, without exception: one building one detail (1B1D). The Web of North-Holland became the landmark project for the revolutionary 1B1D paradigm. The complete pavilion features just one complex detail, mapped on the nodes that populate the surface. The single detail includes the exact geometric description of the structural parts, the skin parts, and the connecting components. The detail includes the exact positions of the holes for the bolts. The 1B1D design strategy integrates structure and skin into one coherent relational system.

A2 Cockpit in Acoustic Barrier

In 1998, I was selected by Nora Hugenholz of the Projectbureau Leidsche Rijn to develop a proposal for a building along the A2 highway. The design task was to communicate the presence of an industrial area behind a 5-m high acoustic earth wall. We proposed to embed a whole building into the acoustic dike. My proposal was positively embraced as an innovative and sustainable solution. It would build a showcase for a company with a front to the busiest highway in the Netherlands. And it would function as a shopping window for the industrial area behind the earthen wall. The concept of the glazed cockpit embedded in the green acoustic dike structure was made part of the master zoning plan that was at that time in the making. It was not before 2004 that the project took off, after some big changes in the requirements for the acoustic performance. Following the new rules of law, the acoustic structure now had to be a whopping 12-m high instead of 5 m. We proposed to add a 7-m high and 1.6-km long structure on top of the earth wall and designed the building in the middle of the Acoustic Barrier in the form of a long stretched glazed cockpit.

Frits Hessing, the owner of Hessing in Bilthoven, was looking for a better location along the A2 highway for his luxury car showroom. Hessing, the once legendary Rolls Royce, and Bentley dealer had acquired dealerships of Lamborghini and Maserati as well. Hessing found out about the unique opportunity to build along the A2 highway and thus create maximum visibility for his business. As a young boy, I planted my nose against the large windows of the Hessing showroom in Bilthoven, where my older brother Peter some 30 years before had been working as a mechanic aka electrician expert. My brother was granted the pleasure to drive the Rolls Royce cars back to their owners when fixed. Hessing trusted my brother that he would not go joyriding. From our first meeting, Hessing was playing tough. After having invited us to sit down in the comfortable armchairs and having served a whiskey, he pointed at the ceramic pistol on the coffee table and told us that if we ever imagined cheating on him, we would have to face a hard time with him. I believe his remark was directed at the project manager Hans Altink, but the tone was set for an adventurous design process.

The fluid design with long elastic lines and smooth transitions in the surfaces from concave to convex featuring fold-lines fading in and out self-challenged us to come up with a unique parametrically designed structure to control the thousands of different components of steel and glass. We carefully looked into a possible generic simplicity of the system, to have simple rules for building the smooth complexity of the Acoustic Barrier and the Cockpit. At that time in 2005, not a single building in the world had been designed, not on this scale and with this level of consistency, according to an uncompromising parametric design to the CNC production process. The Cockpit and, to an even greater extent, the Acoustic Barrier are radical examples of the new paradigm of a component-based, fully scripted architecture, producing the exact data needed for the CNC production of the constituent components instead of 3D models or drawings.

Not Market Conform

Shockingly, it was the last architecture commission that we received in the Netherlands up till today. I am still puzzled why the Acoustic Barrier/Cockpit project, well appreciated by the general public and by my international peers, so extensively published, internationally and nationally, greatly appreciated by Hessing, the client of the A2 Cockpit, by the Projectbureau Leidsche Rijn, the client of the Acoustic Barrier, well appreciated by virtually everyone who passes by on the A2 highway, has not led to follow-up commissions. At the Hessing Cockpit's opening party, the famous Dutch actor Huub Stapel predicted a bright future for ONL; he predicted that it would bring in dozens of commissions and tons of money as what happened to Roberto Meyers and Jeroen van Schooten after they completed the ING House in Amsterdam. In the Netherlands, no new projects were coming in. The 2008 financial crisis certainly has played a role, but that is only part of the story. The building costs for the A2 Cockpit luxury car showroom, with the apparent grandeur of an airport terminal, were extremely competitive, a mere 750 EUR per m^2 all-inclusive. When we submitted the credits and project data of the A2 Cockpit for an authoritative yearbook of building costs, we mentioned the actual building costs. When we saw the yearbook printed, the editorial team had doubled the costs; they mentioned 1500 EUR per m^2, simply because 1500 EUR per m^2 was considered the market-conform price for a building with such a luxurious and complex appearance. The damage was done. No one in the traditional building industry was ready to believe that the Cockpit showroom had been built for such – in their view – unlikely low price. Before Meijers Staalbouw became involved, Hessing's project manager had invited an Amsterdam-based building contractor with whom he had worked before to offer a price. Their price was, as expected, almost twice the budget. They added 20% for the shape, another 20% for the triangulation, and another 20% for the unforeseen, demonstrating their ignorance of complex geometries and new CNC building methods. We turned to Meijers Staalbouw again and managed to fix it within budget, again in a design & build consortium, with Meijer Staalbouw as the main contractor.

Rationalizing the lack of follow-up commissions, I believe that such a complex high-profile architecture simply is not supposed to be that cost-effective. By being so keen on staying within the budget, and by finding innovative solutions to get there, it seems that we effectively doomed ourselves, that we stood ourselves in the way of breaking through. Otherwise respected quantity surveyors who were previously involved had estimated the Cockpit and the Acoustic Barrier to be at least double the cost. In their uninformed view, such a low non-market-conform price is not allowed to be possible; they have to protect their consultancy businesses. According to their consultancy standards, derived from traditional methods of design and construction, a complex and luxury visual appearance simply has to have a high price tag.

Outside the Netherlands, fortunately, new clients found us. The A2 Cockpit did get a lot of international attention, was published widely from the East to the West, and triggered the attention of the Hungarian project developer Imre Márton of Porto Investments kft. Márton was advised by Iván András Bojár, at-the-time director of the Hungarian design magazine Octogon. Iván Bojár had interviewed me some years before for his bilingual magazine. Together with Iván Bojár, Imre Márton drove up nonstop to the Netherlands in his black Audi A8 with a German license plate. Exhausted after a 12-hour drive, they arrived at our ONL office, and from there, we went to visit the A2 Cockpit. They were deeply impressed by the imposing presence, the spaciousness of the showroom, the innovative performance of the structural skin, and the overall cost-effectiveness of the project. For them, the price tag sounded ok while building costs in Hungary were at that time half of the Dutch average building costs. Soon after their visit, we were invited by Porto to join them in a competition for a cultural center along the river Danube in Budapest. In 2007, we convincingly won the competition and established our Hungarian office ONL Hungary kft. in Budapest, where we were forced to develop the design within an even more strict budget than we were used to in the Netherlands. Despite numerous cultural and financial hurdles, we managed to complete the building approval drawings set and the building approval within one year.

Killing

We proposed similar component-based details as in the Cockpit building to build the atrium roof and to wrap the body at the front of the Bálna Budapest, at the time called the CET, Central Europe Time and also meaning "Whale" in Hungarian. We developed the building permit documents, submitted them in 40 boxes with 40 copies, were extensively praised for the design, and were given the building approval instantly. Submitting the documents for building approval is, in Hungary, accompanied by an academic protocol where the submitter has a promoter and an opponent, as in the Ph.D. thesis defense, presenting for the building approval committee. Architect Judit Z. Halmágyi, who at the time was the office manager at Erick van Egeraat's Budapest office, was asked to be the opponent. She was praising the design extensively but was disappointed to see that the escalators, which were fashionably randomly placed in the preliminary design, now were symmetrically positioned along the length of the atrium space. But I was happy that we skipped an accidental deconstructivist element from the design.

In the follow-up phase for the working drawings, the project developer insisted on welding the steel structure instead of the proposed assembly of components, as in the Cockpit. During the last meeting I had with the developer in their apartment office in the Buda side of Budapest, a package was delivered to the room. The package was addressed to Halmágyi who joined the meeting as an architectural advisor for the developer. Halmágyi unpacked the box and showed a big sharp decorated knife, supposedly from her grandfather. I was kind of shocked to see the knife on the table – what sort of message did Halmágyi try to convey? The atmosphere at the small meeting table became even tenser when she started an argument, for which she was most likely invited into the meeting, that we as ONL should hire at least 30 people for the execution phase, as they did at Van Egeraat's office. We had a compact team of five dedicated people in our ONL Hungary kft. office on the project, and another three in our ONL office in Rotterdam, and we were performing well. Besides, the scanty fee that we had to agree upon in the contract did not allow us to hire more people in the current team. Contractually, the next phase of the project, i.e., the construction drawings, was to be commissioned after the successful completion of the previous phase. The developer Márton requested that we hire more people to get the order for the next phase. We explained how we work parametrically and that more people would not make the work any easier, and that traditional welding of the structure would seriously compromise our design. At a certain moment, Márton became very nervous and aggressive and took it out on our project architect Marthijn Pool who made notes on his laptop. With force, Márton closed down Pool's laptop. Things escalated, I banged my fist on the table saying "OK, this is it," walked away, and decided to refrain from further commitments to the project; there was no chance to further build the CET according to our logic, and I did not want to spend years to be subordinate ourselves to an ineffective traditional form of execution, as the selected main contractor requested. I told them that welding the steel structure kills the integrity of the project and that it will cause problems in the precision of the triangulated structure, and hence guarantee a waterproof skin. We had seen the geometrical and waterproofing problems before, in the welded triangulated structure of the MyZeil shopping mall in the PalaisQuartier in Frankfurt, designed by Massimiliano Fuksas. The overall shape and layout as documented in the approved building permit set of the CET/Bálna were to be respected, but the details of the steel and skin structure were changed. The welding corrupted the integrity of the file-to-factory design system. The contractor's/project developer's coup killed the effectiveness, the internal logic, and the performance of the component-based design-to-production system. The costs of the system that we proposed, and the price that we were offered by CIG-Architecture, would have fit nicely within the tight budget. But that was not what the project developer wanted. Márton refused to work with us in a design & build consortium with Centraalstaal. The obvious strategy of developers and contractors alike is to create gray zones for negotiations with subcontractors. Precision, transparency, and integrity as in parametric design to robotic production are not exactly favored in the traditional building industry. In the end, I was offered a sort of a supervising role for the visual appearance only but was not allowed to enter the building site. Years later, after lots of complications in the execution, which I could only observe from a distance, by and large, caused by the traditional way of welding the steel structure, we could only visit the Bálna after the official opening. Their unfortunate choice of a 2D cladding system from Schüco for the double-curved skin inevitably led to a permanent leaking of the skin.

Second Life

Today, the systemic approach that is based on a dry assembly of prefabricated components is instrumental in what is known as the circular economy. Component-based design addresses the two main pillars of the circular economy. Parametric design to robotic production of components reduces waste and allows for a second use. What is typically undervalued by proponents of the circular economy is the role of data-driven design and production. In circular building systems, the constituting components must be identifiable; they are the known knowns. The origin of the materials, the impact on the environment, the performance, and the capacity to be exchanged, taken apart and to be reused, all must be quantified and qualified. The file-to-factory design to production and dry assembly system complies effortlessly with the criteria of circularity. I remain critical, though, of the word circular. Nothing is circular in nature as we know it, and nothing is circular in product life. The word circular implies that the used materials can return to the same initial condition. This can never be the case since the world is a dynamic adaptive system. After one cycle, everything around the product has changed; the components can only be reused in a new setting, a new context, to form part of a new system, in which societal values have changed. An alternative term for circularity would be the cyclic economy. Cyclic as in the water cycle, clouds that generate rain, rains that feed rivers, rivers that fill oceans, and oceans that evaporate into clouds that condensate into rain again. Everything in nature somehow evolves cyclically; nothing comes back to where it started.

The iWEB at the TU Delft meant the second life of the Web of North-Holland, but it was not an exact re-assembly. The Web of North-Holland was taken apart into its constituting parts after the Floriade 2002 show and temporarily stored for possible future reuse at a parcel of unused pasture near the factory of the steel construction company Meijers Staalbouw. Hans Beunderman, at the time the Dean of the Faculty of Architecture of the TU Delft, had shown interest to take over the pavilion for the symbolic value of 1 EUR. It would function as Hyperbody's experimentation lab, for which Beunderman successfully acquired a budget. We formed a design & build team to rethink, redesign and rebuild the structure – cyclic, not circular. While the Web of North-Holland is designed to be a temporary summer pavilion, the iWEB had to be prepared for cold winters. The iWEB is designed to become Hyperbody's protoSPACE 3.0 lab. Hyperbody's experimentation labs protoSPACE 1.0 and 2.0 were located inside the faculty building. Insulation was added to the inside of the structure to maintain the exterior visual openness of the synchronized steel structure with the super lightweight aluminum Hylite skin. File to factory produced by Polyned, a seamless bright red PVC interior skin – basically a balloon – was put under pressure and pushed against the inside of the steel structure. The balloon was sprayed from the inside with polyurethane foam to get the required 10-cm insulation. The triangular pattern of the steel structure was kept visible, with the visual attractiveness of an embossed surface. The original pentagonal single-layer PVC roof was replaced by a double-layered inflatable PVC cushion and was kept under continuous pressure by a ventilator. The details to connect the steel structure with the skin were redesigned, more robust, and esthetically establishing a stronger coherence between structure and skin. All the original steel and aluminum components are reused. The Web of North-Holland evolved into the iWEB, with a new function, improved performance, and with an innovative energy-efficient climatic installation. I never promoted ONL's projects as sustainable because the term sustainability is too vague; it can contain all sorts of conflicting design measures. At one end, there is parametric design to robotic production, and at the other end, there are green roofs, buildings with mud, and other ready-to-use natural materials. It is disturbing to find two radically opposing design strategies under one umbrella. A better word for sustainable would be lean. Lean implies doing as much as possible with the smallest number of resources, being effective, establishing direct links, and finding the shortest route to the desired destination. Digital technology is helpful here, especially in design to production of anything, establishing directly verifiable and quantifiable links between the idea and the execution of the idea. The major difference between digital parametric design to robotic production and analog traditional building methods is scalability. Most of the otherwise sympathetic feel-good small-scale design strategies cannot be scaled up to the size of high-rise buildings, while that is exactly what is needed to have a measurable impact on the global climate crisis.

Parametrically designed and robotically manufactured projects have only little waste in the factory, which is fully recyclable, and zero waste on the building site. The constituent components are produced as and when needed. The dry assembly of the components needs no scaffolding. The production of the components needs no molds. Only a few workers are needed for the assembly of the relatively small structural components that are synchronized with relatively large skin components, assembled without polluting onsite welding. My designs intentionally avoid a secondary structure between the main structure and the skin. The designs typically feature small feet, reducing the size of the foundation. Where possible, the most basic materials are used, which need little or no post-processing. The dry-assembled constructs are fit for future cyclic reuse. They are perhaps not sustainable in the popular sense of the word, but ultimately lean and green.

Doomed

The iWEB, the home for the protoSPACE laboratory of Hyperbody at TU Delft, was standing proudly in front of the faculty. However, the dreamt-of ideal situation of having our own laboratory in a pavilion of our design was not meant to last long, not even a year: the Faculty of Architecture burnt down in 2007 due to a short circuit in a defective coffee machine, demonstrating chaos theory in action. As a by-effect of the fire, the iWEB was disconnected from the infrastructure, depriving the lab of data, electricity, and water. After two years of being temporarily hosted by the Faculty of Industrial Design, without a laboratory to continue Hyperbody's interactive architecture experimentations, protoSPACE Lab 3.0 was partly re-installed in an old, renovated building. After years of negotiating and trying to find a new function for the iWEB, TU Delft Real Estate decided to remove the iWEB from its real estate portfolio, thereby closing the opportunity for a possible third life of the Web of North-Holland /iWEB. TU Delft Real Estate refused to invest in the re-connection of the iWEB to the infrastructure, and the iWEB was doomed. The iWEB was eventually demolished in 2015, by the brutal force of the raptors of the demolition company. Ironically, investing in a third off-grid life for the iWEB would not have cost more than what was spent for the demolition. Although the Web of North-Holland, alias the iWEB, is internationally recognized as an icon for parametric design to robotic production and extensively published internationally, there was not enough interest in saving the iconic pavilion. I still fail to grasp how the respective deans and colleagues of that time could not accept the importance of preserving this prototypical statement for nonstandard architecture, which is groundbreaking even according to international standards. One of my colleagues told me at the opening of the iWEB in 2006 that he considered the iWEB more of a piece of art, not architecture. That was, at the same time, meant as a form of critique and as an appraisal. None of my international nonstandard colleagues had achieved something like this on the scale of a complete building of 250 m^2 yet. The iWEB is an uncompromising example of nonstandard architecture. Respected colleagues have realized indoor and outdoor nonstandard installations, mostly delicate open roof structures but not complete buildings as the iWEB. The Web of North-Holland existed for half a year only – the second incarnation of the iWEB almost for one year until it was brutally torn in pieces after years of being brain dead.

Meme Components

The Dike Houses scheme in Zonland, Groningen (1991) is embedded in an earth wall, thereby substantially reducing the need for heating in winter and cooling in summer, while the earth wall itself has a constant temperature of 6 ºC. We were chosen for the Acoustic Barrier project by Projectbureau Leidsche Rijn because they appreciated the concept of a building that is embedded in an earth wall. The A2 Cockpit is embedded in the Acoustic Barrier along the A2 highway near Leidsche Rijn Utrecht, thereby doubling the use of the surface of the earth – two functions in one structure, adding value and saving money for the stakeholders at the same time. The structure functions both as a showroom and as an acoustic barrier, saving the costs for 160 m of acoustic barrier, capitalized to be estimated at one million Euros while offering the car dealer client a prime location for a substantially reduced price. Because in the master plan, there was no building plot to start with, we created a building plot by merging the building into the Acoustic Barrier. The design strategy meant a typical win–win situation, beneficial for the city council and the private client alike.

Many of ONL's buildings including the A2 Cockpit, the iWEB, the Saltwater Pavilion, and, to a lesser extent, also the Bálna Budapest typically have small feet, featuring large cantilevers, meaning reducing the area that a building claims from the landscape. The iWEB has a projected surface area and internally a usable area of approximately 250 m^2 while the actual footprint is only 150 m^2. Small feet buildings reduce the number of foundations needed, allowing the ground to breathe and absorb the rainwater. We consider lean and mean strategies to be active meme components in the design process. Any strong conceptual meme acts as a leading component in the design process; it frames the process, it processes information in a particular rule-based fashion, and it weighs heavily on the following design decisions. The most important contribution we have made to *de facto* leanness is the efficient and also cost-efficient file-to-factory process combined with the dry assembly of the components to constitute our buildings, indeed decades ahead of time. The digital file-to-factory (F2F) process that we have practiced since the nineties was years ahead of recent BIM-supported projects and, above all, lean and green. The F2F process has not been acknowledged to be lean; LEED and/or BREEAM points were not harvested.

Taking Radical Steps

Since the pandemic, the general public became almost overnight interested again in radical visions for the future. We are not exactly short of visionary design concepts, on the many levels of conceptual design, rethinking society, art, and architecture, lean IT technology, radical green and lean planning, swarm theory, parametric design, robotic production, participatory design, and designing DIY design instruments. I have been exploring what it means to be an information architect sculpting data; as is the famous saying by Marcos Novak, I dived into what is emotive architecture, as explained in my inaugural speech at the TU Delft titled E-motive Architecture. As early as the nineties, I pioneered in the domains of interactivity, adaptivity, and proactivity.

As from the Web of North-Holland, the components in the built constructs are identifiable, performative, and interchangeable. What would be the next game changer? As I told my former graduation supervisor Rem Koolhaas at a celebration party in 2001 where he was awarded the French Legion d'Honneur for his theory of bigness, and for his Lille project in particular, in a lucid moment of self-analysis: "I like quietly working in the periphery, where I can take radical steps forcing myself into ever more remote corners of the architectural discourse, as to eventually come out into the open and pop up in the center of it," referring to the direction of interactive architecture that I had embarked upon. The Web of North-Holland was built in a remote location in the Haarlemmermeerpolder, the Saltwater Pavilion on a man-made island in the remote South-West of the Netherlands (1998), and the Elhorst/Vloedbelt Waste Transfer Station in the far eastern regions of the Netherlands. It was at that same occasion that Koolhaas, by a slip of the tongue, told me that of the Dutch architects, he respected me the most, which I found hard to believe because I never was a docile follower of his, but I had chosen my own way. Eventually, I gained international recognition and was invited to become a professor from practice in digital architecture at the TU Delft in the year 2000. During the interview at TU Delft for the professorship with the committee, presided by the dean at that time Cees Dam, together with the professors Leen van Duin and Umberto Barbieri, Barbieri asked me his closing question about how I see the role of traditional pencil drawing in relation to digital architecture. I answered that I see digital architecture as a new instrument in addition to pencil drawing, as an extra layer, but not replacing it. Barbieri and the committee members were relieved, and I was hired. The part-time 0,4 fte university position inspired me to take even more radical steps, both in practice and in academia. I renamed the chair of digital architecture into Hyperbody. The new challenge was to design building bodies that change shape and content in real time, a real game changer, merging the physical world with the virtual. I explored my fascination for interactivity through the Sculpture City event, since the Waterpavilion and since Trans_Ports 2000, and chose it as the leading theme for Hyperbody. Interactive architecture builds upon the earlier notion of the comprehensive building component. Comprehensive component-based design is a prerequisite for interactive architecture. In component-based fluid architecture, the components are parametrically related to each in the design process, and in interactive architecture, the related components "talk" to each other. The building components communicate in real time with their immediate neighbors. Designing parametrically means the start of a process, a procedural way of working. In the execution phase, that process is normally frozen; when built, the components no longer change shape and performance. I asked myself if there was a way to continue the process in the execution phase. The concept of interactive architecture is driven by my discomfort with stopping the excitement of the process. As every computer program is a running process, I became more and more intrigued by the idea that we could keep the process running by keeping the programs running and that we could keep the components processing incoming data, and keep them talking to each other, as birds do in the swarm.

Hyperbody

Since 1989, our innovation studio ONL has been a digital platform for the fusion of art, architecture, and technology. Since 2000, Hyperbody acted at the forefront of interactive architecture until my retirement in

2016. The brand name Hyperbody is an extrapolation from the term hypertext and <u>hypersurface,</u> coined by Stephen Perella for linked virtual worlds that are projected on surfaces, into a hyperbody. Hyperbodies are building bodies that transform not only the content in real time but also the shape, denying Vitrivius' "firmitas." Hyperbodies are vectorial bodies that live simultaneously in physical reality and digital hyper-reality, in real time. Embedding real-time processes into the constituent components of architecture is the game changer. A hyperbody/hyperbuilding gradually evolves by receiving input data, processing data, and streaming output data. The major changes are performed by switching from one mode of operation to another.

As from the end of the nineties, I felt – more and more – that we need to keep the process going. Games being played on computers are executing processes; the games are interaction instruments. The participants in the game access a running script, and they join worlds that evolve in real time. A charming example of early interactive games is the handheld egg-shaped digital pet <u>Tamagotchi,</u> the wonderful little virtual creature inside a little pink body that one had to feed and interact with to keep it alive. Tamagotchi was easy to get addicted to, lovingly feeding the baby to become an adolescent, an adult, caring for Tamagotchi up to its death. One developed an emotional relationship with the virtual Tamagotchi creature. Those cute computer games were harvesting my attention, and I could not resist bringing that new interactive technology into architectural design concepts. The idea of serious gaming in architecture was born, later leading to participatory design. Tamagotchi urged us to reset our thinking about what exactly is a component in an interactive digital environment. These new living components have to be played with in real time, like Tamagotchi begging for attention.

Attractor Game

Back in 1998, I was invited to participate in a competition to develop a master plan for a new area called Reitdiep for 1500 homes in Groningen, in the North of the Netherlands. Inspired by handheld computer games and building upon our earlier efforts in programming and scripting, as in the Sculpture City CD-ROM and the Genes of Architecture CD-ROM, we developed the Attractor Game. The Attractor Game is programmed in Visual Basic by the then ONL architect Menno Rubbens. We defined four major design components, each of them identifiable in terms of quantity and quality. We defined the tree component, the water component, the house component, and retail as a component. The components are defined as attractors; negative values turn the attractors into repellers. The attractors are dragged into the playing field, and the players set the size of their sphere of influence and their strength, using one pair of sliders for each attractor. A maximum of 12 attractors can be dragged into the playing field. Apart from the parametric design techniques, one needs strong conceptual ideas about why and where to position the attractors, choose their quantities, and define the reach of their influence. Each move with one of the sliders means a design decision. Playing the Attractor Game equals designing the master plan, like playing an instrument to compose a musical piece. After the placement of the attractors and setting the values for one particular component at a time, the resulting image is transposed into a DXF file. By an in-house written AutoLISP routine, the image is scanned and translated into a population of referenced Autocad blocks, functioning as quantifiable components for the bill of quantities. The blocks form the basis for further three-dimensional representations of the points. Blocks can be in a later stage replaced by a three-dimensional tree, the 3D model for a house, a body of water, or the units of the retail area. The integrity of the quantitative data and their qualitative performance remains intact. A dense population of trees becomes a forest, clusters of homes a residential domain, and clusters of water a lake. To constitute the master plan the different ingredients are superimposed to define the rules of the game for each plot, in the form of a graphic representation. Thanks to the serendipity of the superposition, a house can be on the water, and a tree in the house. Any combination of components is possible and challenges the skills of future architectural designers. In the master plan for Reitdiep rules, statistics, chance, and serendipity are applied to create a pleasant living and working environment. Looking back from the here and now, Attractor Game already had many of the ingredients of the emerging theory of component-based swarm architecture, participatory design, and parametric design to robotic production. The key components for future projects were about to crystallize. The Attractor Game represents a new direction in architecture, writing algorithms running the design instruments while simultaneously playing the design instruments. Design is redefined as the design of the design instrument with which the design is played as in a game. The goal of the game is to define the data construct that informs the production machines.

Trans_Ports

The next disruptive step, after having defined the constituting components as identifiable, performative, and interacting components, is to equip the components with sensors and actuators, to prepare them to act in real time. I presented this idea for the first time at the first Archilab Conference in 1999 in Orléans, organized by Frédéric Migayrou and Marie-Ange Brayer of the Centre FRAC in Orléans. I showed a video of a soft bluish loaf-shaped volume, the Trans_Ports project, which breathes in and out, twists, and talks while changing sections of its well-rounded body into alternating concave and convex surfaces. Showing Trans_Ports 1.0 was highly provocative, and none of my conference peers thought of such living objects before, and I asked my colleagues to respond. It was the French punk-style architect Odile Decq who made a meaningful comment, saying that she liked the video but wondered what would happen inside such a structure. That was a sensitive remark. I had not developed the inside yet. I decided to have an LED-based light-emitting skin for the interior. There was an urgency to make the Trans_Ports video, while I did not find it logical to freeze the computational process of animated blobs. I was annoyed by Greg Lynn's notion of the separation between design space and contractor space, which to me is a weakness, a subordination to the traditional building industry. Lynn is also responsible for coining the term blob – that terrible B-word. Lynn, who worked for Eisenmann, was highly influenced by the theories of deconstructivism, and I think that he could not come to terms with the synthetic nature of constructing complex shapes; he gave priority to deformation over formation. While discussing some angular-shaped designs like the Erasmus Bridge by Ben van Berkel (1996) in his book <u>Folds, Bodies & Blobs</u> (1998), he omitted writing about the Waterpavilion (1997), nor did he mention earlier complex geometry designs like the Waste Transfer Station (1994) or Mecanoo's ING penthouse in Budapest (1994). Together with Douglas Garofalo and Michael McInturf Lynn was at that time building the Korean Presbyterian church, a disappointingly low-resolution attempt to design the complex. The complexity in Lynn's church design is a series of frozen instances of a moving object, more like a deconstruction of movement than of a synthesis of movement, probably inspired by the cubist-futurist style <u>Nude Descending a Staircase</u> by Marcel Duchamp (1912). At the time, Lynn experimented with supersmooth meta-balls that merge but did not manage to translate the smooth shapes into building components that are designed to be manufactured by computer numerical controlled production. I straightforwardly dislike the word "blob," knowing that a blob in programming language means a Binary Large Object. There is something inherently wrong about the word blob architecture, demonstrated by the fact that in popular language, blobs are equated with organic undefined <u>Barbapapa</u>-style shapes. Binary Large Objects contain data that are exact and known; there is nothing about it that is vague or arbitrary. Only from the outside, it is contained as a bag full of data. The term blob refers to formlessness, while my work features a precise formation of interacting components.

I had to find a way out of the frozen blob frame that somehow got stuck to my designs. The apparent solution was to continue the life that resides inside the virtual design process into the real-world existence of the project. The "live" acting components in Trans_Ports are electronic pistons that can extend their length with a factor of 1.5. Twelve actuators of 2 m in the default position and 3 m in the extended position length placed at both sides inside the double skin of the volume are programmed to move individually. The actuating components change the shape of the pavilion in real time. The name Trans_Ports popped up during a meeting with Marcos Novak in 1998 in Los Angeles. Novak had invited me to speak at the Getty Center, where I witnessed a weird gender discussion. Some speakers argued that 3D space was for boys, while 2D graphics were for girls. Later, we sat down in the Mondrian bar on Sunset Boulevard, designed by Philippe Starck, then owned by Paul Simon, whom we spotted hanging at the bar. Novak and I came up with the idea to design two pavilions that would interact with one another, one in LA and one in Rotterdam. Putting a dent in the pavilion in Los Angeles would immediately be seen and felt in Rotterdam, and vice versa, like spooky action at a distance. There would be an ongoing dialogue between the two pavilions, communicating with the speed of light, entangled. Like the telephone and the fax machine, the encoded information is sent over long distances to the drivers of the pistons, causing instant physical movement. Marcos and I aimed at the integral merge of the real and the virtual, a scheme of things that has never left us since. The Archilab Trans_Ports video developed further into the Trans_Ports 2000 project that I showed at the Venice Biennale in 2000, themed Less Aesthetics More Ethics. Invited by curator Massimiliano Fuksas, we got a prominent central location of 15 × 15 m² in the Italian pavilion, with a window to the International Space Station that Fuksas had hung from the ceiling in the adjacent room. We designed an arena for interaction with the projected 3D models of the Web of North-Holland, of Trans_Ports (in self-explaining mode), and of the art project Handdrawspace. The public navigates and interacts with the projected 3D models simply by changing their positions in space. Trans_Ports 2000 is a collective public instrument. Moving around in the interactive area triggers the proximity sensors that send their signals to the script that configures the 3D models in real time. Moving around tweaks the geometry of the 3D model and refreshes the projected hypersurfaces. The immersive interactive installation Trans_ Ports 2000 is an instant choreography of the public testing the effect of their movements on the unfolding virtual worlds. There are three main acting components in the Trans_Ports ecology loop: the people, the arena, and the virtual objects. The people components inform the arena, and the arena, in its turn, informs the projected virtual worlds, while the video projections trigger the people to move around.

After the Biennale 2000, the next step we took in our quest for interactive architecture was to drive physical components in real time. We went from interacting with projected worlds to interacting with physical constructs. Toward the end of 2002, I was invited to propose an interactive installation for the game-changing Nonstandard Architectures (NSA) exhibition curated by Frédéric Migayrou and Zeynep Mennan at the Centre Pompidou in Paris. Together with interaction designer and sensor expert Bert Bongers and Hyperbody students Chris Kievid, Sven Blokker, and Remko Siemerink, ONL designed and produced the NSA Muscle, a programmable inflatable sculpture that moves, dances, and twists. NSA Muscle is triggered by sensors that are attached to the nodal points at the sides. An actuator network of 72 Festo muscles is wrapped around the inflated blue balloon. The muscles are constantly kneading the soft shape; they provide for the tension forces and the balloon for the expansive counterforce. Despite the limited behavior due to material restrictions and safety measures, the NSA Muscle project has set rules for the future of interactive architecture. Interactive architecture has a life of its own, proactively proposing new configurations that change the behavior of the public.

Since the Corona crisis has made us more aware of the interaction between humans and our natural environment, also our interaction with the built environment will become a natural aspect of future design tasks. The ecology of things and people is a dynamic adaptive system, supported by digital technology rapidly evolving toward a more and more immediately interacting environment.

Game Set and Match Conferences

Assisted by my then assistants Misja van Veen and Hans Hubers, we organized the first international Game Set and Match (GSM I) conference at the Faculty of Architecture at the TU Delft, in December 2001. We played interactive games during the breaks. In style with my inaugural speech at the TU Delft, in November 2001, in the format of a live discussion with my virtual friend Marlon, I had an <u>interactive conversation with my new virtual friend Cindy</u>. Both real and virtual people inhabit cyber architecture. We invited Paul Verschure and Kynan Eng who just completed the intelligent <u>ADA space</u> at the Swiss national exhibition Expo.02. We invited Maia Engeli, who wrote the book Bits and Spaces, computer game designer Xavier de Boissarie, who worked with the game development program NEMO, Kinetic architect Michael Fox, and interaction designer Ted Krueger. In the GSM I program, we had an introduction by Ole Bouman, intermezzos by composer Edwin van der Heide, the young kids of Hans Hubers playing real-time games, and a real-time structural analysis by student Nils Addink. The motto of Game Set and Match I is provocative and forward-looking based on state-of-the-art digital technology. While architecture is played like a GAME, users *set* the parameters for interaction, and *match* spatial conditions in real time.

The key components in building up theory and praxis of emotive, nonstandard, and interactive architecture are adequately described by the headers of the paragraphs that I have written as an introduction to the proceedings of the <u>Game Set and Match II international conference</u> (GSM II) organized by Hyperbody at the TU Delft in 2006:

1) Space is a computation,
2) the building becomes the installation,
3) quantum theory,
4) real-time behavior,
5) swarms of building components,
6) personal history from synthetic architecture to swarm architecture,
7) implications for the daily practice of architecture,
8) swarm architecture from research to practice,
9) uncertainty and unpredictability, and
10) top-down styling interventions and bottom-up swarm behavior.

These titles leave no doubt in what direction architecture is going, namely bidirectional relationships between real-time interacting components. Much of the interactive projects are atomic in their initial condition, building dynamic universes from a point cloud of reference points. My designs are based on a coherent swarm of reference points in space, and the basic materials of my nanoscale, mesoscale, and macroscale thinking are the thousands of mutually communicating data points. The core principles of bidirectional relationships form the basis of the collaborative participatory design instruments that were developed at Hyperbody in the years between 2000 and 2016. Every component in the design game is parametrically connected to its neighboring component, while families of components are connected to families at a higher level. I took the principles of complex adaptive systems and applied them to architecture, to be formed and performed in real time, as actors in a dynamic scheme of things, on the Internet of People and Things. In the merged world of reality and hyper-reality, people and things are actors on a level playing field, playing by the rules.

Ten years later, the <u>Game Set and Match III</u> conference marked my retirement from TU Delft. The Hyperbody team organized a three-day international conference, Day One programmed by me. I invited partners, collaborators, and Ph.D. candidates from the last two decades to tell their stories, tell how they started at ONL and/or at Hyperbody, and how their careers went from there. I invited Frits van Dongen, my first business partner (1978–1980), many former ONL collaborators, and previous and then current Ph.D. candidates at Hyperbody. I asked them to reflect on their stay at ONL and/or Hyperbody and elaborate on the work they do as of today. Many of them made it into a thriving design consultancy practice of their own. For Day Two and Day Three, I gave the floor to my assistant professors Henriette Bier and Nimish Biloria to put together their program. During my 16 years at the TU Delft, I worked for 2 days a week, as a professor from practice.

After I retired from TU Delft, I chose to work for some years as a full-time professor at Qatar University in Doha, Qatar. In the meantime, the ONL portfolio was running dry; so I negotiated a transfer of our experts in parametric design Gijs Joosen and Pieter Schreurs to Royal Haskoning, intending to collaborate in future projects. The years at Qatar University are among the most rewarding experiences in our life. Ilona energetically took up painting again and exhibited a series of large acrylic paintings in the renowned Sheikh Faisal Bin Qassim al Thani Museum and at the Doha Fire Station cultural center in Doha. Besides teaching and research, I was hired to advise on the curriculum innovation of the Department of Architecture and Urban Planning (DAUP) and to secure international NAAB accreditation. I initiated and acquired big research projects for the department, and taught design courses for students, from the first year to the fifth senior year. Together with Ilona, I did a parametric painting studio in the course Design Methods. We introduced the concept of parametric thinking through exercises in painting and sculpture. One does not need computers to get acquainted with the principles of parametric design. The key idea of parametric design is to build relations between the constituent components, and that can be achieved in an analog manner as well. Subsequently, I introduced the students to digital parametric design and to the importance of connecting parametric designs to robotic production methods of the building components. The Qatari students were keen to pick up parametric component-based design concepts. Fifty percent of the exclusively female students of the Department of Architecture and Urban Planning are native Qatari girls and the others are from neighboring countries in the Middle East, except Saudi Arabia, Bahrain, the UAE, and Egypt, which countries started the four-year-long blockade on Qatar a few months before we moved in. In February 2019, I organized at the Library of Qatar University (QU) a sequel to the earlier GSM conferences, Game Set and Match IV Qatar (GSM4Q), a two-day international conference with scientific paper presentations. I invited architect friends Antonino Saggio, Marcos Novak, Lyudmila and Vladislav Kirpichev, and Philippe Morel, along with the Dutch natural physicist Vincent Icke, architects Philippe Block, Shajay Bhooshan, Nimish Biloria, Ali Mangera, Rolf van Boxmeer/Tessa Peters, and Fatima Fawzy of Msheireb Properties, and local architects Hani Hawamdeh of AEB and Hafid Rakem of the

Doha branch of the Ateliers Jean Nouvel. QU students were finally able to meet major league players from the international architectural arena in person. The GSM4Q lectures and papers are well documented in the <u>proceedings of Game Set and Match IV Qatar</u>, published by Qatar University Press.

3

Another Normal

Quantum Architecture

From early on in my studies, I have been interested in quantum theory and what implications it may have for architecture. If it would not have implications, I postulated that architecture would miss the connection to science. To keep up with actual developments in technology, trying to grasp what is buzzing in the minds of natural physicists is one thing, and applying their findings to one's field of expertise is quite another thing. In one of my first publications dated 1986, featuring three of my XYZ projects, Polder XYZ, Conservatory XYZ, and Lego XYZ, in the bespoke Dutch magazine Wiederhall, invited by the editor-in-chief Joost Meuwissen, I explained my relationship to science and general relativity, in particular, in 10 points:

1) The XYZ system of coordinates materializes as the basic relational framework for further signification and complexity.
2) The XYZ system of coordinates attaches the projects to their context creating a self- imposed fixed center point related to the site.
3) The XYZ system of coordinates defines the exact position of the points, the planes, and the volumes in a continuous space.
4) The center point of the XYZ system of coordinates is a floating point in space, either vaporous, liquid, or solid, from which the design develops in all six axial directions.
5) The six directions describe the project as dynamic volumes/ planes on a continuous scale, above ground and underground are

equivalent, in front of and behind are equally important, and left and right are decentralized and asymmetric.

6) The major events which are set apart from the materialized XYZ system of coordinates are mathematically described as elements whirling in space, following their own self-sufficient concentric rules.

7) The events are sublimated into tangible elements, mind elements or function elements, form elements, or any possible mutual superposition.

8) The major events are never experienced simultaneously; they represent the general relativity of the system.

9) The XYZ series propagates an intuitive scientific style, testing permanently existing images on recent social and technical developments.

10) The designs of the XYZ series are scientifically international, geologically national, or even local, and intuitively personal.

Although I did not mention the Q-word, quantum uncertainty was exactly what I had in mind to create a quantifiable generic framework that allows for the qualitative unknown, the alien, the unexpected, the unforeseen, and the specific. I took a bolder approach in another article I wrote titled "Space Time Volume" for Wiederhall 12, trying to link general relativity to quantum theory and applying the findings to my emerging theory of temporal and local centers of observation, preceding my interest in interacting swarming components as developed toward the end of the nineties. I wrote back then, 30 years ago:

"In a microscopical sense the object is fully transparent, in a macroscopical sense the visible empty space is tangible like matter. Depending on the imaginary position we like to take on the scale of minus infinity to plus infinity, a section from this total space, whether very large or very small, may be represented as 'matter' [visually and physically impermeable] or as 'void' [intangible]. Or in architectural terms, as objects or as space. What we call objects in our everyday life is related to the biological body structure. Our body is a delicate spatial and temporary balance of cells, atoms, quarks, of information. Our human perception is implicitly related to this fragile balance and thus our direct perception enables us to perceive only similar balances as dead or living matter. In daily life, we are fixed at a certain place and a certain moment in time-space. But in our imagination time and place are free. We can move ourselves into different scales, microscopically small or macroscopically large, we may compress time as we choose, or stretch it infinitely, which makes it possible after all to visualize a non-observable coherence between apparently elementary particles on a much smaller or much larger scale. Currently, scientists are working on a theory of unification, which aims to link the theory of relativity of space-time with the quantum theory of the small particles."

Can we, 30 years later, establish such a link in architecture? Can we dig deeper than the periodic table of elements and find the underlying force fields that architecture is made of? Can we reduce the 118 atomic elements to more elementary forces? Can we come to a better understanding of fundamentals than RK's 15 supposedly fundamental elements? And are we talking about a reduction at all, or should we rather see the world as an emergent property of information? What are the smallest particle-field siblings, and how can they be considered components rather than elements? Asking the question is answering it. Components are defined as elements that are inclusive of how they are tied together, in that particular position, at that specific point in time, from that particular vantage point of the universe. Components are inclusive of their relations with their immediate neighbors, while elements are autonomous generic parts, produced with at best a generic relationship to some other element that is not identified yet. Since Rutherford, the schoolbook image of an atom is that of a solid core with electrons orbiting around it as if it were an isolated element. That image is problematic since we know that an atom has no solid core, and electrons are not particles that are simply rotating around it. There are no particles at all but only

97

tension fields. Based on insights from quantum mechanics, what we, as humans, see and feel as matter is nothing else but a specific form of an entangled tension field of information. Each atom can be described as a specific interference pattern. Numerous atoms bind together as components to form matter. At the subatomic scale, tension fields are the components that constitute an atom. At the super-atomic scale, tension fields bind the atoms together to form matter. Atoms are containers of hidden information in much the same way as a sea container contains its content. Atoms tie together to form matter; components tie together to form a built structure. A component is considered matter when it contains many atoms bonded together to build specific properties, tangible for humans living their temporally and locally restricted lives at an arbitrary point in between the galaxies and the quarks. When one would single out building parts as elements, the inherent consequence is that they are denied relational properties. Nothing and nobody can exist without their relations, and no brain would even work without being related to other brains. No computer would mean much when not connected to an intranet or internet; they form part of a vast ecology of computation that has emerged out of the digital revolution. Elements as such have no value in a theory of architecture, but components do because they contain the information on how to form the swarm. Buildings are an intentional assembly of a variety of mutually related parts, not a random stack of unrelated parts. Indeed, there must be something much more fundamental to chemistry than a list of elements – whatever choice of elements one would make, and from whatever scale one observes the world: macroscopic, human, or microscopic scale. What is more fundamental resides in the dynamic relationships between the components. Components are informed elements. There must be a theory of everything that we have only lightly touched upon as of yet. We could think of a possible quantum character of the information fields, from which component-based matter emerges, constituting the spatial constructs and eventually the organisms (i.e., ourselves) that feel and experience such constructs. It might be fluid and discrete at the same time, like Schrödinger's cat who is both dead and alive.

Making Ends Meet

Recent scientific discoveries reveal that in each galaxy resides a supermassive black hole, where gravitation has collapsed, warp-holing atomic matter and even subatomic information towards unknown dimensions. Black holes suck up the event horizon, and everything that comes close to the wormhole drain is vehemently tossed around and stretched into endlessness, that is, from our human vantage point. It is impossible to imagine yourself being inside the black hole physically since one would be stretched into billions of kilometers. Can we even think when stretched into infinity? In our imagination, we travel through the powers of 10, but inside the black hole, we would be infinitely small and infinitely long stretched at the same time. Matter will be diluted into a form of information that is not yet understandable by humans. There is a limitless expansion in the discernable universe on the one hand and super strong implosions in the centers of an unlimited number of galaxies in the universe on the other hand. A fundamental postulate of the Copenhagen interpretation of quantum mechanics is that complete information about a system is encoded in its wave function up to when the wave function collapses, known as the black hole information paradox. Once you get inside the black hole, it is believed that the laws of physics, as we know it ceases to exist, and new laws are put into place. Black holes are believed to be connected via wormholes, representing a sort of inverted universe. Space and time as we know it ceases to exist, and quantum information is probably what rules are inside.

My layman's speculation is that the origin of matter is pure information, whereas matter is an emergent property from that immaterial and non-spatial information. Further speculation is that the big-scale immaterial information content must be seamlessly connected to or be the same as that other universal infiniteness, that is, the world smaller than the Planck length of 10^{-35} m. Assuming that the space inside the black holes is equally vast and deep and similar to the subatomic world containing information in its purest form, would it be not possible that the two endless worlds actually are connected or even could be the same? Like two sides of the same coin. Could it imply that there is nothing like a big bang, but the closing of the loop from the endlessly small inside the sub-quark world to the origin of our universe? It means that the information ruling the universe and the information ruling the quantum world are the same. The same information ruling the very big and the very small would mean nothing less than a possible alternative unification theory. That gives another dimension to the phrase "making ends meet." I speculated on this idea before and illustrated the concept in a diagram published in Wiederhall 12, modeled on my Atari PC in the late eighties, the "hourglass" diagram. It could also explain the number of atoms in our universe, estimated to be a whopping 10^{78} atoms. I named the graphic representation of my hypothetical unification theory the hourglass of space-time. In the diagram, we humans occupy a unique temporal and local position in the powers of 10, from where the very big and the very small are looked at. The idea is as simple as it is elegant. Assuming that the space inside an atom is endless and emerges from a pure form of information, and assuming that the same endlessness applies to all 10^{78} atoms in the measurable universe, then pure information could be the invisible, intangible, and immeasurable force that is underlying all equations, perhaps even responsible for what today is called dark matter. Pure information is neither an element nor a component. Pure information has neither content nor meaning. Pure information is probably more something like potential energy, forming the intelligence of underlying non-temporal, non-spatial, immaterial, inside–out universe, out of which our observable universe has emerged and out of which numerous other universes may emerge. According to this worldview, matter, as we know it, is an emergent property of information, as we are ourselves.

Our former general practitioner Mattees van Dijk coined the word holoconsciousness in a cute book Little Big Think that he wrote after his retirement in 2019. He believes that holoconsciousness is a state of universal consciousness that life, as we know it, is tapping into. Perhaps universal consciousness is congruent to my assumed state of pure information. Both of us assume an underlying system to the world as we know it.

As a former practicing doctor, van Dijk concentrates on human life, while I focus on the totality of the natural world including the world-making by humans. The state of pure information supposedly has no awareness of itself; it has no meaning to perform in one way or another specific way. However, pure information must be an as-of-now unknowable form of potential energy that can transform via phase transitions to become a universe like the one we are acting part of. Phase transitions are not conscious acts; they just happen when a certain condition has been reached, and they emerge. We, humans, are not at the center of the universe, nor is any other form of life. Either everyone or everything is at the center of whatever universe, or no one is. It is the delusive illusion of the human condition that we think self-centered. Information is the basic non-spatial, immaterial, and non-temporal stuff that serendipitously generates all we know and all we see around us – information is all-inclusive. Everything in our universe is informed in one way or another. The universe is governed by a simple set of rules, creating an – in our eyes – immense complexity as an emergent outcome. The role of black holes in this picture could potentially represent the missing link between the super small and the super big, linked by information. Only information in its purest form, which is unbound to matter, lives in the black hole – the connecting link between the very small and the very big. It is a very speculative assumption indeed but one that works for me and helps me in choosing directions in art and architecture.

Quantum physics tells us that, from that default state onward, multiple universes are possible and could exist simultaneously, like the Other Now in Yanis Varoufakis' streetwise social fiction fantasy Another Now. The Other Now is a benign parallel world, tapped into by an ultrasmart device called the Freedom Machine constructed by the author's fictional brilliant tech friend Costa in Silicon Valley. The machine establishes communication between the Now and the Other Now, the world as it could have been if other political choices were made after the crash of 2008. The downside of the Other Now as described by Varoufakis is that the intended perfect society unintentionally has inherited some of the flaws of our current society, notably that of gender inequality.

101

Varoufakis also warns of the risk that the advanced technology of connecting directly to one's alter ego in the Other Now would fall into the hands of malicious dictators, using the same "<u>walled garden</u>" techniques to divide and rule forever, referring to giants like Google, Facebook, and Amazon. In Varoufakis' Other Now, the Freedom Machine is eventually destroyed by its maker, while the inventor chooses to live in the imperfect Now, but not without taking a female Other Now friend with him. The open question that sticks is how can one be uncompromisingly forward-looking, and take advantage of the most recent technologies, without falling into the trap of a socio-technocratic tunnel vision that favors the rich and famous.

It may seem a bit far-fetched to discuss quantum mechanics in a book on components in art and architecture, but when digging deeper, there is a correlation. Understanding the underlying principles of emergent components shapes my view on art and architecture, with a special focus on design concepts that are based on simple rules from which complexity emerges. As a conceptual thinker and world-maker, I have developed a clear preference for abstract procedural art and inclusive architecture, which is the architecture of diversity, complexity, interaction, and ultimately the architecture of participation. Within the realm of complexity, I am interested in a design approach that is based on simple rules that create a critical mass for future diversity and eventually allow for immersive participation in the design process. Speculating on the nature of a non-spatial sub-quark world and of all-absorbing black holes helps me in reformulating art and architecture, by finding underlying principles that lead to their apparent states. The clue is in the notion of information; information that rules the universe and information that rules the design-to-manufacturing process. The designer must choose a clear direction in the endless sea of possible directions that simple rules may bring about and subsequently test that choice. When dropped in an ocean, with no land in sight in any direction, how do you choose a direction? That pretty much describes the condition we found ourselves in when starting to work with digital media. Choosing direction is a combination of rational thinking and intuitive acting – super-controlled and out of control at the same time, meaning nothing less than searching for a unification principle for art and architecture.

Synthetic Architecture

While the world synthesizes from pure information into informed atoms, from atoms into molecules, from molecules into matter, and from matter into building components, architecture defines itself as synthetic architecture. The synthesis takes place not only in the design process, but subsequently in the production process, and the procedures of the assembly. Designs are developed by a force that works from within, defining the project-specific components, and by listening to external forces from the environment where the design will be embedded into. The inner workings and their relations with their nearest neighbors are merged with imposed external forces, with governmental rules of law, performative criteria for climate control, and forces that are demanding minimum requirements for stability and strength, often with imposed urban bounding boxes, within which the design is allowed to develop. The environmental constraints subsequently form part of the design code for the future building components. The design code of the component is informed by both external and internal forces, both top-down and bottom-up. In the production process, a component of a typical ONL project is produced by CNC machines informed by digital strings of code. The machines of the designers are directly short-circuited to the production machines: machine-to-machine (M2M) communication. The synthetic production process ideally takes place in a factory that is close to the building site, but in the practice of the mass production building industry, the parts may come from anywhere in the world, transported over long distances in sea containers. Parametric design to robotic production (PDRP) forms the prerequisite for the shortest possible route from design to manufacturing, when possible, in a field factory on the building site itself using mobile robots. Synthetic assembly procedures combine the parts to form a whole. With the ubiquitous use of robots on the building site, the building self-assembles. In a time-lapse movie, the people and the robots are disappearing, and only the assembled parts constitute the image, and the structure self-builds, emerging from its informed basic building components as if it "grows."

The production facilities of objects of synthetic growth are located outside the body. An organic body produces the building blocks for growth from its internal production units, fueled by nutrients from its immediate environments. Although referring to growth from the time-lapse point of view, I typically refrain from too literal biological references, which have become popular in the last decade as in the bio-digital, coined by Alberto Estevez, and material ecology, a term coined by Neri Oxman. They are inspired by natural growth and aim to find ways to grow architecture. However, their efforts remain limited to small scales, failing the skyscraper test. The skyscraper test poses one simple question: can you build a skyscraper with the proposed technology? While ONL's component-based parametric design to robotic production easily passes the skyscraper test, none of the proposals of organically growing building materials and building parts comes even remotely close to passing the test. A skyscraper that produces its own materials and self-assembles has not been seen yet, and I do not expect to see that in the near future. Even if this biodynamic skyscraper would perform a form of self-growth, it still would need tons of traditional building materials and hence carbon-based fuel to materialize the load-bearing structure. According to biodynamic theory, the fuel must come from its immediate environment, meaning that the skyscraper would need a large area to feed upon. Of course, we can replace the skyscraper with any other large structure in the equation. The bio-digital advocates would probably argue that they do not want to build skyscrapers at all but only organically shaped low-rise structures in a low-density environment, in balance with what fuel the immediate environment offers for organic synthesis.

While we can educate ourselves with the logic of biology, it does not directly provide us with guidelines on how to design. An airplane is not using the same technique and logic as that of a bird. It does not mimic its flight. If airplanes would want to take off, fly, and land like birds, they would need a dynamic connection between the wings and the body that is a thousand times stronger, and the complex flapping of the wings themselves would be the motor. The bird and the airplane are fundamentally different types of vehicles. Birds are grown into shape from one single cell by cell division and cell specification, securing a coherent whole in every stage of development. Buildings are assembled into their shape from discrete components that are produced as separate entities, only forming the coherent whole after the assembly, which is a completely different process of coming into being. For buildings, the whole world comes into action to put the components together, and the world acts as the distributed factory. Even when parametric design to robotic production aims at using materials and machines as close as possible to the building site, still it depends on the connections between the minds and the machines of the whole world to get things done. Biomimicry may result in structures that look organic, but their methods of building are very different, mostly even using outdated traditional building techniques. Mimicking biology simply is the wrong starting point. One should start a design process that is natural in itself and hence will lead to natural outcomes. Natural in the sense of a synthetic nature, based on digital design techniques and digitally controlled production processes that have a logic of their own. The synthetic logic of a digital operation is different from a biological operation. A digital process facilitated by complex pulses through silicon cannot directly inform a biological process that is facilitated by complex molecules and their interactions. The bio-digital is at best a partner in a larger complex of building techniques. One may use genetic algorithms and multi-material 3D printing machines to produce a complex structure that is the spitting image of a natural structure. One may build a digital twin that is alive, in sync with the original biological organism. But it will not be anything else than a simulation, inspired by nature, but certainly not a biological structure that is grown from informed cells. To reach the point of a truly bio-digital production process, one should program the very DNA string that informs the cells to divide and specialize and feed the cells the right nutrients to gain weight. Oxman has coined the phrase "mothering nature," thereby rightfully rejecting the idea of mimicking nature, and pleads for design-inspired nature as opposed to nature-inspired design. But Oxman also rejects the synthetic assembly of thousands of parts in favor of organic design from a single source. Her experiments with silkworms are partly convincing, partly misleading, and partly cruel to the captivated worms themselves. It leads away from the synthetic, which is an emergent phase change in the building industry we have just touched upon. By rejecting the phase of parametric design to robotic production of informed components, which is extremely efficient in their means of design, production, and operation, and fit to build large structures and skyscrapers, the bio-digital and the concept of mothering nature leads to a form of escapist small-scale architecture, totally unfit to pass the skyscraper test. Instead of the mean and lean design to production strategies, which are fit to build the affordable for billions of people, one would end up facilitating a small and privileged elite.

Time will tell, but as of now, I predict only small-scale applications for working with organic design-inspired nature, not for bigger constructs, and certainly not for skyscrapers, not even potentially soon. To claim that design-inspired nature would be a game changer for the building industry is as of now charmingly naive, making the designers drift even further away from where the action is, the voluntary prisoners of nature as we know it. The essential difference in thinking between synthetic assembly and natural growth is that I consider synthetic architecture as a new form of nature, not as something opposed to it. We are living inside new nature, and we are an acting part of it. There is nothing organic about an architectural construct; the process of synthesis means new nature in action, putting informed parts together as an assembly of components. After a synthetic process, a building may be compared to organic bodies, like the public refers to our Bálna Budapest building as a whale, whereas "bálna" means whale in Hungarian because the shape vaguely reminds them of a whale. Buildings may look organic, but they are not, and cannot be. Although I use words like bodies, cells, and behavior in my design vocabulary, I do not mimic organic forms, and I do not grow materials or simulate growth and decay. Rather I design systems that emerge from their own logical rules of play, based on available materials, available technologies, and, above all, driven by abstract conceptual thinking. I work from a simple inner drive to come to a possible complex outcome. Neither the synthetic architectural designs nor the abstract paintings of Ilona Lénárd, which I discussed in the Ubiquitous Symmetry paper, have the intention to resemble something that is known from nature. Eventually, our work may look like something familiar, something biological even, but that is purely the interpretation of the beholder. It is only logical that the public wants to refer to something they know, while there are no references for the alien. Often my designs are labeled as futuristic, which they are not, but they are alien indeed. We use available materials and technology, we invent new ways of using those materials, and that technology, by design, and we inform the components that make up the whole. But we do not refer to a possible future look of things. It is what it is, based on what is available to us. From a clear aversion to the notion of the organic, I have chosen from the beginning to describe our work as synthetic. The first international exhibition of my work was in Gallery Aedes in Berlin in 1990, located Unter Den Bogen, directed by Kristin Feireiss, curated by our architect friend, the late Konrad Wohlhage, and I titled the show Synthetic Architecture. The notion of synthetic architecture I have nurtured ever since. During that same exhibition, Ilona and I organized a workshop titled Artificial Intuition. We worked with invited artists and architects to sketch intuitively in 3D cyberspace with the explicit aim to construct interpretations of how that subconscious sketch could be consciously synthesized into a spatial structure. Synthesis is the combination of components to form a connected whole. The term synthetic covers accurately how we see architecture moving away from the organic, learning from the developments in the arts at the beginning of the 20th century. I am more interested in the concreteness of Piet Mondrian than in his earlier abstractions. Mondrian ended up making concrete art, a term coined by Theo van Doesburg. Concrete art goes beyond the abstract. While organic equals the abstract, i.e., abstracted from nature, synthetic equals the concrete, synthesized from the most elementary components, constructing the hitherto unknown, welcoming the alien, ultimately the Other Normal, by combining rational thinking with intuitive acting.

Science Inspires Art Inspires Architecture

Frank Gehry wants us to believe that architecture is the Master of Arts. How so? Art, and especially the early 20th-century avant-garde visual art, has paved the way for the progressive development of architecture. So, who is leading? It was natural physics that inspired artists to rethink art. Now, who is the Master of Arts, when science informs art, and art informs architecture? Without Ernest Rutherford who described electrons orbiting the nucleus and without Albert Einstein who developed the theory of general relativity (1907–1915), there would not have abstract/concrete painting, and without Norbert Wiener, who formulated the theory of cybernetics and feedback in 1943 and published the book with the same title in 1948, there would not have been interactive art. The scientists constructed atomic models and models of the universe that are a total abstraction from reality, bottom-up constructed from abstract mathematical knowledge, proven through laboratory tests. As we know, and as they knew at the time as well, their models are static illustrations of otherwise dynamic processes. The dynamic nature of atoms and the universe must have been buzzing inside the heads of scientists and artists. The dynamic nature of the very small and the very big had a viral influence on the avant-garde artists at the time, the early 20th century. Inside the heads of the artists, abstract universes of pure form and color were taking over their worldview. No longer was art meant to be an individual interpretation of nature, but it was to become a new universal form of nature in itself, a new nature – nature 2.0. Scientific developments foreshadow technological innovations. New scientific insights in chemistry and physics paved the way for engineers to invent technological instruments and devices that trigger new developments in the arts. The inventions of photography, electricity, the automobile, the telephone, the radio, and, later on in the 20th century, television, the computer, the internet, artificial intelligence, and the Metaverse, led to new art forms that wanted to capture the dynamic nature of the new technologies. No longer did the

artists go into the fields to be inspired by nature, artists started to create a new form of nature. The artistic challenge has shifted from the natural and the human toward the synthetic and the super-human, toward the cyborg, the merge of the natural and the virtual, and toward a virtualized reality and a realized virtuality. Revelations by scientists and engineers of the until-then-unknown, aka the alien, are so strong that the front-runners of the arts inevitably have to reconsider their way of thinking. At the end of the 19th century, photography spawned impressionism and hence expressionism. While scientists and mathematicians were digging into the underlying principles of mass, energy, and gravity, artists started to rethink the underlying, fundamental principles of painting. In the first and second decades of the twentieth century, a group of artists rejected the idea of painting after nature completely and built up abstract universes from scratch, based on abstract notions of color, points, lines, surfaces, volumes, relations, force fields, movement, direction, infinity, and energy. The main actors were Wassily Kandinsky, Vladimir Malevich, Piet Mondrian, and Theo van Doesburg in painting, and El Lissitzky and Vladimir Tatlin in spatial constructs; they leaped from a two-dimensional way of thinking, from paintings, plans, and sections, toward three-dimensional constructs, applying the new principles on abstract sculpture and abstract spatial assemblies. Abstract thinking in science infected technology and the state-of-the-art of the technology inspires the arts. Dutch architect Gerrit Rietveld was one of the first to follow, stimulated by Mrs. Schräder; he designed what is now known as the Rietveld Schröder House, and then the Bauhaus architects took inspiration from the Bauhaus artists. Typically, architects are slow in applying the deeper meanings of what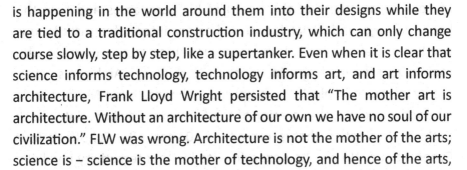
is happening in the world around them into their designs while they are tied to a traditional construction industry, which can only change course slowly, step by step, like a supertanker. Even when it is clear that science informs technology, technology informs art, and art informs architecture, Frank Lloyd Wright persisted that "The mother art is architecture. Without an architecture of our own we have no soul of our civilization." FLW was wrong. Architecture is not the mother of the arts; science is – science is the mother of technology, and hence of the arts, and hence of architecture. Only science digs deep enough to advance

technology and the arts and thus is entitled to the title of the mother of all arts. Without science, artists would have no clue where our civilization is going. Without science, technology, and the arts, architecture would have no idea in which direction to go.

When we have a look at our era, notably the developments that have led to the inventions of the personal computer and the digitalization of almost everything, we see the same process unfolding again: scientific breakthroughs, notably in mathematics, led to the invention of the first electronic computer. The groundbreaking paper titled On Computable Numbers by Alan Turing paved the way to build Colossus, the first digital programmable computing device that eventually was able to break the codes of the German Enigma encryption machines. The theory of cybernetics of the American mathematician Norbert Wiener inspired the Hungarian- French artist Nicolas Schöffer, born in Kalocsa in Hungary, to develop the Tour Cybernétique, realized in 1961 in Liege, Belgium. At the base of the Tour Cybernétique Schöffer had installed number-crunching computers that you could hear switching while performing the calculations. The Tour Cybernétique marks the rise of interactive art. Art would become dynamic, cybernetic, and interactive with the public. The cybernetic artists were the first to develop ideas for interactive architecture on the grand scale of architecture, not architects. Nicolas Schöffer wrote his groundbreaking book La Ville Cybernétique, which offers a concise blueprint of how creative citizens would claim control of their cities. Schöffer imagined the artists to become members of the board of big companies and turn the knobs of large industrial facilities. For my interactive installations and larger-scale projects, I am deeply indebted to those artists who took revolutionary leaps in conceptual thinking. Architects, even the avant-garde, typically take up the challenges later. My first truly interactive project, the interior of the Saltwater Pavilion, dates back to the early nineties, that is, more than 20 years after Schöffer's Tour Cybernétique. No one had experimented with interactivity involving digital technology in architecture before we (ONL and NOX) did the

Water Pavilion. The only examples came from the arts, promoted by institutes like the ZKM in Frankfurt and V2 in Rotterdam. I considered the whole building as an environment to interact with. In the early nineties, I took the speculative jump from interactive installations, where one establishes a relationship with the object from the outside, toward immersive interactive architecture, where one interacts with the environment from the inside. From the early nineties, I saw Another Normal coming – Another Normal, which is based on dynamic relationships between the constituent components including the people.

We Cannot Eat Coal

When the very small meets the very big, how would this apply to architecture? When the very small meets the very big, how would this apply to politics? In the field of architecture, my take on it is that we must find those simple rules that create the conditions for a wide variety of components. As seen from the outer helicopter view, it would mean introducing simple universal rules that govern the behavior of the smallest parts. As seen from the inner atomic view, it would lead to emergent behavior of the components that communicate with their nearest neighbors. The top-down imposed rules of play of Another Normal frame the bottom-up emergent behavior of the swarm of components. Information feeds both the very small and the very big. Information connects the components that are identified to be unique individuals to the swarm. Now, how is the unification between the very big and the very small played out in architecture and in politics?

The design of the constituent components of Another Normal has a political dimension as well. How would the global meet the local in an informed loop? In politics, I would argue that simple globally imposed rules of law must be designed to improve the conditions of local individuals on Earth. Simple rules create complexity, diversity, and opportunities. Not only for the privileged and not only for the poor but for each identified individual who communicates with the nearest neighbors. People move freely around while building components are typically fixed to one position. For both types of swarms, only simple rules may create diversity. In the time-lapse movie, the swarm of people behaves like a liquid, communicating with many different individuals, while the swarm of building components behaves like a solid that hardly moves at all, and the components typically communicate with the same neighbors during their lifetime. People and building components represent two distinct emergent phases that act and interact on the Internet of Things and people. My assumption that there is a common underlying information base for both the very small and the very big has consequences for the process of world-making of artifacts as well as for community building.

How do we get from our here and now to that Another Normal, that other possible world that is equally possible as the current now? When we continue walking along the same beaten paths, the best we can achieve is more of the same. Addressing climate change the way it is currently done only reinforces the bankruptcy of the current capitalist system which prevails almost everywhere. Unfettered capitalism inevitably leads to ever-increasing inequality, and thus to unsustainable poverty, which is the main cause of the rapid growth of the world population. Climate change and poverty are two sides of the same coin. Climate change hits the poor hardest. In the words of climate activist Vanessa Nakate: "We cannot eat coal, we cannot drink oil, we cannot breathe gas." Africa is responsible for less than 5% of global warming yet feels the devastating effects stronger than any other continent. A fair and just world must imply that Africans are entitled to a similarly high standard of living as the people in the Western world. And that would mean a drastic increase in global warming if the Western world does not change course drastically. As long as economic growth is propagated at the expense of a sound ecological balance, we may only achieve small improvements when we follow that doomed path of doing too little too late. Defending capitalism or promoting a more extreme form of anarcho-libertarian capitalism feels like blindly appeasing what has gone awry, looking away from what caused climate change and excessive inequality in the first place, and closing one's eyes to what needs to be done to switch to a possible alternative. Not everyone would agree on the causes, but everyone experiences that the global atmospheric conditions are gradually going out of control, and everyone should be appalled by the ever-increasing inequality. There is no common agreement on what the alternative to unfettered capitalism might be. Not many want to lose the sumptuous abundance of food, transported from all over the world, not many want to lose their ample choice of fashionable gadgets, not many want to refrain from their holidays abroad, their flights around the world and their luxury cars, and not many want to lose their comfortable homes, their five-star hotel stays, and their second homes. Soon, more and more people will want the same privileges as the people of the Western world, which will inevitably lead to even harsher climate change when the decision-makers do not change course. I propose a possible new normality, a possible society 2.0, named Another Normal, paradoxically facilitated by the ubiquitous availability of almost everything, based on the digital connectivity of almost everything and everyone. The main players to achieve this are well-informed techno-social components to form well-informed synthetic bodies, thriving in a well-informed societal fabric, inhabited by well-informed and well-connected inhabitants, thriving on a digital platform that includes both the things and the people, often referred to as the internet of everything. In Another Normal, everything and everyone are connected digitally, like in Nature 1.0, everything and everyone are connected analogously. An internet that is not dominated by a few private multinationals, but an internet in which people have control over their data, Web 3.0, is based on blockchain technology; a bidirectional internet, which is not a super global surveillance institute but a widespread distributed being a global connectivity machine. What kind of proactive thinking on the constituent components for a fair and just society is needed to find an alternative to the current looming situation, an alternative that will lead toward a more fair, resilient, and ecologically balanced Another Normal? Another Normal means that things are designed in participation and only produced on demand as and when needed. Another Normal is where the production of food, energy, and water happens as close as possible to where it is consumed. How can we as artists and architects contribute to such a fair and just world, without becoming an escapist? How can we make modern technology, computers, robotics, machine learning, and artificial intelligence work for all?

Personal Universe

For decades, I have worked on the fusion of art and architecture on a digital platform. With Ilona, I established a specific relationship between art and architecture, which cannot easily be generalized, as there are many different types of artists and architects. But there is one thing that nearly all artists share, which is the way of working that is inherent in their profession – their natural attitude to take full responsibility for what they do. Typically, artists are not working for someone but follow an internal drive to materialize the unknown, to visualize the unseen. Artists in general are self-propelled, proactive, and emphatic to scientific discoveries and societal issues, not serving anyone else but themselves. In general, artists have developed an ability to internalize a wide variety of external stimuli, transform that into a vision, and eventually make a work of art. While many of their fellow citizens are consumers and/or performers or executing predefined operations, artists are producers of new content. There are many different ways of being an artist, but they all share some of the above. I am interested in those artists who internalize the world around them and express themselves in a way that opens gates to the hitherto unknown, to the unseen; in short, to the alien. These artists are creating new personal universes, with their unique internal logic, existing in parallel to other universes. These artists do not paint after nature, they do not reinterpret an existing reality, they do not comment on societal situations, and they are not producing more of the same. On the contrary, artists intuitively create new realities that are equally real, and they create that other normal, equally concrete as the hitherto known, and likely more challenging and surprising than what was considered normal before. In their modus operandi, abstract painters, sculptors, writers, composers, and choreographers are closer to the informed speculations of natural physics theorists than any other group of professionals. Artists are not waiting for a client to work; they have this internal drive to express themselves and they take risks without the certainty of a financial reward. Naturally, stipends and income from selling the work are more than welcome, but the absence of such income does not stop the artist from working. Self-research, and eventually recognition of the self, drive independent artists. Even without recognition, many artists keep pursuing their ideas. Many artists hesitate to talk about their work because it is personal. How do you share your intimate personal universe with a stranger? Typically, artists find it difficult to explain what they intuitively confess to their canvases. Artists like: "Why don't you just look for yourself and find your own meaning." Artists typically do not want to educate others; they just offer new universes for them to explore. How to explain a universe of thoughts and a complex of deliberate and intuitive acts? Learning from an abstract artist's mind intimately near, I strive for a similar level of autonomy and self-driven research in architecture. An architecture that is, on the one hand, autonomous and alien, while, on the other hand, fully aware of what is going on in society at large, trying to synthesize the two extremes in bold yet coherent proposals. Sometimes I refer to our buildings as spaceships that are built in weightless digital space and successfully landed on Earth, after an intense communication between the two different environments that until then did not know each other. It is a key factor for successful communication that the communicating parties start from a sufficient degree of autonomy, a character, and logic of their own. Communication needs that tension field to open up the road to Another Normal.

The other thing that virtually all artists have in common is that they work per definition on a one-to-one scale. Art is what it is – art is not a model; in art, it is the work itself that counts. Moreover, individual artists are 100% responsible for their work; they cannot blame others for failures or mistakes – no one to hide behind. Taking full responsibility for one's work is currently mission impossible for architects. Only when architects are directly involved in all subsequent stages of a design, from financing to concept to modeling to the production of parts to the assembly and eventually to the daily operation of the construct, can architects take their share in the responsibility. Taking full responsibility for a larger part of the building chain is what happens when a designer establishes a direct link to the production of the parts, when they adopt an in-house parametric design to robotic production procedure, as today numerous start-up companies are experimenting with. Design-to-fabrication designers take more responsibility and gain greater freedom in charting their path in their productive life. Independent digital file-to-factory start-ups partly have a similar position in society as independent artists. They produce their work on a one-to-one scale, and they take full responsibility for the production.

Reversal of Values

While science and technology liberate the arts, the arts are the liberating force for architecture. In the last century, several radical technical innovations were introduced, building up a critical mass that is necessary to trigger further evolutionary effects. Industrialization was breeding new production techniques. Photography urged painting to reinvent itself. From impressionism to expressionism to abstract/concrete painting, all happened in a period of only a few decades. Both widespread industrialization and hence abstraction in the arts urged architecture to redefine its esthetics and the production of its constituent components. First, decoration was elaborately synchronized with structure, as in Art Nouveau, mainly as a secondary addition to the load-bearing structure. Later, decoration was abandoned completely in favor of mass production methods for the essential parts, as in the Bauhaus. The industry produced common goods for the many, not only for a privileged elite, and architects had to adapt to this new reality and were commissioned to design the factories and more healthy housing for staff and workers. Adapting to socio-technical modernism meant a new method of design thinking, inevitably leading to an until-then unknown esthetic. A series of mass-produced columns, windows, doors, and complete buildings that are the same soon became the dominant esthetic. Mass production of parts meant a reversal of values in architecture, emerging from technical innovations in the first place and subsequently given a boost by new ways of thinking in the arts.

The currently ongoing revolution is the digital revolution. Quantum physics has led to the advent of number-crunching computers, which in turn led to a range of new technical innovations. After World War II, the idea of ubiquitous computing inspired artists, and then architects, to indulge in utopian visions. Artists started imagining kinetic and cybernetic structures in the fifties, and in our times, artists are the first to design interactive installations. Computer games have become the most popular art form. As in interactive art performances, popular games request the active participation of the players, while the plot is unfolding in real time. Games are an art form that lives in the virtual real. Virtual reality is the frontier to Another Normal, which is shaped by the fusion of the real and the virtual. While digital technology informs art, digital art informs architecture. I realized back in the year 2000 when I was appointed professor from practice at the TU Delft that the IT revolution would liberate architecture from its very foundations. The IT revolution disenfranchises mass production to form the foundations of architecture; it reverses all existing values, and it reverses the paradigm of mass production into that of mass customization. Mass customization is the digital version of creating the unique; this time not for the few but for the many.

Performance art foreshadowed what was to become interactive architecture, which is unfortunately still in its infancy today. Back in 2012, I compiled and edited the thick-as-a-brick Hyperbody, First Decade of Interactive Architecture. In the first decade of its existence, Hyperbody staff and students developed many interactive prototypes, built to act on a one-to-one scale, establishing bidirectional relationships between the people and their environments. What interactive art has done to architecture is liberate architecture from Vitruvius' Firmitas, Utilitas, and Venustas. Finally, we can leave Vitrivius behind us, without second thoughts. In my Another Normal, architecture no longer needs to be strong, functional, and graceful. Strength (firmitas) is substituted by being agile and resilient in real time. Functionality (utilitas) is a useful byproduct of the alien that favors exploration and curiosity over predictability. Gracefulness (venustas) is an emergent property that is not imposed on the structure but happens by the bottom-up interactions between the constituent components. Firmitas, Utilitas, and Venustas are no longer architecture's primary goals, but at best possible byproducts of the agenda of Another Normal, which is to be there for the users when, where, and as needed, just there, just that, and just then. The 2020–2021 Corona pandemic accelerated society to turn toward inclusiveness and resilience, and toward the production of components only when, where, and as needed. The current pandemic stimulated the digital economy, which boosts general interconnectedness, augmented virtual experiences, and the blur of work and free time. Add to this universal basic income, omnipresent self-driving vehicles, and wireless payment systems, and we are well underway toward Another Normal, which is a well-informed interactive digitized environment.

Language, then tools, then instruments, and as of today real-time computation and personalized apps further evolve humans into augmented beings – augmented in the form of smart wearables, embodied in clothing, mobile phones, homes, basically containing miniature wireless transaction RFID chips, performing streaming identity verification checks; perhaps in the form of <u>subdermal implants</u>. In the era of the internet of people and things, there is no need for having a physical purse in your pocket.

Paying no longer requires queuing before a cash desk; one just collects the items and walks out of the shop, paying on the fly, while the embedded chip takes care of the instant communication with the scanner. The tagged products are probably scanned by putting them in the smart shopping cart and are automatically paid for when leaving the store. Virtual reality glasses or contact lenses will augment people's vision with an information overlay, to inform them about virtually anything they need to know. Users might want to take their pictures in a subtle blink of the eye, with no need to carry a physical smartphone. The AI revolution compresses daily routines into a continuous bidirectional flow of information, paying credits and getting credit points as rewards for certain activities. The credit points are stored in your digital wallet on the decentral blockchain and can be harvested without the interference of the centralized authority of the banks. VR allows you to be informed on virtually anything that is around you, including products in the store, but also the public information on people and things, of which the content is controlled by the people themselves. The people decide for themselves to what extent they will be visible in the public domain, as no big brother is managing their profiles. For such self-controlled management of one's profile, ubiquitous blockchain technology is needed, by which the people own full control over their data, verifiable by blockchain. The digital revolution is science at work. Artists jump in to create NFTs. Architects will follow and build the digital twins that, in real time, co-evolve with their physical counterparts. Digital twins will emerge soon in climate science, healthcare, product life, and then in urban planning and buildings. Such technological developments that inevitably take place in the immediate future are a prerequisite for the further advancement of socially fair and just politics, and, in my profession, for the democratization of parametric design to robotic building technologies. As there is no escape from the digital revolution of Industry 4.0, we might just as well claim ownership.

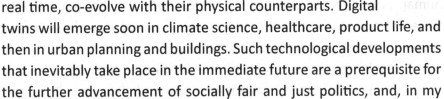

The Swarm

The analogy I typically use in lectures to describe Another Normal is that of the swarm of birds. Birds are the constituent components of the swarm. In the swarm, the birds keep a close eye on their immediate neighbors and respond in fractions of seconds to even the smallest changes in their neighbor's trajectories. Living in the swarm is a condition that we, as humans, experience daily as well, for example, when we drive our car in the seemingly chaotic traffic around the Arc de Triomphe in Paris. If in such dense city traffic without the indication of lanes on the street one car that is close to us changes course a little bit, we immediately adjust to that small change. We take attention to our nearest neighbors in the bustle, in a similar way as birds in a swarm. When we walk into a busy department store, we cleverly avoid bumping into each other. The smallest change in people's trajectories is immediately noticed by our built-in positioning system. Like we avoid falling by walking, we avoid bumping into each other by estimating in real time our distance from moving people and immobile things around us. Birds constantly measure the distance to their neighbors, and they minutely adjust their speed according to the speed of their neighbors. It is impossible to notice who acts first, the bird or one of its neighbors. The swarm feels like a smart fluid knowing what to do and where to go. There is an evident logic to it, yet utterly unpredictable in the long run. The patterns of the swarm and their destinations are full of surprises, at least from a vantage point outside the swarm. Birds move relatively fast when seen from our human perspective. We, humans, observe and think slower than birds move about in a swarm since the complexity and the multi-tasking capabilities of our brains come at the cost of the speed of observing.

The reference to swarms of birds is not randomly chosen. As early as 1986, Craig Reynolds programmed the groundbreaking computer simulation Boids, applying a limited number of simple rules to the behavior of simple triangular graphic objects in the computer. The three basic rules are as follows:

1) separation: steer to avoid crowding local boids and objects;
2) alignment: steer toward the average heading of local boids; and
3) cohesion: steer to move toward the average position (center of mass) of local boids.

The rules worked surprisingly well on a two-dimensional screen; the simulation of the movements looked quite accurately like a swarm. In the eighties, the exponential rise of the personal computer for the consumer market meant a game changer. I got my Sinclair ZX81 home computer in 1981; this prototype of the future PC could do only texts and perform calculations, as it came without a graphic screen. My designer instinct told me that this would become something big for architects. Typically, conceptual designers are per definition interested in rule-based design. I was into analog conceptual thinking and programming, soon to be enriched with digital binary programming running on the computer. Rule-based design and conceptual design thinking are close allies. Having worked with number crunchers like the Intergraph workstations mainly for 3D visualization of otherwise traditionally designed projects and having the personal Atari 1024 ST computer close at hand, also used by Ilona for art experiments of free sketching with the mouse of the computer, I began applying the newly acquired knowledge of swarm behavior on our conceptual design thinking. That is where and when the notion of the point cloud comes in, i.e., the digital twin of the analog swarm. Around 2000, I initiated to establish a group named The Swarm with the elder well-established Rotterdam-based architect Jan Hoogstad († 2018), with publicist Piet Vollaard of Archined, and with architecture historian, critic, and curator Ole Bouman, at the time editor-in-chief of Archis. We readily agreed on the common source of inspiration, but, unfortunately, we did not manage to find suitable projects or acquire commissions to put our teeth into in due time, and the initial enthusiasm to establish the Swarm drifted away. However, in ONL and Hyperbody projects, the concept of the point cloud, the referential basis for dynamic calculations between the components of any construct, stood firm. We developed the new normal of dynamic swarm behavior of constituent components further to the limits of practical applications. Since then, a dynamic point cloud of interacting components forms the referential basis for my Another Normal.

The Point Cloud

In the point cloud of reference points, the data points are behaving like boids, moving in space, occupying volatile positions described by floating XYZ coordinates. In the design process, points are dropped in digital space, and the points are connected to define a curve or a volume. The earliest 3D modeling program that I used was the affordable DesignCAD 3D software, which was launched in 1988 and still exists after some upgrades, essentially the same software architecture. This software still can be bought for a mere 100+ USD and has been developed further to communicate with Autocad and 3D MAX. The feature in DesignCAD that we used for intuitive sketching is that you can sketch from viewport to viewport, switching from top view to front view to right and left views to perspective view, as is now also possible with the Rhino Grasshopper software that we are currently using. The then-advanced Mechanical Engineering (ME) software of Intergraph that we used for the Artificial Intuition workshops in Galerie Aedes in Berlin in 1990 also offered the possibility to draw continuous polylines by subsequently clicking on the different viewports. Architects used to be framed by their university educators to draw plans and sections, which effectively promotes two-dimensional thinking. The design software that architects could use back then was programmed by engineers who developed the software for draftsmen rather than for designers. It was an eye-opener to discover the possibility of freely sketching in a virtual, weightless, three-dimensional space. It allowed jumping from Flatland to Spaceland.

We were sponsored by Intergraph to use their ME software and their table-sized workstations for the Artificial Intuition workshops that Ilona and I held in the Aedes Gallery in Berlin and a later sequel at the TU Delft. In an intentional intuitive fashion, the participants were asked to click points in space to trace free-form curves and surfaces. The participants in the workshop created alien constructs that none had imagined before, indulging in a spatial complexity that had not been seen yet. It was undecided whether the traced and assembled components should be looked upon as art or as architecture, which was left open for interpretation. For the workshops we initiated in the early nineties, Artificial Intuition in 1990, Sculpture City in 1994, and The Genes of Architecture in 1996, we invited both artists and architects. The sketches were to be intuitive, that is, without intention or meaning, value-free, weightlessly hovering in cyberspace, open for further interpretations into works of art or an object of architecture. The conceptual idea behind the workshop is that, only when one is free from preconceived ideas, the participants would be more open to accepting their sketches as raw material to work with. A line could become a structure, a surface could become a skin, and several skins combined into a volume, eventually identified as building components. Sketching in 3D virtual space is at the same time intuitive and exact. Per definition, everything in virtual reality is known, meaning that the data points positioning the components are fit for further development, without information getting lost in translation, which sets the precondition for a lean and mean design to production process. The initiated process is, to some extent, similar to the concept of the blind sketching method as practiced for a short period by <u>Coop Himmelblau</u> and their subsequent interpretation of their analog sketches into a built structure: "Created from an explosive sketch. Drawn with eyes closed in intense concentration; the hand acts as a seismograph, recording the feelings that the space will evoke." Their projects remained by and large clashes of loosely connected shapes, literally decomposed boxes, and while I was looking for more self-contained shapes, I wanted to use the sketch to disclose hitherto unexplored volumetric shapes. It is instrumental for our joint art-to-architecture shape development that Ilona is trained as a sculptor at the Willem de Kooning Academy in the early eighties. Ilona had long ago abandoned the use of platonic volumes, looking for more challenging shapes based on intuitive sketches. She intentionally wants to surprise herself. One of the key projects based on intuitive computer sketches is the sculpture titled <u>Folded Volume</u>. After analyzing a choice of 3D sketches, eventually, 20 data points in space are identified, a point cloud floating in space, without predefined meaning. Using the 20 points, triangles, rectangles, pentagons, and even one hexagonal surface are formed. Following the wordplay of Antonino Saggio, in his book <u>IT Revolution in Architecture</u> (2013), "Information also means in-formation, to bring into formation." Certainly, this applies to the Folded Volume; the data points are informed to become the folded volume. Folded Volume is not an autonomous sculpture but, intentionally, a sculpture that also can be interpreted as a 1:20 scale model for a sculpture building. Function and shape are disconnected, and the volume is set free, thanks to the new digital instruments – a volume that is set free from its function I referred to earlier as an "open volume." The 1:20 model is built by pre-fabricating and assembling strong wooden frames covered with fiber-reinforced PVC in green color. The constituent components form the whole by simply being bolted together. The idea of the ubiquitous unique component took shape by designing and building the sculpture Folded Volume, a crucial moment in the development of my view on architecture. One of the structural challenges of Folded Volume is the large cantilever that just does not touch the ground; it hovers a few centimeters above the floor. Since then, building on that intriguing spatial effect, I introduced large cantilevers in virtually all ONL projects, pushing the structural concept of a unibody, a structurally strong three-dimensional bodywork, as in airplanes, boats, and cars.

Reference Points

In *Towards a New Kind of Building* (pages 130–131), I introduced the notion of personal time. I stated that everyone and everything in principle must be considered to be the local and temporal center of the universe. Everyone carries with him/her the highest resolution "real" time, as opposed to the time zones of Universal Time Coordination (UTC). The communication between the personal time of a person and the personal time of another person needs instant bidirectional time conversion. Each person carries his/her personal time on the mobile phone. Calling someone would display the other's personal time. Everyone is at a local and global center of the universe. When one has agreed to meet at a certain place, then the real time of that place is also a local and temporal center of the universe, the reference point to which the personal time of the invitees is related while moving around. While the Earth spins toward the East, moving toward that meeting place from the West shortens the minutes on the invitees' watch, while coming from the East stretches the seconds and minutes the time the person has on his mobile phone. The concept of personal time makes the time zones redundant. Personal time is ultra-high resolution, exact to the fractions of a second, while time zones are an extremely rough division of time into 24 one-hour parts. According to the theory of general relativity, time is elastic and not divided into time zones. Personal time is general relativity in action on the face of the Earth. The direction one takes and the speed with which one travels determines the pace of one's personal time. To abandon fixed time zones in favor of super-local versatile personal time is a radical concept that is feasible when all devices with a clock adjust their speed of time in real time via a high-resolution GPS with their nearest neighbors and with those they are communicating with over longer distances. The very far and the very near are versatile parameters in the concept of personal time. As in the theory of everything, I turn the very big and the very small, the very far and the very near, and the very short and the very long into actors in one inclusive system. Nonstandard architecture is inclusive of geometrically simple standard architecture, while the other way around is an ultra-low-resolution approximation at best.

There is no such thing as one single center of anything; there are trillions of centers, as many as there are things and people and as many as there are stars and atoms, fields, and waves. Not even atoms have a center; they are varying constellations of dynamic interaction fields, where there is nothing fixed and everything in flux. Centers appear in the abstract virtuality of mathematics, phantom centers of gravity, functioning as temporal and local reference points. Imagine one *XYZ* coordinate system on the surface of the Earth. Now imagine another *XYZ* coordinate system randomly placed further down the surface. The three axes will point in different directions; the *Z*-axes point toward the center of the Earth, and none of the *Z*-axes are parallel, meaning that the zero point of each *XYZ* coordinate system is super-local, which is only valid for that single immaterial, mathematical, and abstract reference point. There is not one single Euclidian *XYZ* system for Earth, nor for any constituent component of anything. Each local center, wherever on Earth or inside the entire Universe, has its super-local coordinate system. The *X*-, *Y*-, and *Z*-axis system is a spatial referential framework, not a representation of three dimensions. One framework with three axes, positioning the reference points of the point cloud in space. I always found the concept of three-dimensional space misleading, as I feel uncomfortable with the concept of time as the fourth dimension. Concerning earlier conceptual ideas about information as the source from which matter emerges, that is, matter as seen through the eyes of other materialized beings like us, information may be the only dimension there is, while matter only can be measured by another form of matter, would be an emergent property of information, and time, that can only be measured by movement, an emergent property of informed matter. Time will tell whether exceptional personal time will become the rule for Another Normal.

In the dynamic tension fields of the universal flux, both things and people are the acting components, well aware of their immediate neighbors, whether interacting at a short distance in physical space or over long distances, wirelessly connected or wired via the internet. People and things behave like birds in a swarm. Birds keep a negotiated distance from each other, whether flying in formation in the air or settling down on an electric powerline. While sitting down on a powerline, the idea of the swarm still is maintained, while the birds still respect an agreed social distance. In the current Corona times, we are more than ever aware of the social rules of keeping distance since people have to adapt to new rules of social distancing, as from February to March 2020, imposed by the pandemic. More than ever, people and things are aware of each other, estimating the distance to each other, checking their speed changes, and changes in direction, and spatially negotiating which path to follow to avoid close contact. In pre-Corona times, people were just unconsciously flocking the streets, in the department stores, in rallies, and at home. People always have followed unwritten social rules to keep a proper distance. Due to the pandemic, there is a growing awareness of social distancing; the parameters have changed, and will change again in Another Normal, while the rules by and large remain the same.

The adaptation to new parameters for the social rules during the pandemic is the equivalence of a culture shock. This culture shock is not something that happens when we visit other countries; this culture shock comes from within. What was believed to be normal is no longer normal; one readily adapted to another normal. Another normal that is based on scientific data is effectuated by digital applications. The new normal highlights the super-locality and super-temporality of our positions in time and space, administered by apps on mobile phones. One's position, measured by a mobile phone, comes with a verifiable relationship with neighboring people with mobile phones. Each phone component has its local coordinates and, as a consequence thereof, its local private time, living in the here and now. Components, people, and things inform each other about their position and time. Time is not absolute; real time is not divided into time zones. According to quantum theory, time is not in a super-fluid flux but moves on by tiny Planck scale steps from instance to instance in Einstein's space-time continuum. Each component is relentlessly repeating: "these are my coordinates, and this is my time." Like birds do when singing, like dogs do when barking, they communicate a simple message: "I am here, now, I am here, now," with the purpose of either attracting attention or scaring away. Reference points for complex nonstandard constructs do the same as birds and dogs: they inform their neighboring points where they are and when they are.

Points in space are not fixed points but dynamic fields; they are not static but always on the move, like the birds in the swarm, the dogs barking, and the automated vehicles finding their way on the road and changing position in space and time all the time. In principle, components in a building body behave similarly to the birds in a swarm. Each component has a floating neighboring component that it is connected to, by maintaining the rules of behavior: distance, direction, and speed. Gravity spoils the freedom of movement of building components across the swarm. Building components are looking to their nearest neighbors to transfer their gravitational loads, to components below, like the components in the Tetris game. 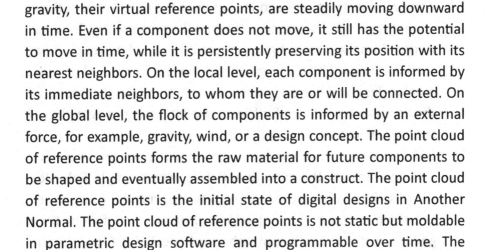 I endlessly played such gravity games during our long flights to the USA and Japan in the early eighties. Their centers of gravity, their virtual reference points, are steadily moving downward in time. Even if a component does not move, it still has the potential to move in time, while it is persistently preserving its position with its nearest neighbors. On the local level, each component is informed by its immediate neighbors, to whom they are or will be connected. On the global level, the flock of components is informed by an external force, for example, gravity, wind, or a design concept. The point cloud of reference points forms the raw material for future components to be shaped and eventually assembled into a construct. The point cloud of reference points is the initial state of digital designs in Another Normal. The point cloud of reference points is not static but moldable in parametric design software and programmable over time. The point zero of each building component refers to at least one reference point. There is nothing more elementary in design development than these referential data points. Every detail and every specification can be ultimately traced back to its initial reference point. By moving the position of one or more reference points, everything related to those points will change with it, and the shape of the component will adapt to the new configuration of points. As hair follicles are inward folds of the surface of the human skin, thereby specifying the topology of the skin in greater detail, in architectural design, we add new reference points, parametrically related to the initial reference point, for further specification of the design, i.e., the points defining the shape of the component and the connection details. In reverse, data points also may be deleted to simplify a design.

From Point Cloud to Building Body

In 2009, I summarized my vision of architecture in 10 compact points. These are my design guidelines. As of today, I would not change any of these points. The body parts are what I refer to now as the components.

1) Start with a Point Cloud. Start imagining your Personal Universe of free-floating immaterial dots in endless space. The immaterial points of the Point Cloud form your basic design material. Your brain will construct connections between the dots. Some of these connections will be strong; others will be connected by weak forces. Some will form tight groups; others will be much more loosely related. All are constructed from this endless, shapeless, meaningless, and dimensionless Personal Universe. It contains all the necessary substance for the imagination and the evolution of conceptual ideas.

2) Define internal drive. Impose an internal drive on your Personal Universe. Inform the individual Points with a simple behavior. Write scripts to have them communicate with their neighbors. Let them execute the behavioral program and follow their genetic instinct. Let the members of the Point Cloud swarm interact to configure your unique architectural concept. Let them work for you and let them play your game, but take care

that they strictly play according to the rules. Architecture is a rule-based game.

3) Define external forces. Each Point Cloud is a temporary and local densification related to other Point Clouds/Universes. For each new design project, your Personal Universe is affected by external forces, redirecting the size, position, shape, and meaning of the Points in the Point Cloud. The external forces represent the climatic and physical context in which your Personal Universe is embedded. The interaction between external and internal forces upon the swarm of points of the Point Cloud forms the basis for the evolution of the architectural concept.

4) Create powerlines. Put attractors and repellents in and around your swarm of behavioral points. These attractors/repellents can be points, splines, or surfaces. We call the linear attractors powerlines. The attractors attract the points of the Point Cloud to populate the lines or surfaces. These points can be further used to form the nodes of a tessellation, for the structure and/or for the skin. Repellents can chase away points to clear the space from points, leaving room to move in between the swarm of points. These areas can be later used as functional space, for linear spaces (passages), and enclosed spaces (rooms).

5) Shape the building body. A set of powerlines organizes the positions of the main modal points of the structure, like birds on a wire, which represents a special configuration of the swarm. Typically, we advise using the powerlines as folds in the building body. The simplest form to create the fold is to span rules surfaces between the powerlines. But we often use more complex outward bulging doubly curved surfaces as well surfaces. Look at car body designs to understand the communicative power of the powerlines. You will never be bored again in a traffic jam.

6) Specify body parts. Each body plan evolves through a process called specification. Certain parts of the building body specialize to be the structure; other parts specialize to be the skin, again others the internal empty spaces, and others the arteries aka the MEP installation. There is a specific instruction in the evolution of the genes for the body plan for each of those functions. At the same

time, all points/cells of the system keep communicating as members in a swarm. They become members of a specialized sub-swarm.

7) Develop the parametric detail. Develop each specialized swarm of nodes of the structure or the internal/external skin, either enveloping spaces or enclosing gas/water/electricity/airflow, according to a parametric detail. Make all details one big family, where all family members share the same detail in principle but with local and temporal different values. They share the same formula but from point to point changing values for the parameters. Aim at limiting the number of different parametric details, and try to integrate as many details as possible into one complex systemic parametric detail. One building, one detail.

8) Build your personal BIM. Build your personal building information model. Do not import doors and windows and such from a library but develop your personal library. Reinvent all parametric details from scratch, based on the availability of basic materials. Create your own added value to the basic materials, and do not accept added value as created by others. Create your own building families of components and inform them with great exactness. Your personal BIM labels in painstakingly precise detail what information is needed to build your design. This information you must communicate directly with the manufacturer, not the contractor.

9) Write scripts for CNC fabrication. To be sure of a direct relation between your BIM and the actual fabrication, you must write your scripts to link your machine to their machine; this is called machine-to-machine (M2M) communication and file-to-factory [F2F] fabrication. Design such as fabricating only by CNC (computer numerical control) machines. Avoid bypasses but make sure the manufacturer imports your data directly, without rebuilding 3D models and rewriting scripts. Talk with the manufacturers and prepare your data in such a way that they can be used unconditionally.

10) Assemble the body parts. Finally, customize your body parts to form the complete 3D puzzle of your building body. Assemble them following the principle of dry montage, using bolts or even without. Prepare your components to avoid local adjustments

on the building site. Do not use molds, scaffolding, or other temporary support structures. Build your building only once. Make sure they can be disassembled after they have fulfilled their purpose. Design your building components to be recycled or to be used as basic material for other designs. Think C2C.

Parameters

Typically, at the start of a new project, I do some basic calculations. It may start with one figure indicating the total number of required square meters. This single number is represented by one single reference point, which typically resides in the center of gravity of the volume, or in a point zero. Each project is, in its initial phase, visualized as a quantified volume connected to a reference point. That volume is like a single cell that can be manipulated and divided into smaller cells over and over again, whereas certain groups of cells specify into one direction, and others into another direction, developing into the different functional parts of the whole body. The initial volume can take any shape; it is only the quantified volume and its space and time coordinate that count. Any complex shape described by more reference points than the eight points that describe a simple cube that contains the same number of cubic meters will do. It must be guaranteed though that molding the complex shape does not alter the given volume when that is considered a temporarily fixed target value, to keep the integrity of the given task to build the target amount of m^3. It means that if we pull at one side, the shape as a whole has to adjust its shape to keep the target value. In the background runs an algorithm that calculates the volume instantly and adjusts accordingly relative to the reference point. The target number of cubic meters is a global parameter, while the adjustment of the shape is driven by local parameters. Either the global parameter or the local parameter can be kept fixed or volatile. When the global parameters are fixed, the local parameters will adjust, and vice versa.

Following my guidelines, Hyperbody's Ph.D. candidate at the time Tomasz Jaskiewicz programmed a parametric design instrument Swarm Toolkit (2004), a design game written in the game development software platform Virtools. The Swarm Toolkit has in the left side menu the global and the local parameters that quantify the components that are placed in the action field. The global parameters quantifying the components are total volume, the global floor space index (FSI), local FSIs, footprint area, attractor strength, influence area, the position of the component in the vertical Z-axis, color, type of shape, and transparency. Components can be placed along the X-, Y-, and Z-coordinates in the playing field. The parameters can be typed numerically or by using sliders to inform the active components in the field quantitatively. The components are addressed either as individual ones or by a global FSI that applies to all components. Everything from a building component to a house or an office building, a car or a carport, and even people and a tree can be a component. A group of components can be assigned a unique representation in either cylindrical or spherical shape, in a choice of color and level of transparency. It is advised to have the components semi-transparent to indicate the conceptual character. The outcome of playing the Swarm Toolkit design game is not the architectural design itself but the quantified framework for the design. The design game starts with placing one single data point, which is positioned at the bottom center of one single component that encompasses the target amount of m^3 in one single volume. As with lifeforms on Earth, the design game starts with one single cell that subsequently splits, couples, specifies, groups, and aligns. The volume is represented in the form of a semi-transparent cylinder or sphere, with the same volume. To represent the initial volume as a sphere avoids making the standard assumption that a building is a cubical block, as all sorts of shapes are possible in the further growth into a building. In the conceptual design phase, arbitrary design decisions must be avoided as far ahead in the design process as possible. At ONL, we have used the Swarm Toolkit in some projects, to quantify the programmatic content spatially. The initial component can be freely placed and dragged anywhere in the Swarm Toolkit playing field. In the menu bar at the right side of the screen, there is a set of operators. One of the operators splits the volume into two equal volumes that together represent the same volume as the initial one. The split cells can divide again, and a selection of divided cells can merge into a conglomeration, such that we can create any number of smaller components, any conglomeration of constituent components, and any variation in the respective volumes. The individual component or group of components is qualified to represent a function. When we change the size of one of the smaller components, all other components will become slightly smaller, when the value of the global parameter for the FSI has been fixed. Thus, the overall volume of the whole project is, at any stage of the design, guaranteed. Whatever operation is performed on the components, the total volume remains always the same as the initial value that has been chosen. Each volume can be moved around and interact with neighboring volumes. Components can either attract or repel their neighboring ones, whereas the area of influence can be set by one of the sliders in the left menu. Components can be linked to any other component, whatever the position of that component is in the field. Two components can be linked to having a special relationship with each other, while the distance between any of the two components can be set at will. This feature is used, for example, to fix the proximity of objects of different nature like a house and a car, or a tower and a pavilion containing services like restaurants and cafés. If the position of the house moves, the car moves with it, and when the tower is moved, the facilities move with it. Any function can be assigned to the components for its qualification and represented by a color. One of the strongest features of the Swarm Toolkit is the linear attractor or repeller that selects and magnetically aligns dispersed components by attracting the components to the line along its length or repelling them from the line. The strength of the attractor, which becomes a repeller when we choose a negative value, and the area of influence are chosen by using sliders, a powerful function that is often used. Components can be split endlessly and rejoined into any conglomeration of components and in any desired level of detail. The Swarm Toolkit design game works perfectly for complex urban design tasks. At ONL, it is used for the 2,000,000 m^2 development of the Manhal Oasis masterplan for the abandoned palace of the former sheik and founder of the UAE, Sheikh Zayed bin Sultan Al Nahyan, in the heart of the peninsula of Abu Dhabi. Every component is quantified from its initial condition to the smallest subdivisions in its subsequent steps. Each component is informed by setting the quantitative and qualitative parameters and dragging the mutually related components around to visualize the design concept. The Swarm Toolkit is a design instrument that guarantees that in each stage of the concept design, each component and group of components are quantified and qualified.

The elementary building block of the swarm is the informed point in space, within the program or interactively in real time communicating with its immediate neighbors. The swarm is the default initial condition of my Another Normal, whereas the constituent components are precisely known, traceable, verifiable, quantifiable, and qualifiable. The outcome of the design game forms the quantified and qualified basis for further modeling and scripting in commercial design software. Hyperbody did not bother to develop the Swarm Toolkit into a commercial product; Hyperbody was primarily research-driven, i.e., not focused on commercial exploitation. If Hyperbody still would be active today, we would probably have set our minds on the commercialization of the unique interactive, participatory Swarm Toolkit. Bob McNeil of the 3D modeling program Rhinoceros had the commercial insight that a graphical interface such as Virtools would have a potential for architecture. Probably inspired by Hyperbody's state-of-the-art research output using game development software that is based on a graphic interface, McNeil came in contact with David Rutten of the Faculty of Architecture of the TU Delft to program a graphic interface on top of Rhino to parametrically manipulate the relations between the 3D components, using operators, sliders, and modifiers. The Grasshopper interface is more user-friendly than the Virtools interface, but there is also a crucial difference, which makes the two pieces of software incomparable. Virtools is a time-based software, developed for game design and interaction design, while Grasshopper does not have that option. Virtools is a perfect fit for programming real-time behavior, to simulate and drive interactive installations. Hyperbody and ONL use Virtools to act in that other real-time, transparent, and verifiable Another Normal that is bound to become a new normal.

Inclusive

How do we go from points to lines, from line to surfaces, and from surfaces to volumes, which make up the spatial dimension, and to time, that other dimension that has emerged from the spatial dimension? Time is a dimension that defines anyone's or anything's historic position in space-time. There is one spatial dimension with three axes to describe a position in space and one-time dimension that measures how it changes position. On Earth, two of the three axes are the longitude and the latitude meridians, and perpendicular to the surface, crossing the local *XZ* position on Earth, the third vertical *Z*-axis. As a consequence, the *XYZ* system is super-local and in a different position for everyone and everything. Everyone and everything have its local verifiable coordinate system, its identifiable spatial dimension. On top of that, people carry with them their personal local time, while time is different in relation to the planetary system for every spot on the surface of the Earth as well. I became aware of the super-locality of the *XYZ* system of coordinates when modeling the space stations around the globe for the Global Satellite workshop we held in 1991 at Museum De Zonnehof in Amersfoort during the Synthetic Dimension exhibition. None of these space stations shared the same *XYZ* in the digital model. Projected on the surface of the Earth, the *X* and *Y* axes are curved. In a discussion that I had with Henk Meijers of Meijers Staalbouw when detailing the Cockpit in the 1.6km long Acoustic Barrier, we figured out that even in the relatively short length of the barrier, the deviation from a straight laser line, as straight as the line in our digital space, would be a whopping 3.2 cm. You can visualize the curvature of the globe with your own eyes. Go to the beach, take a long ruler with you, and hold it horizontally against the horizon. You will see the difference between the straight ruler and the indeed quite strong curvature of the Earth. When objects are constructed, they are constructed in themselves, in their self-contained spatial dimension, defined by their *XYZ* sub-dimensions, in weightless digital space, referring to a specific point on the surface of the Earth.

The Russian painter Wassily Kandinsky has written a book about the three sub-dimensions of the spatial dimension titled, Point and Line to Plane (1926), setting the tone for a scientific approach toward art. Kandinsky did not go beyond the notion of the Plane; as a painter, he voluntarily restricted himself to the flat surface of the painting. Architects typically restrict themselves to the dimensionality of the volume to describe spatial designs. Architects work from point to line to surface to volume. To describe my interactive architecture projects, I add time to the equation as a temporal dimension – point to line to surface to volume in time. From reference points to curves to double curved surfaces to open volumes to real-time behavior, paraphrase the jump from the rigid painter's canvas to Another Normal's versatile space-time. I prefer the word curve instead of the word line, while a straight line is a poorly informed curve. A curve connects a multitude of points in space, while a straight line is the shortest connection between two points. A straight line can never become a curve, while a curve can be straight. Curves are inclusive of straight lines. When all points are aligned, a curve with many points on its trajectory appears as a straight line. But then again, it is only to be distorted by the relativity of space-time beyond our earthbound vision. I typically use the inclusiveness of graceful curves to orchestrate the population of reference points. In the 160-m-long A2 Cockpit, embedded in the 1.6-km-long Acoustic Barrier, I described the inclusive curves as long elastic lines. The curve is inclusive for all possible positions arranged along this elastic line. Similarly, double-curved surfaces are inclusive of all other kinds of surfaces, simply by varying the coordinates and the parameters for the handles on the vertexes, defining the amount of topological curvature from one point to the other. A smooth double-curved surface is a "non-uniform rational basis spline" (NURBS), a mathematical model commonly used in computer graphics for generating and representing curves and surfaces, whereas the shape of the surface is determined by control points positioned exterior to the generated surface. The roundness of the surface can change from sharp to smooth just by manipulating the control points. In the time dimension, real-time behavior is the default state in the Another Now of space-time, whereas fixed objects are per definition frozen instances from a dynamic process. Everything moves, even rocks move, albeit very slowly to our human perception. Real-time behavior is inclusive of all possible changes in the shape of the constituent components. Real-time behavior is the inclusive informed state of any organic or inorganic body living between the subatomic quantum fields and the cosmic event horizon, including our bodily selves. Our daily reality unfolds somewhere in the thick of it, at a particular spatial and temporal position in space-time, out of trillions of other possible space-time constructs.

Just There, Just Then, and Just That

The logic of inclusiveness applied to architecture naturally leads to the "just there, just then, and just that" paradigm of Another Normal. A paradigm shift is an emergent outcome of proliferating scientific insights and technologies, usually arriving via the arts into the domain of architecture. The paradigm shift triggered by quantum physics leads to an architecture that is process-oriented, cross-disciplinary, participatory, and ultimately unpredictable for everyone who is involved in the design process. The paradigm shift caused by the implementation of swarm logic inevitably leads to an architecture that acknowledges the active role of the smallest building blocks, dynamically connected to constitute the bigger picture by following a simple set of basic rules. The paradigm shift from the traditional manual modeling of floors, columns, walls, and roofs into a generative field of elementary point clouds of reference points leads to an inclusive design practice that connects the phases of the design process dynamically. The paradigm shift from static elements to components that behave and interact leads to an architecture that adapts to changes in real time in exterior and interior conditions and changes in user-driven requirements. Buildings are complex adaptive systems. The paradigm shift from static 3D modeling to dynamic parametric design and scripting leads to an architecture that naturally deals with gradual transformations of nonstandard complex geometries. We are witnessing the paradigm shift from the traditional design methods that lean on complex rules with many exceptions, toward an approach that starts from an inclusive nonstandard complexity based on simple rules without exceptions to the rule. Nonstandard architecture is an inclusive architecture that is visually diverse and functionally multimodal and adaptive – an architecture that facilitates full integration of structural, climatic, and esthetic aspects. Time-based architecture leads to design to production to operation strategies, whereby the constituting components are designed and produced in just that unique moment in the process of conception, for that unique position in the larger whole, assembling to just that unique shape, for that unique performance at that particular moment in time. When needed and as needed, just that, there, and then.

Measuring is Knowing

Adaptive architecture senses its environment in real time, processes the information, and shares the processed information with its immediate neighbors, whether in its immediate space or with the speed of light in space-time. Amid the COVID-19 pandemic, it strikes me how similar the basic condition for a truly adaptive architecture is to the spread of the virus. Managing the spread of the virus comes down to a social condition where every single person, be it the bird, the virus, the people, or the things, measures and processes its position in real time to adjust their distance from other people. To maintain the integrity of the swarm, the members of the swarm never stop measuring, processing, and acting. As we will see, this non-stop process applies just as well to the adaptivity of a building, and the resilience of society at large. Dr. Tedros Adhanom Ghebreyesus, the Director General of the World Health Organization (WHO), clearly pointed out in his regular press conferences in February–March 2020 that to combat the virus,

one must acquire knowledge, and that means "test, test, test." Testing is knowing who has the virus, when s/he was where, whom s/he has been interacting with, or who already has had the virus. "The moment you know you know you know," sings David Bowie (2014) in his swan song "Where Are We Now." When you do not test, you will not know, and you do not know how to work together to combat the virus.

Surprisingly, many countries have responded late to WHO's essential guidelines and are still facing the consequences as I write. The urgency to do as many tests as possible has been largely denied. Alternative facts have been fabricated, as the most adopted practice was to test only those who have severe COVID-19 symptoms. But one needs to know before the symptoms have developed. To be able to act as one coherent swarm, everyone must be tested and inoculated. Everyone must be included as a member of the swarm, with no exceptions. Either everyone is in the swarm, or there is no swarm. Being informed is one thing to build the swarm, processing that information must be the immediate follow-up step. Processing in real time means acting in real time, keeping a distance, self-isolating, avoiding becoming infected, avoiding infecting others, wearing masks, and vaccinating. Processing information in real time means staying an active member of the swarm or dropping out, like a bird falling from the rooftop, as sings Anouk (2013) in "Birds"; there is nothing in-between. Self-tests must be repeated regularly, as there is always a possibility that one has caught the virus after the test. Living in a dynamic swarm means adjusting one's course in real time, based on incoming data, by processing that information and acting accordingly. Measuring is knowing, as is the well-known wisdom of engineers. Not wanting to measure equals refusing to know and obstructs an informed collective process. What is true for adequately responding to a deadly virus spread is equally true for the agility of complex adaptive systems like a fleet of autonomous vehicles and like an ecology of programmable constructs. To combat the virus, precise and correct information is needed, distributed to the participating constituents, and precise and correct information is equally needed to build up the resilient structures of Another Normal. The building components of the hyperbodies of Industry 4.0 are per definition well-informed; they process incoming information following some simple rules and act accordingly. Hyperbodies behave proactively in the eyes of their immediate neighbors, while *de facto* this input > processing > output process is happening in real time. Hyperbodies respond in a fraction of a second to incoming information in Another Normal. The emergent behavior of ubiquitously interconnected people and things is by definition surprising, not predefined; the outcome of any emergent process is inherently unpredictable, and it may feel irrational, and even bizarre. This is what we see and feel when watching a swarm of starlings; we cannot predict where they will go next, when they will settle down in the tree, neatly align on a powerline, or when they fly out again in mesmerizing formations.

Informed Distancing

My takeaway from the COVID-19 crisis is how beneficial it would be if we could vary the social distance toward other people in real time, for example, by using smart wearables in a smart environment. Informed distancing is when the environment informs you, when you inform the environment, when you inform your nearest neighbors, and when they inform you. As in the swarm of birds, birds supposedly are looking at their seven nearest neighbors to plan their course. We, as informed humans, do the same in natural circumstances. The human condition of standing and walking is one of not falling and not bumping into each other. We balance our bodies, and we balance our position towards neighboring bodies. If one person steps forward, the other automatically steps backward, as if they have magnetic fields around them. One keeps proper distance from others in shopping centers, just enough not to bump into each other. People sit in a highly organized manner in cinemas and theaters, like the birds that are neatly arranged on a powerline. Unconsciously, one knows perfectly well how to behave in the swarm, informed by one's natural senses. Now, what happens if natural senses are augmented with an additional set of senses, for example, with RFID chips in combination with miniature force feedback actuators, invisibly built-in into our clothes? The distance toward the nearest neighbors can be programmed when another pandemic strikes again, to protect oneself against too close contact. A benign force feedback signal might inform us when we are taking unnecessary risks. Red alerts on your mobile might warn you when entering an area that is not safe. Individuals may set their personal preferences on how much distance they want to keep, bottom-up style, to be negotiated with top-down interference, when the government imposes a minimum advised distance to protect its citizens from getting infected with the nasty virus. The values would dynamically vary depending on whether one is walking the street, sitting in an audience, or taking the train. In Another Normal, there is nothing like one size fits all. Individual people set the bandwidth for the parameters to fit their preferences and capacities. Distances would vary over time, in real time calculating one's position to other people and other things for the many different situations one maneuvers oneself in. My smart car is already doing that in relation to other vehicles, so why don't we? My car has sensors in the front, in the back, and at the sides, which warn me if I tend to come too close, and correct proactively the direction that the car, aka me as the driver, is taking. The force feedback is hardly noticeable, but just enough to make me aware of the potential danger of the situation. I can override the suggestion with my grip on the steering wheel. My car gives me a hint, a bit of well-meant advice. In the automotive world, this is already the new normal, and all new cars have it. When driving, I have become one with the car, in a fusion between the human body and the car; so why wouldn't I want to become one with my digital self?

Such a new normal informed behavior would have a huge effect on spatial planning and architecture. In general, we will require much more space for pedestrians, which is facilitated by the fact that autonomous cars do not need the same space as human-driven cars. Cars that are augmented with sensors and flocking algorithms take up much less space in the streets, while the algorithms distribute the cars in an optimal variable density and thus guarantee a maximum smooth traffic flow. Traffic lights will become a relic from the past. The space that currently is reserved for cars only will be reclaimed by pedestrians and cyclists, who will need more space for dynamic distancing. Pedestrians will not just walk and sit on terraces but move themselves using small electric scooters and bicycles as we see already thriving in the streets of bigger cities. As I write, there are thousands of electric bicycles and step-on rollers in Budapest like from Tier, which you can use from anywhere and leave anywhere. There are different competing brands providing the same service. The GPS coordinates of the e-steps, e-mopeds, e-bikes, e-scooters, and e-rollers are known, and a service task force takes care of maintenance and relocates the vehicles when and as needed. The electric vehicles are active components, actors in the network, merged into one with the person it carries. Each acting component has a unique number, is traceable, and is known by other components. All is based on knowing, being known, and being informed by simple rules. Rollers and bikes will soon be equipped with sensors and algorithms to guarantee that no clashes will happen, as we have sensors in the cars and, as is already widely used, clash detection software for building information modeling. This new form of small-scale electrified transportation, which is a direct extension of our bodies, claims space on the streets. At the same time, automated car traffic will inevitably reduce their claim on the available space. The only barrier to getting the new normal implemented sooner and at a larger scale is subject to politics and commercial considerations; it is not a question of available technology. As for the design of buildings, they will claim the freed-up space as well. The informal constructs in front of the buildings, small pavilions, terraces, bushes, and trees will reclaim the streetscape; the space in front of buildings will become an extension of the interior of the buildings. People in the new normal, Another Normal, move at individually varying speeds, in real time exchanging data with their immediate neighbors, informed by their smart wearable devices. It is only a matter of time before the environment itself becomes more responsive to constantly varying circumstances, proactively self-proposing changes in the environment, in real time adapting to the fluctuations in density, the changing behavior, and the varying health conditions of the users. Buildings and the urban spaces in front of them will become hyper-adaptive to changing circumstances; the pavements and the streets will merge to become livable. Streets and buildings will merge to form a fluid urban continuity, a shared space inhabited by the real and the virtual, as in Another Normal.

No More Queuing

A special form of distancing is queuing. When many people come together at a certain time at the same address, they have to wait in line. Queuing is a special form of swarming, an informed form of slow swarming, with one neighbor in front of you, and one neighbor behind you. It is similar to the cars waiting in line before the traffic lights, similar to the birds sitting beside each other on the powerline, and similar to waiting for a telephone connection through an automated call center. At a Business Exchange Forum in 2016 in my then hometown Rotterdam, I asked the innovation manager of Dubai World Expo 2020 how she would manage crowd control, which is a form of top-down managing the swarm. To my pleasant surprise, she had a challenging point of view, apparently prompted by the recent Dubai policy to promote happiness as an important societal factor. The innovation manager straightforwardly said that her ideal picture is to avoid any queues at Expo 2020. Remembering well the long tiresome queues we got stuck in at previous world expos of Milan and Shanghai, it would be just a stroke of genius if that could be realized. I do not know what solutions she had in mind, but the solution is of course to have personal time slots registered on an accessibility app for the entrance to the Expo site and for the individual pavilions. If only one could just walk into the Expo site and the pavilions without waiting for hours in the heat of the day. Only that person gets in at that particular time, through that particular gate – just that person, just then, and just there. It is not a matter of the daily capacity of the facilities, but ultimately a matter of personal time management, to avoid peak moments. Customized time management would facilitate the public to be identified, be guided to walk fluently to their destinations, and be alerted by the app in due time, while the app knows your location coordinates and the time needed to go to the pavilion. People would be smoothly swarming while securing COVID-proof distance to each other. People, space, and time need to be tagged. Everything and everyone have to be bottom-up connected to the nearest neighboring things and the nearest neighboring people while receiving top-down information on the amount of time expected before entering the pavilion. On the scale of the city, the Expo visitor will be fetched by an automated taxi service at the right time from any place in the city, or even in the world, delivered at the personally assigned time slot at the right porch at the Expo site, and then the visitor just walks unhindered by fellow visitors into the target venue. If one might arrive later or earlier than originally planned, due to unforeseen circumstances, the order of visitors getting in may adapt automatically. The application knows the exact location and the speed of all people who have registered for a visit and share that new information with you in real time. Think of Tickets, Booking, Google Maps, Uber, Amazon, Meetup, Whatsapp, and Instagram combined. Thanks to the new standards for interoperability in Another Normal, your broadband smartphone connects all platforms on which you are registered to deliver you there and then.

Delivery services are an absolute convenience and have become increasingly precise in their scheduling of delivery. We order online food, books, clothes, and anything we need to live our daily life. During the pandemic, home delivery services have proven to be essential for upholding economic activities. At first sight, delivering to each customer individually might seem an unsustainable luxury, a waste of energy, but the opposite is true. It takes less energy to distribute the products to the customers at their homes by one delivery car taking the fastest and shortest route, using a traveling agent algorithm, than when individual customers would take their car to the shopping center, often dozens of miles away. The total amount of kilometers driven by the delivery car is substantially less than the total kilometers traveled by the individual cars. Home delivery is more efficient and cheaper for both the producer and the customer. On-demand produced products may be delivered directly to the customer, i.e., not brokered by the Lidls, Auchans, and Tescos. Ubiquitous delivery of almost anything builds an open framework for the digitally augmented nomadic international citizen, who wants to go places where one has never been before, and still be personally served with anything that is needed for living one's customized life at any time at any place. I repeat, just that, just there, and just then, when needed and as needed.

Real-time Banking

In the new normal, the inevitable Another Normal, payments will be handled in real time. Checking your bank account on your phone means seeing the meter running, like your electricity, water, and gas meters at home. You would see money coming into your account in a steady flow of millions of discrete small amounts, probably measured in cents per second, and you would see your expenses subtracted from your account in an equally smooth fashion. Incoming and outgoing payments will be spread over agreed periods, and one would not get or spend the whole amount in one go. Thus, the flow of money can be managed to run smoothly. The subtracted taxes and interests will be spread over longer periods, to avoid sudden disruptions and greedy speculations on the value of the product on stock markets. In your digital wallet app, two graphs will be shown next to each other: one graph of money flowing out, and the other of money coming in. Expenses and income will be leveled out in a smooth steady process, and not the total amount in one go. The payment will be executed in millions of small packages – small containers of information that distribute themselves over the global network and reassemble when and as needed, probably verified in a blockchain. The future of banking is real-time personal banking, where all payments are spread over hours, days, weeks, months, or even years depending on the amount sent or earned, to stabilize the capital flow, to avoid casino-capitalist disruptions. Your money will flow as money flows in the stock exchange, with meters going up and down in real time, measuring the millions of small transactions in real time. Every person on Earth will have their own super-local super-personal stock exchange; all personal stock exchanges will operate in a global scale-free network of hubs and connections. Shareholders should only own, sell, and buy shares in the companies they work with or work for, related to the amount of time they dedicate their time to the company, and related to the projects in which they have shared responsibility, as described in Varoufakis' Another Now. For your digital wallet, you are the stock exchange and the trader in one. You might want to build red alerts when your expenses are running faster than what is coming in. The personal banking app will feature a dashboard with parameters you can manipulate by using sliders, to speed the money flow up or to slow down the flow when becoming overheated, in real time, restoring the balance. If your account is in danger of running dry, you will be automatically advised to slow down your spending; perhaps, you are urged to skip some transactions completely. It is not the government that imposes such advice but the personal settings of the app that the user has set in the preferences menu. The role of the local government certainly will be in the sphere of setting values for the parameters of the interest rates and taxes. New values for interest rates and taxes will be introduced as smoothly as the meters on your banking app will go up and down, stretched over years. One would not even notice the change, while all changes will be gradual, allowing you to have ample time to adjust your settings and parameters. Would one feel real-time personal banking as harassment, as interfering in one's customized sphere, or would one welcome it as useful information? I think the second, while I am sure that being well-informed is mandatory for a sound personal financial regime. I would rather know in what financial waters I am than remain ignorant, or find out when it is too late. Current tax systems are such harassment; it is hard to predict when and how much the tax authorities will charge you, and the real flaw in the tax system is that one feels accused of a crime rather than that one feels happy to contribute to the advancement of society. You do not have a choice in how your tax money is spent. In the personal banking part of the spending should be administered as tax money, or rather your contribution to society, and it should be a choice of the citizen to set preferences in what direction the tax money flows. Fewer people would check the box "warfare" when offered such a choice, and more people would choose "climate crisis control."

 Augmented personal digital banking is one of the tools for high-resolution direct e-democracy. Real-time democracy would involve continuous voting for actual socio-political issues. When a majority is consistently voting for or against a rule of law, that rule of law will come into action. The role of the people is to bottom-up put rules of law on the agenda, and the role of the scientists and technicians is to validate, qualify, and quantify the effects of the rules of law, based on which information the informed people make their choice, while the government executes the chosen rules of law, which are then imposed on those affected by the rule of law, in a continuous feedback loop. Rules of law would not be black or white, but formulas with parameters to adjust the effects in real time, according to real-time refreshing data. Like real-time banking, real-time voting must be spread over longer periods to get a stable consistent picture of the commonly agreed opinion.

Real-time computation of almost everything will bring back the nomad in us. People will live in the here and now, and you will be able to make instant choices about where, when, and how to go places, and how to organize your life. Permanent private ownership will make way for temporary forms of ownership, lease, and rent, or long-term shared ownership. You own it as long as you use it. Known forms of shared ownership are car-sharing and time-sharing for homes. Almost anything can be shared and hence be used more effectively. The IT revolution, intricately interweaving top-down and bottom-up information, will free us from exclusive ownership in favor of shared forms of ownership. "Sharing is the new having" in the new normal of the information economy, which will be an active part of Another Normal.

The New Nomad

The information economy according to Wikibooks: "An information economy is where the productivity and competitiveness of units or agents in the economy [be they firms, regions or nations] depend mainly on their capacity to generate, process, and apply knowledge-based information." After centuries of feudalism and consumer capitalism, the world economy entered a new era 30 years ago, the era of abundant information, cheap production costs, the era of embodiment of computation, do-it-yourself markets, and lots of free time to spend in sports clubs, restaurants, theaters, and museums. The Homo Ludens (1938) of Johan Huizinga, the sort of mankind that plays, has become the default situation in our times. Constant Nieuwenhuys based his monumental project New Babylon (1956–1974) on the idea of the homo ludens, which he remodeled into the new nomad. The new nomad is a digitally informed gypsy, always on the move, and always going places. Constant's architectural models give shape to an alternative society that consists of linear structures that find their way zigzagging through the Dutch polder landscape, structures that are elevated from the ground, clustered into highly urbanized networks, leaving the ground floor level to nature, not owned by anyone, and respected as collective property. As architect Oscar Niemeyer stated, the ground belongs to the people. Niemeyer sculpted the ground floor level of his projects as public playgrounds, as in his cultural center in Le Havre (1982), nicknamed Le Vulcan, on account of the shape of the main auditorium. In New Babylon, productive labor has become superfluous, and robots have taken over the mind-numbing work; the people are free to go and live and be creative wherever they fancy, for whatever reason. Constant imagined that when the nomadic citizens accidentally would return to their point of departure, they would not even recognize the place, and so much would have changed. For the new informed nomad, home is where you are at one particular moment, and home no longer is that place where you return to all the time. The nomadic international citizen feels at home anywhere, while any place in any country will adapt to his/her personal preferences as set by the ubiquitous booking app, as I have described in my paper Multimodal Accommodations for the Nomadic International Citizen (MANIC) (2018). Sixty years after New Babylon, it is striking to see how adequately Constant predicted the Industry 4.0 society, and how the architectural structures of New Babylon were the forebodes of 24-hour airport cities like Schiphol Airport. International airports have become cities that never sleep; their inhabitants are always on the move, to destinations where they feel at home for a shorter or longer time.

Thanks to the evolutionary success of digital technology, the production and distribution of almost anything are cheaper than ever. Mainstream economists use the term post-capitalism, while architect Patrick Schumacher, the self-declared libertarian advocating extreme liberal anarcho-capitalism, refers to our current time as the post-Fordist society, thereby marking the end of an economy that is based on mass production of goods and services. In the post-Fordist society, there is a reduced need for work, the edges between work and free time are blurred, and the relationship between work and wages has been loosened. Automation drastically diminishes the amount of human labor needed. Our society is no longer dependent on mass production; our current society is step by step entering the era of mass customization, potentially providing an abundance of almost everything for everyone. More and more digitally literate people work from anywhere, and employees are no longer confined to one working place, one city, or one citizenship; they are the new nomads, the nomadic international citizens. The pandemic accelerated the nomadic process in unprecedented ways. More and more people work from home or their favorite coffee shop, instead of daily commuting over tiresome boring long distances to the centralized place of their employer, which represents a huge loss of time and energy. Experts expect that after the pandemic, perhaps only half of the work will remain to be executed in the firm's offices; employees will spend a maximum of two days per week at the central office. More and more, the office becomes a social meeting place, a place to exchange stories and ideas. Skype, Zoom, Meet, Teams, and Webex tie people together in new forms of cooperation, where the participants in the video conference can act on a level playing field. Boss and employee, teacher, and student are digitally coupled in a non-authoritarian way; the teacher does not stand before the class but is part of the class. The boss comes out of his oversized room, decorated with leather sofas and designer furniture, by and large invisible to his employees. The pandemic has changed productive life in a revolutionary way; without even pushing for it, it just emerged. Online groups are formed, non-discriminatory linking people from different nationalities and different disciplines, merging the physical with the virtual. Social media like Whatsapp, WeChat, Telegram, Instagram, Facebook, and LinkedIn are extensively used to share stories, opinions, and insights, post about private lives and ambitions, and tell each other about ongoing projects. They order books online, clothes, groceries, instruments, tickets, travel, hotel bookings, and complete holidays; shares are traded, and one does their exercises via fitness apps, one uses apps to search, find and hire craftsmen for small-scale construction works. Different nationalities communicate via real-time translating apps. In Another Normal, everyone speaks their preferred language yet perfectly understands the other. In Another Normal, people live in their customized time zone, and yet perfectly communicate with others in their time zones. In Another Normal, there is an app for everything.

The central element in the MANIC research is the merge between a ubiquitous booking app, basically facilitating booking anything, anywhere, and anytime, and the programmable Pop-Up Loft spatial unit that I developed some years earlier. The <u>Pop-Up Loft</u> is a classroom-sized space programmed to transform into anything from a bedroom, dining room, workshop, meeting room, bathroom, dance floor, home cinema, or classroom, to an empty Zen-like space. When designing the concept for the Pop-Up Apartment, I did not realize that the <u>House of the Future</u>, designed by Alison and Peter Smithson back in 1955, has similar features, such as tables, beds, and bathtubs that can sink into the floor to make space for other activities like reading and listening to music. That kind of versatility of a compact space is exactly what the programmable Pop-Up Loft is capable of, in its most radical form, using the ubiquitous booking app to manage the configurations, in a more neutral design language than the plastic pop style of the late fifties. The Pop-Up Loft is the ultimate temporary home for the new nomads. I prefer the term nomadic to catch the essence of today's dominant architectural style, representing the actual state of the information economy, representing our era of tagging, networked in ubiquitous connectivity between things and people. The term nomadic adequately refers to societal trends, characterized by digitally informed swarming behavior. Informed swarming behavior is the default condition of the nomadic international citizen of Another Normal.

Hive Mind

The images of the swarm as in numerous YouTube videos are local temporary configurations of flocks of birds. Most of the time, birds do not murmurate. Most of the time, they fly individually from one branch of a tree to the other, demarcate their territories, reach out to others, sing their unique tunes, beg for attention when preparing to mate, build nests, collect fodder for their offspring, send out signals to warn of danger, and migrate to other parts of the world, guided by their built-in sensors to read and follow magnetic fields. The song is designed to let the nearest neighbors know the bird's position, mood, and state of mind, at the same time revealing its physical condition. Although most of the time living at larger distances from each other, they still communicate from neighbor to neighbor, and their behavior is still considered swarm behavior, but in another spatial situation, at another pace, caused by different external circumstances and unfolding internal biological processes. Murmuration at dusk is a specific instance of the swarm. Birds that position themselves on electric power lines represent another specific form of swarming. On the powerline, they still calculate their mutual distances, which are substantially closer than during murmuration. On the powerline, they keep a small distance, just enough that another bird would fit in between when the population on that power line would become denser; still the swarm, but in a less dynamic form. On the powerline, most birds look in the same direction, with only a few exceptions, ready to take off when the immediate neighbor takes off. Birds mainly murmurate in the fall season when there are fewer leaves on the trees. Why are they murmuring at all? To enhance their safety? To protect themselves from predators? To warm up for the night? To play? To exercise to find out who is fit enough to migrate? No one knows for sure, perhaps a combination of the above. Birds and virtually all other organisms, including humans, are by definition operating in swarming mode, in any possible instance of the dynamic of the swarm, in any instance from slow to fast, from distributed to compact. Either operating as an individual or a collective, as a nomadic individual, as a couple building their home, as a member of a team, as one out of many in the crowd, but always communicating with the nearest neighbor.

The default condition of swarming also applies to things; it is not a privilege of living bodies. Houses that populate a city are arranged according to swarming principles, and houses are looking to their nearest neighbors, following the rules of master planning. Objects in a house are arranged according to swarming principles, and the pieces of furniture and paintings on the wall are looking at each other, to avoid clashing with each other, following the rules of interior decoration. Cars are arranged according to swarming principles on the streets, keeping a safe distance from each other, or packing as close as possible in parking garages, following the rules of play of traffic. Things avoid similarly bumping into each other as birds and people do, albeit most of the time not in real time. Things do have a hive mind that is active in the design process, mostly passive after being positioned. Things and people are permanently checking on each other, checking positions, speed, and direction. A chair that is positioned in a room has a direction; it has a forward-looking vector that requires a specific relationship to other pieces of furniture. In the same way, I refer to my designs as vectorial bodies; many functional household objects have such a vector too. Houses do have a vector; they are oriented toward a view, toward the sun, and the connection with the street. Cities can be considered to have a vector, as they are oriented toward the sea, toward a river, or aligned with a street. Swarming is a ubiquitous organizational principle, governing movements, and positions of virtually everything and everyone on the surface of the Earth. Swarming is a collective built-in adaptive system of artifacts for reading, processing, and sending out information, organizing themselves, and being capable of adjusting their behavior when circumstances change. The hive mind governs how people and things move and are moved, and how they bottom-up distribute themselves or are top-down distributed in their respective territories, mostly following simple rules that create visual diversity. The real-time acting swarms of Another Normal emerge from the digital platform that was established by the fourth industrial revolution. The constituent components of the constructs of Another Normal inform themselves in real time to be fit to respond, physically and virtually, to changing circumstances.

Another Normal

In the alternative present of Varoufakis' Another Now, there are no bosses and employees; companies are spontaneously ordered, based on personal responsibility, through a non-hierarchical management structure. In Varoufakis' Another Now, everyone is a co-owner of the company to which one is tied, and shares are only earned by, given to, or sold to persons inside of the company. Decision-making is decentralized according to the one-person-one-vote system. Companies are ruled by a democratized form of inequality, based on a person's quantified and qualified performance as acknowledged and voted for by colleagues. In Varoufakis' Another Now, there are no private banks since a bottom-up activist movement called the Crowdsourcers has effectively defunded the until-then omnipresent capitalist banking system. Another rebellious movement called the Bladerunners broke down the monopoly of the big tech companies, while the Environs broke down the hegemony of the fossil fuel and cement industries.

In Varoufakis' Another Now, targeted activism proved to be an effective means for the rebels to transform society into a fair and just anarcho-syndicalist model. Varoufakis'

alternative present could very well emerge naturally from my techno-social Another Normal, which is based on real-time accountability, verifiability, and quantifiability of all things and people, as postulated above. In both Varoufakis' Another Now and my Another Normal, every bit of input and output information is transparent, and the processes are verifiable. Every transaction can be traced back, as in a blockchain. Design decisions are micro-transactions as well. All steps in design development are transactions of the players, acting faithful to well-defined rules of play. Another Normal is since decades already our daily normal, when it comes to parametric design and robotic production. In our daily lives, everything and everyone are digitally related to each other in a logical, transparent, and at the same token highly intuitive and unpredictable network of decisions. Design decisions behave like members of an open-source version of the swarm, relentlessly checking each other out, exchanging information, building relations along the design journey, and immediately responding to changing circumstances. Millions of micro-decisions of the evolving design are virtually connected in the cloud, eventually physically tied to each other in the built

assembly. The data of the design decisions are published in a distributed Open Data Commons database, securing the integrity of the design process. After the fatal fire at

the Faculty of Architecture and the subsequent brutal destruction of the iWEB, we shared the data of the iWEB on Github, a web portal for open-source software development, data that is needed to reanimate the iWEB into its third life.

4

Ubiquitous Components

Precogs and Cyborgs

A *component* is a smaller, self-contained part of a larger entity, often referring to a manufactured object that is part of a larger device. An *element* is an entity that does not necessarily need such a context. As there is a periodic table of elements and as there are the building elements that constitute a catalog of parts, an element is an autonomous thing without a predefined context, without a unique identity, and without defined relations to neighboring elements. An element as such does not form part of an organism or a building. A component does have a unique identity, and it does have predefined connections to its immediate neighboring components. The relations between components are an integral part of the component. The notion of the component and the connections between the components appears in language, science, the universe, art, nature, societies, products, and, of course, built environments. Although it is self-evident that the world is built-up from interacting components in a huge variety of associations, not much research is being carried out on the bidirectional relationship between the constituent components with architecture. We must specify relations between components that belong to the same family, between components from different species, and between groups of components from different species. I aim to identify a universal design language, based on my experience in teaching at the AA London, at Hyperbody TU Delft, and subsequently at Qatar University, and based on my experience in realizing radical design projects in our joint art and architecture practice ONL (Oosterhuis_Lénárd). My design language is by and large founded on the design, production, and assembly of interacting components.

155

The design-to-production process may lead to fixed end-products or, when the constituent components are programmable in real time, to interactive and proactive constructs, of which I have built many at Hyperbody and ONL along the way. Technological extensions to the human body, the exo-brains, the exo-tools, and the exo-instruments, make up the human condition of today. We are one with our technical extensions, one co-evolving ecology of things and people. We have been evolving into the techno-human, the homo arte, the cyborg, from the moment that we started to use physical and virtual tools and instruments to extend the reach of our bodies. Buildings, and the built environment at large, are extensions to our physical bodies; they are our exoskeletons, or rather our exo-shells. Our homes are physical exo-shells and social interfaces at the same time, protecting us from and connecting us to the world around us. Similarly, the internet is a collective exo-brain, selectively connecting us to the world around us while comforting us in our virtual social bubbles. Our human extensions, from the bricks and mortar of cities to the bits and bytes of the internet of things and people, are a carefully assembled global ecology built up from their constituent components. Raw ore is mined by human bodily extensions, i.e., our exo-arms, and transformed into materials by our exo-processing plants, the materials are transformed into components by our exo-factories and exo-workshops, and the components are assembled into our exo-buildings. These human bodily extensions are not grown from within as snails grow their shells from the mantle, which is that part of their skin that connects to the shell and continues to produce the shell while getting older. A snail's food intake is processed by its gland into a calcium-rich mass that hardens to form the shell. The shell grows spirally with the growth of the snail's body. Building components in synthetic human-driven processes are designed to connect to their neighboring parts in one particular place; their raw materials are mined in another often-remote place, and their half-products are processed into components in yet another place called a factory and assembled at their final destination by our exo-machines to form the whole construct. The information of where they are coming from, how they are processed, and where they are going is in the identity card of the component. Components must be considered "precogs," as in the film <u>Minority Report</u>, when it comes to the future assembly because they already know where, when, and how to connect. Components are precogs when it comes to forming the whole construct; they have foreknowledge. They know what their future place is, and they know where and when they should act.

Components are informed actors in a dynamic network of components and ecologies. This applies to any ecology, to the ecology of languages, science, art, the universe, music, architecture, industrial products, programming, and generally speaking the real-time behavior of things and people. Component-based ecologies are interacting on different scale levels with each other, they are feeding upon each other, peer-to-peer processing the information from their neighboring components, and top-down being informed by neighboring groups of different components while giving feedback to connected ecologies of a different nature. Language has been developed by animals and humans as an instrument to exchange information with other beings. Language is a virtual extension of the sense organs, to the mouth, the ear, and to the eyes. Language is built up from clearly defined components developed for communication from person to person to their nearest neighbors. Other than animals, humans have found a way to vastly diversify the components of language, compiling layer upon layer of meaning and intention, through body language and spoken language. Spoken language in the first place, later augmented by written language, became the ultimate instrument to evolve complex relationships between people and things. Spoken and written languages are extensions of our body language. We have found sounds, then words, then stories, and then treaties and books to expand our bodily gestures. Printing of books revolutionized the distribution of knowledge and stories among larger groups of people, not only to one's physically nearest neighbors. Readers are virtual neighbors, establishing long-distance connections between people and things. Eventually, we humans got hooked on playing the instrument of language. We may think it is natural now, but it must have taken a long road of synthesizing sounds into the richness of language to get to this point. Language and humans together have co-evolved into something that is an analog cyborg, an entity that has a natural body that is augmented with synthetically constructed immaterial components, subsequently augmented with physical components later on in the process of becoming the cyborg, represented by the material grammar of languages, and the physical materiality of books and computers. Our cyborg-ish extensions are grown from our bodies and our minds. Our cyborg extensions consist of raw material that is extracted from our immediate environments, synthetically processed, designed, produced, and interwoven with our daily lives. From where we are now in our ongoing development as cyborgs, we would not even survive without language.

Beyond Elements

 As suggested in the movie <u>The Quest for Fire</u> (1981) by the French film director Jean-Jacques Annaud, humans started to point at things and make specific sounds to indicate a specific object. "Oh" means this and "Ah" means that, as an identification of an object or action. Those "ohs" and "ahs" are specified over time using different and more complicated sounds and diversified into a series of vowels and consonants. Vowels and consonants eventually constituted an alphabet of sounds. A language-specific alphabet of sounds is what the periodic table of elements is for chemistry. The elements of the periodic table represent the many possible configurations of atomic tension fields with increasing density, the theoretically isolated building blocks of nature. The first 94 of the elements are found in nature, and the other 24 are synthesized in the laboratory. These elements cannot be broken down into smaller "elements." When having engaged in a specific context-related relationship with other elements of the same kind or other kinds, elements have become components. Elements are customized to form a component. H_2O molecules are the constituent components of what we perceive as water. Only very few elements exist in their pure form and bond to their peers to form a pure material, like gold, or a pure gas like hydrogen. Components are an emergent property of elements, beyond the elementary.

The vibrating sounds and the written representation of the 26 letters of the English alphabet are for language what the atoms are for matter. Some languages have a larger number of letters, like the Hungarian alphabet, which has a whopping 44 letters. Quantum physically, spoken language is a pattern of waves, while written language has been crystallized into particles, as the elements of the periodic table. The alphabet is the breeding ground for a virtually endless number of combinations of letters, eventually leading to an extensive lexicon of a hundred thousand words per language. A global variety of differently structured languages emerge from the alphabet, in an endless number of combinations of words, formalized from local dialects. Words are an emergent property of guttural sounds, framed by the pool of letters. Words consist of a minimum of one letter, of which the most elementary is probably the I, up to dozens of letters. Potentially, there are 676 possible two-letter word combinations in a 26-letter alphabet. In English, there are only 24 two-letter words commonly used, limited by preferred combinations of vowels and consonants. The pool of three-letter words in a 26-letter alphabet contains a whopping 17.576 possible words, of which only around one thousand are used. While some words contain a dozen or more letters, the number of possible letter combinations for words is virtually endless, a limitless universe in itself.

The language of our era is the digital language that is made up of zeroes and ones only. Digital language has another wealth of possible combinations while 64-bit computers can process integer values up to $2^{64} = 18,446,744,073,709,551,615$ at a time. The sheer number of possible combinations boosts new inventions and applications. Digital computation has warped us humans into a society of digital cyborgs in no time, within half a century. Letters are the basic elements to form words, words to form sentences, sentences to form paragraphs, stories, laws, and theories. While individual letters and words may be considered elements, words that form a sentence are considered the constituent components of that sentence. Words in the context of a sentence will only fit there and then, making up the meaning of the message. Words are customized to become a constituent component of a sentence. Similarly, the sentence is a more complex component in the context of a comprehensive text. Words are used in certain rule-based connections to each other. The order of the different words and the connections between the words make up the meaning. A simple sentence containing one subject, one verb, and one object may describe a complex action. In a three-word sentence like "I build [a] house," the word components form a sentence that accurately describes a complete process that involves the actor (the I), the processor (to build), and the output (the house). That sentence is an input-process-output vehicle, in which the words are the acting components. The sentence is a conglomerate cluster of word components in language, featuring a subject or pronoun, a verb, and a subject put in a meaningful way together. The sentence is informed by the connections between the basic word components. The verb adapts both to the pronoun and to the subject, thereby creating a mutual bond, in a dynamic relationship. One level up, sentences are tied together to form opinions, stories, documents, reports, and laws. Language is a fascinating instrument, especially in our digital era, where many new digital applications are emerging, driven by algorithms, thriving in their own synthetically constructed universes. One sentence or line of code builds upon a previous sentence and foreshadows the following sentence. Each sentence somehow connects to the previous and the next sentence. Sentences are active players in a dynamic swarm. Sentences are components that are looking to their nearest neighboring sentences, informing themselves of the meanings, intentions, and implications. If one changes one sentence or line of code, the logical structure may be broken, and the previous and the following sentences will have to adapt to restore the logic of the storyline/program. Similarly, when we map the concept of language onto the building blocks of the building industry, raw materials are customized to form elements, while elements are customized to form components. Elementary materials, wooden plywood plates, sheets of steel, and glass plates that come directly from the raw materials processing industry are manipulated to become components or subcomponents of larger, more complex building components. While elements are the most reduced and abstracted form of a base material, components are intentionally manipulated, customized, and informed to form a more complex component. The order of the customized elements and their mutual connections make up the meaning of the component, while the order of the different components and the connections between the components make up the meaning of (parts of) the building.

Spaceships

Numbers and letters, integers, alphabet, calculations, words, formulas, sentences, theories, and stories together form the ecology of language-based entities. Each entity lives in its preferential bubble and upholds its relations with other entities of the same kind. Words relate to other words, not to single letters. Sentences relate to sentences, not to words, and stories to stories, not to sentences. A word is a sequence of letters, a sentence, a string of words, a story, or a series of sentences. The dynamics inside sentences are ruled by syntax and grammar, whereby syntax describes the rules of the word order in sentences in a specific language, and grammar prescribes the word class, conjugations, functions, and relations in a specific sentence of a specific language. The dynamic interplay between words inside stories is bottom-up ruled by the intentions and top-down by the style of the writer, by the personal scheme of things of the author. When mapping the hierarchical structure of languages to the structure of the built environment, one could consider the smallest generic elements to be the equivalent of letters, the screws, and bolts of words, and manipulated informed building components the equivalent of sentences, the spatial arrangement, and mutual relationships of thousands of components to form buildings equivalent to stories. Words are the components forming sentences, sentences as the interacting components shaping the story, and collections of stories are the components to build a culture. In analogy, the city is a library of stories, built up over time, and re-edited over and over again. Syntax in language shares similarities to the rules of physics that govern the connections between building components to form a coherent larger structure. Grammar in a language is the equivalent of the guidelines in master planning. The interventions of the author are the equivalent of the design choices of the designer, imposing shape, style, and meaning on the constituent components of a built construct. The semantics of the chosen design decisions make up the characteristics of a culture.

Our current society cannot exist without the elementary bits and bytes or without all the more complex components that keep emerging from it. Society would not work without naming things and situations, categorizing and counting, identifying things and people, and without quantifying and qualifying. In the digital era, new kinds of buildings are emerging from digital syntax, grammar, and semantics. As there are many languages, and still counting due to the rise of programming languages, there are many unique ways to define a component and to compose components into larger structures. Individual design languages vary wildly, based on different component libraries and a different logic connecting the constituting components, based on different semantics altogether. New digital languages facilitate new design paradigms and new production paradigms. Parametric design and CNC-driven robotic production methods have created a new logic to define the components and for the assembly of components into a building. Digital design

 instruments generate alternative esthetics that fall to Earth out of their weightless habitat, which is cyberspace. "Yes, we build spaceships" has been one of my provocative statements from the early zeroes. Spaceships are not easily recognized as a serious alternative to prevailing types of buildings. Alternative versions of what is normal emerged from the digital soup, leading to a renewed definition of what is a component, assembled into buildings that look like alien spaceships. Digital technologies, and

 parametric design in particular, lead to a new appreciation of the notion of beauty. As I proclaimed in my TEDx lecture at the TU Delft in 2011: "We are changing your view on what is beautiful and what's not."

Design Language

Language is used as a ubiquitous instrument, which is played by its users to arrange its components into a meaningful whole. The constituent components are the swarming building blocks of communication. The components of language together form a greater whole, whereas each identified component is content-wise strongly related to their neighboring components, locally and temporal. People invent words, construct sentences, use common expressions, tell stories, write letters, essays, and complete books. A fully developed language is one of the main instruments for us humans to communicate. Humans play language as an instrument, some of us as professional writers, some of us as scientists for the exchange of expert knowledge, while others use it for day-to-day small talk. That is the beauty of a ubiquitously available instrument. Everyone finds their unique way to play the language as an instrument. Everyone naturally enjoys the complexity and richness of the billions of possible letter/word/sentence combinations. Everyone accepts the emergent complexity of the grammar and the syntax of language as something beautiful, especially of that language one is familiar with. Performers in the theater recreate existing constellations of components; writers put together new combinations of words and sentences, sometimes inventing new words. Thinking and talking are playing the instrument of language, as to create a personal universe. The spatial design language of my design universe is built upon an active role of the constituent components, upon their relationships with neighboring components, and their assemblies into an as yet unseen yet consistent whole. Creators, writers, scientists, composers, and designers imagine new building blocks to compose their unique personal universes. My design universe has inherent values, its integrity, its own (in)consistency, and its form of beauty. By the transparency of my rule-based design principles, my design language has the potential for use by others. The simple rules and methods of my design language are designed to make a pleasant form of complexity, to define the components that give shape to that new form of beauty, the logic of which I believe is relevant to share with students and colleagues.

Living Inside Evolution

In science, there is no such thing as an atom, and yet we use it to explain the properties of the basic elements of the periodic table of elements. Atoms are not physical things; they do not have any materiality, and they are delusive petrified simplifications of what they in immaterial reality are. It is not known exactly as of yet, but that does not undermine this statement. Matter is something we only see, feel, and understand because we the people are built the way we are built. Our bodies are an illusion of materiality, while, in reality, we are just a complex formation of tension fields. Our bodily tension fields interact with other seemingly materialized objects, and these tension fields cannot penetrate each other. My hand cannot penetrate a table, while all the immaterial waves in the tension fields that the hand and the table are structured by can move freely through quantum space. There is no such thing as a tangible electron, a neutron, or even a quark particle. The illustrational concept of tangible particles is just a mental model intended to help us understand the world, but it is fooling us at the same time. The mental image of the particle prevents us from a further deeper understanding of the world. We may use it as a simplified model of reality but certainly not as a material construct. How is it that we as humans experience the complex arrangements of molecules as something tangible and real? How come we see and feel an immaterial construct of immaterial components as something real and solid? My answer is equally simple as it is self-referential. It is because we are built the way we are built, based on the same immaterial molecular constructs. We are in the same boat as the rest of organic life, including rocks and thin air; yet, we cannot see and feel any other construct that is outside of our common boat. It seems perfectly logical that the material illusion of our universe is an instrument that we as molecular organisms play, playing by the rules of that universal instrument. We are caught in the limitations of what our universal instrument is capable of. I am open to accepting the idea of many other possible universal instruments for a completely different type of reality. We have different instruments already inside our reality bubble like we have language, music, and computation, and like there is the built environment that we use as an interface to the world around us. But outside this particular virtually augmented real bond with our universe, there might be nothing to see and/or feel for other types of universal constructs. We are trapped inside the game of life, while any other alien culture is excluded from it. The beauty of it is that we are living inside the evolutionary bubble of the game of life, unfolding before our own eyes, co-created by ourselves.

Instead of seeing atoms as physical particles, it will be more accurate to consider one specific atom to represent a specific instance of the ubiquitous quantum tension field; vibrations that are not made of matter, but vibrations of pure immaterial information, out of which the world, as we know it, emerges. Material worlds and immaterial worlds are structured, informed, and emergent manifestations of the quantum-universe information continuum. Information is transformed into matter in our local and temporal material worlds, as noticed by our material senses and felt by our materialized bodies. Our ears, eyes, noses, and skins intercept the quantum field vibrations and translate them into physically felt sensations. We are built up from molecules that are tuned by quantum vibrations, and we feel matter that is tuned by quantum vibrations. We internalize the informative tension fields surrounding ourselves, resonating with the world around us.

Infoland

A closer look at the constituting components of the universe as we know it teaches us that there are still huge amounts of components missing to get an accurate understanding of the universe and the subatomic world. Concepts like antimatter and black holes are credited for balancing the missing components. To imagine how things might work at the minuscule scale of the quantum world, it helps to dive deep down, by downscaling ourselves and imagining how to interact with the tension fields inside the quantum universe. We must leave the idea that molecules, atoms, electrons, and quarks are the vehicles for interaction. Quarks are smaller than any current instrument can measure, yet still firmly rooted in theoretical natural physics. A known number of quarks make up electrons, protons, and neutrons. We do not know the measured size of any of these subatomic particles as of yet. Suppose there are no such subatomic vehicles at all, but just vibrations, fluctuations in interference patterns, and formations emerging out of the primordial information soup. Hence quarks are not seen as particles but as a tension field. A certain fluctuation in the fields might be measured, seen, and eventually felt by ourselves as something tangible, only because we are built up from similar quantum fluctuations, as I pointed out before since we are in the same boat. What we measure, see, and feel kills the dynamic of universal interacting tension fields of the information universe – we may name it the Infoland – and flattens into a lesser dimensional Spaceland, out of which there is no way back to Infoland, similarly as there is no road from Flatland into Spaceland. A spatial building body cannot be constructed out of plans and sections, since it lacks information on how the components are working together to form the whole. Members of the Flatland cult who know the secret coded language of two-dimensional drawings may keep up the illusion, but it does not work for space inhabitants, like myself. The reductive methods of drawing plans and sections rely on an intentionally poorly informed representation by nature, as opposed to the inclusive nature of 3D parametric design, as is the common methods in Spaceland, and, on level up, opposed to the inclusive nature of interactive architecture, as is the common language of Space-Time Land.

Our former, now retired, general practitioner the Dutch Mattees van Dijk, who appears just as hooked to the popular scientific insights in natural physics as I am myself, is a strong advocate of something speculative as "pure information," which is a form of information that is indivisible into smaller chunks, to rule the quantum universe. He is convinced that our consciousness is part of something much greater than we can imagine. He sees consciousness as a fundamental universal phenomenon. That something greater my former general practitioner is thinking a "little big" about is outside of our space-time territories. I do not attribute the information structure of our quantum universe to the phenomenon of something like a universal consciousness. I tend to think that consciousness is an emergent property of quantum field interactions on the molecular scale. Consciousness needs a body to feel it, which means that consciousness comes with the emergence of molecules into full-fledged organic bodies. Due to our organic material limitations, we can only measure, see, and feel shadows of that comprehensive other-dimensional deep information universe, i.e., Infoland. In common language, the term info-verse refers to the earth-spanning information universe that is embedded on the internet. My Infoland lives in the one-level-up unknown dimension that embraces all possible universes, macroscale, and microscale, beyond the spatial dimension and the temporal dimension, the unknown dimension that is inclusive of all real and imagined universes, all of the ones of our past, future, and present. To shape such an informed omniverse, what would be the essential building blocks? What are the building blocks of pure information? Information must be processed to create the different atomic and molecular formations that we measure, see, and feel. There must be an input of data, processing of data, the output of data, and feedback loops in the universal Infoland, also referred to as the omniverse. So, one of the main components of the information omniverse must be the numerous distributed processors that process pieces of information to build a structured output. We might consider quarks, electrons, atoms, and molecules – in a progressive sequence of complexity – as processors of pure information. Those pieces of information are, as of now, written in a language that we fail to understand. Watching the intriguing sci-fi film <u>Arrival</u>, it is simply fascinating to see that possible alien language, incomprehensible to us humans, projected on a huge fluid screen. In the film, we empathize with the efforts of humans to decipher that alien language, that unknown form of communication that in itself has a consistent logical information structure, with language components, a syntax, and a grammar of its own. Likewise, the basic components of architecture are not the physical appearances of matter, but rather a multitude of interacting processors creating the different informed formative structures that we call matter, organisms, and objects, and eventually congregating into building components. Due to my dire lack of scientific knowledge, I cannot dive deeper than this, while I cannot perform any of the natural physics calculations to prove any of my points that thus remain speculative and non-verifiable, much against my principles for the much-needed verifiability of all individual building components. What remains though is the compelling idea that in the end, all that matters are the immaterial interactions between the performative components, at any scale of our self-contained reality. Those information processing components are the programmable building blocks to build our unique personalized universes.

The Amazing Simplicity of Life

 I enjoy watching scientific lectures like "The astonishing simplicity of everything" by Neil Turok, where he explains that the universe is not complicated but a complex computable outcome of very simple mathematical rules. But we do not know what the universe has done to come to this particular congregation, and we do not know where we are going, but we sort of know where we are now, locally and temporarily. In his book "Reisbureau Einstein," Dutch for Travel Agency Einstein, astrophysicist Vincent Icke explores the possibilities of alien life forms in our universe. One of the many takeaways from his book is the notion that our DNA structure is based on many variations and different sequences of a few simple molecules, the basic components of the DNA of all life forms. Vincent Icke describes what he – as does Neil Turok – calls the amazing simplicity of life, played out as complex aggregations based on just a few essential building blocks. DNA is composed of the base ring's guanine, cytosine, thymine, and adenine between the backbones of sugar-phosphate. On a molecular level, these are the components life is playing with as if it were an instrument. Life is an emergent property of the abundance of these elementary nucleobases; life thrives by copying, combining, and rearranging the basic components of the book of life. How this could happen is still unknown. Why would the quantum tension fields, the fields of the interaction of pure information, eventually become so tangled as to form relatively stable molecules, which are very specific instances of all possible spatial arrangements? Is entanglement an emergent property of the pool of nucleobases? When taking pictures of the quantum states of a hydrogen atom, a large yet limited number of possible patterns are observed. Further up the molecular scale, Icke notices that simple molecules are found to be omnipresent in our universe. On interstellar dust particles, scientists have found the methane CH_4 and HNCO, molecules that are thought to be processed in a chemical process to have eventually become the basic building blocks of life as we know it on Earth. Based on the omnipresence of these molecules, Icke believes that the emergence of life must be, in principle, possible anywhere in the Universe. Icke adds that, elsewhere, life may look quite different, even impossible for us to identify as life; yet, all forms of life must be based on that omnipresent process of tension fields aggregating into a kind of molecular architecture. Atoms do not live as independent entities; there is no such thing as a single atom, a single molecule, a single component, or a single person. It is misleading to even consider anything as a single element that exists in itself or by itself, while everything is linked to something else, virtually by exchanging information and physically by their material connections. Our brains are connected to other brains to form social groups and ultimately a world of knowledge. A single brain would not know how to function, while it feeds in its connections upon other brains. What are known as atoms are deep inside local and temporal information fields, forming simple patterns that are interpreted as matter by super-complex patterns such as human beings. Atomic oscillations have different information and hence energy content, which produce different wave patterns. From a dozen different patterns at the most basic level, emergent combinations synthesize to something as complex as a steady molecule. Once a jump into the next level upward has been made, from an atomic-sized quantum state to a molecular-sized quantum state, the molecules become the basic components on that newly reached level. Out of specific simple rule-based combinations of ubiquitously available molecules, the DNA ledger emerges, and from the ubiquitous presence of DNA strands, our animal existence has emerged. On the DNA level, the DNA strings are the basic components on top of which the next level builds, the level of cells that form tissues. The DNA ledger is read by mobile RNA molecules to make up living bodies, step by step, not in a flow, but in billions of discrete steps. Jumping from scale to scale, the components change in size and performance, and on the scale that they inhabit, they communicate with their level playing field peers. A progressive series of ubiquitously available configurations of increasing complexity forms the basis for the astonishing simplicity of everything, including life, including nonstandard architecture. The emergence of informed building components has undergone an evolution similar to the emergence of life. Out of a ubiquitous presence of immaterial bits and bytes, a new design language has emerged that eventually fosters the notion of the informed component. From the ubiquitous presence of digitally defined components, nonstandard, built structures emerge. Inevitably, out of the locally ubiquitous presence of the nonstandard interactive architecture quickly popped up. Interactive architecture is an emergent property of the nonstandard.

Components for Life

Scientists are usually not very prolific designers. They typically observe, measure, and analyze, and especially when they are theoretical natural physicists, they speculate on the possible arrangement of atoms, molecules, and sub-atomic tension fields. Designing and constructing new configurations of atoms, molecules, materials, and organisms only took a big flight after the rise of the digital. Merging computer knowledge and natural physics opened up new directions in science. Once I asked our architect friend Marcos Novak back in 1998 what he would do if he were a biologist. Novak said, as published in our monography 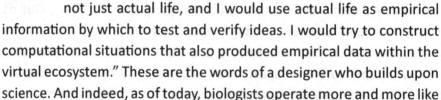 "Kas Oosterhuis, Architect - Ilona Lénárd, Visual Artist": "If I were a biologist today, I would be studying possible life, not just actual life, and I would use actual life as empirical information by which to test and verify ideas. I would try to construct computational situations that also produced empirical data within the virtual ecosystem." These are the words of a designer who builds upon science. And indeed, as of today, biologists operate more and more like designers, creating new organic substances. Chemists are synthesizing new materials, like the buckminsterfullerene (C_{60}), nicknamed the Bucky Ball. Buckminsterfullerenes are new formations of atoms that have not occurred in nature before. More and more, scientists are becoming designers nowadays, crossing the boundaries of their disciplines by and large facilitated by digital techniques. The digital is a common language familiar to different disciplines, thereby facilitating cross-disciplinary work. Yet, the design aspect is almost always completely neglected in the (popular) scientific books that I have been reading for the last decades. Typically, scientists analyze data and evaluate along the lines of predefined experiments and are acting as voluntary prisoners of their experiments. Designers have been educated differently; they invent new components, new building blocks for matter, products, buildings, and cities, bringing to life new aggregations of matter, technological inventions, new products, new building systems, and ultimately master plans for new cities. The designed components for making up society eventually form a critical mass out of which a new actual state of life evolves. Designers are the co-creators of new life forms. There is nothing more spectacular than to realize that we are living, acting, and designing inside evolution, inside a natural habitat that is constantly evolving.

Night views from the international space station ISS show how Earth is lighting up. The atmosphere of the Earth is heating up as well, mainly due to the burning of fossil fuels and the exhaust of methane gases. Climate change – or rather the climate crisis, to avoid the euphemism "climate change" that has been coined by moderate scientists – is an inevitable byproduct of the ever-increasing global consumer activity, which in its turn is the inevitable byproduct of profit-driven mass production. When not changing course, we will see a steadily increasing consumer activity of the billions of people that inhabit Earth and a further acceleration of the pace of climate change, extreme weather, and increasing global warming. No one will deny Asians and Africans to go through a similar evolution and thus increase the consumerist activity that we as Europeans and Americans have lived through, and which is continuing to increase further. There is no sign that fossil fuel companies will let go of their heavily subsidized privileged position. As an inevitable consequence of evolution, the globe is bound to continue to heat up faster than ever before, at least when we continue burning fossil fuels. While we tell ourselves that we want a cleaner world, the new digital technologies, especially the huge data centers that are needed to sustain the data-hungry metaverse, require a lot more energy than today. The further digitalization and ubiquitous availability of almost everything for almost everyone inevitably lead in the short term to an exponential increase in the surface temperature of the earth, and hence to more drastic climatic disruptions. More and more satellites and spacecraft will be sent into the skies from the Earth. In a timelapse movie, it will look like popcorn popping up from the hot plate from a distance. There is no way back to a less active world, we cannot reverse evolution unless a global disaster destroys almost everything; so we must put our efforts into the design of an attractive non-polluting active future that lies ahead of us. The question remains whether we can design our way out from polluting production and consumption toward a non-polluting version of society. When design causes a problem, can design solve that problem? Can the same political system that has supported the exponential growth of everything at the cost of health and justice change course and redirect itself toward a non-polluting future? A drastic regime change in the way we design, produce, and consume commodities will be needed to reach that goal of a just and fair, non-polluting society. A simple solution that comes to mind directly is that every production facility and every home must provide its own energy needs, using clean and networked energy production methods. Is that asked too much? Why would we not require data centers to produce at least the energy they need to operate? To do that, factories and homes will double or triple in size to support the food, energy, and water production installations, but is that a problem? On the contrary, it will mean increased economic activity and offer huge economic benefits. Not growth and profit per se but increasing benefits and social values for society at large must be the currency that leads the activities.

One of the key social questions is whether we can design for less polluting activity and increased fairness and equality at the same time. As long as we remain curious about what the origin of life is, what is beyond the galaxies, and what is inside the quantum fields, we will not stop increasing our collective activities. Such time and capital-consuming research is only possible with the help of piled-up capital. When we would slow down accumulating capital in big cities, global companies, and leading countries, the deep scientific research in natural physics, chemistry, and biology would come to a halt. But who would want to slow down research on virus protection, which simply requires dense capital concentration in research institutes? As the pandemic has shown, the development of vaccines has been much faster than before, not only in the USA. All developers were performing well, made possible by more capital investment. Capital needs to be created first, and that is only possible through increased economic activity. It looks like a devil's dilemma, but I do not see how slowing down economic activities would help us escape the actual climate crisis. It would mean denying the majority of the world population that now lives in relative poverty to develop their societies into a digital techno-social world. And according to Kevin Kelly, the digital world is simply inevitable, as he writes in his book that bears the same title The Inevitable. The digital is ubiquitous, the digital will be immersed in any product including food, energy, and water, and the digital will invade the very minds and bodies of people. AI will assist in the operation of almost anything, including the design of synthetic chemicals, organisms, and buildings. Robots will produce almost anything on demand, everything that can be copied will be almost free, information will be expanding faster than anything else, and science will be expanding exponentially. Your body will be your password, connecting yourself to the cloud. And this will apply to all billions of inhabitants. The obvious downside of an overly naive optimistic view is that George Orwell's 1984 will be seen as a social design guide instead of a warning. The slogan "1984 is not an instruction manual" is shared on Twitter many times. The current rise of populist, nationalist democracy-manipulating autocrats in all parts of the world makes me worry that their supporters are fooled into full denial and chose to find short-term benefits for their nation before anything else. The combat of climate change can be done, but the narcissist autocrats will definitely make it nothing short of a challenge to get things straight. More likely, after having survived the autocrats, the future 10 billion inhabitants of Earth will become designers of some kind, also known as prosumers, a term coined by futurist Alvin Toffler. Prosumers are individuals who consume and produce value, either for self-consumption or consumption by others;

173

prosumers are actively involved in the creation of the unknown. It is in the very nature of the exponential growth of knowledge and information content that it can only speed up even more.

We will need all of our inventiveness as designers to survive as a species. The emergent outcome of the surge of the digital is that all of Earth's inhabitants will become a designer of some sort, co-designing the new components for a fair, just, and non-polluting life. Components for life that are lean and mean designed, produced, consumed, assembled, and cycled up. The future is in synthesizing new materials, rather than in brutally mining and extracting minerals from Earth. Based on an understanding of chemical and biological processes, the building blocks of the future are well-informed customized components, put together in a mean and lean process. Mass production makes way for ubiquitous customization, producing on demand when and as needed. Static worlds make way for dynamic worlds, both real and hyper-real, developing in real-time as unfolding games. Components for the constructs of the future are inevitably real-time informed, and thus in principle prepared for interaction with the users. Constructs of

 the immediate will be "designed on the fly," as we did when designing the Digital Pavilion (2007) for a big-tech client in South Korea, as will be the default situation in the emergent metaverse.

Hyper-real

The very act of designing, producing, and assembling effectively means the inverse of the second law of thermodynamics. According to the second law, everything tends to strive for a maximum entropy, i.e., for a thermodynamic equilibrium, also known as the arrow of time. The very act of making things from base materials seems to be the exact opposite. Magically, the constituent components implode to give shape and performance to the new entity. To make things, energy is added locally and temporarily, reversing the second law. Making things, as we do here on Earth, actually accelerates the entropic processes that are governed by the second law. This applies to the making of atoms from quantum fields, to the making of molecules from atoms, cells from molecules, organisms from cells, and ecologies from organisms. In other words, life is counter-attacking the second law of thermodynamics. The act of designing, processing, and assembling is a form of life, not organic, but synthetic. According to the young scientist Jeremy England, the simple fact that the sun bombards the earth with solar energy inevitably brings the otherwise dumb matter to life.

Likewise, the proactive act of making plans and designs is an emergent property of matter receiving lots of energy. Atoms emerge from information fields, molecules emerge from atomic interactions, organisms emerge from the molecules, humans emerge from molecules, and products emerge from humans. Everyone and everything live in the same energized boat of life, and people and things are speaking the same molecular language. Objects, products, and buildings should not be considered dead forms of life but as exo-parts of living organisms. Like snails build their shells, humans build homes, factories, hospitals, and infrastructures. This means that the totality of the built environment is an inevitable natural phenomenon, emerging from the simple fact that the sun keeps providing the earth with energy. As long as it does, the evolution will continue, probably at an ever-increasing speed. The built environment is a natural extension of nature, and the design of the built environment is an emergent property of human life. We as humans have created our exo-tools like the hammer and the drill, exo-instruments like the piano, the computer, the telescope, microscope, and subsequently larger structures like our exo-habitats, exo-infrastructure, exo-vehicles, exo-cities, exo-particle accelerators aka supercolliders, exo-electricity networks, exo-television networks, and the world wide web as an extension to our collective brains. We have woven an extensive delicate web of interacting exo-structures around the globe. The exo-structures are direct extensions of our bodies, meaning that we have been cyborgs for hundreds of centuries, without labeling them as such. Now that we have entered the digital era, it becomes more obvious that we live in a deep fusion of the corporal internal and the corporal external, of the real and the hyper-real. We have done so since the rise and shine of animals and hence humans. We bipeds are processing information to create physical and mental constructs as actual and future extensions to the prehistoric natural world, in which world-making games we, at first sight, seem to be the primary participating players. Or are we? We should consider the cars, the homes, the factories, and the global networks of our exo-verse as conscious players as well, working with us instead of being created by us. We should leave our egocentric anthropocentric view of the world and replace it with an empathic ecological view. As a species, we are only one player among thousands of other players, including the thing-players. Especially now, we have entered the digital era, which so intricately entangles people with things.

The essence of design is to invert the second law of thermodynamics locally and temporarily. Processing information to create something more complex than the constituting parts *de facto* means reversing the second law of thermodynamics. When no action is taken inside a closed environment, the second law of thermodynamics tells us that any substance is aiming to reach a balanced equilibrium, leveling out differences, and distributing matter evenly throughout the complete closed environment. At the same time, more and more complex molecules, organisms, and physical and mental products like language, music, and the internet are emerging in an ever-faster tempo and ever greater complexities from the self-leveling soup. New components for an extended life are invented by the minute. Design moves upstream of the second law of thermodynamics, like the fictional Baron von Münchhausen who reportedly pulls himself and his horse out of the swamp by his tail. Where does design feed upon? As per definition, the act of design and execution of the design means an increase in complexity, which equals an increase in information content, thanks to the continuous, measured addition of energy to the system. It means an inverse relationship between information content and the second law. If no information is added to the system, everything in the system levels out and equalizes. And when information keeps being tunneled into the system, which is now considered to be inevitable, the system diversifies, stratifies, disrupts, intensifies, and increases complexity. New layers of meaning, intention, purpose, and function are constructed on top of existing ones – roads on top of paths, railway network on top of roads, aviation networks on top of tracks, and internet on top of aviation networks. Old structures do not disappear; they are overgrown and superimposed and overpowered by new layers of matter and meaning. The new structures increasingly feed on information. More and more advanced structures are built according to ever more complex plans, based on an increasing amount of structured information. Design is an emergent property of the incoming flow of energy and information, highly unlikely as seen from the perspective of the second law of thermodynamics, yet inevitable. Interacting tension fields of information form the basic configurations for more and more ubiquitous, immaterial, and material components of our Infoland. Interacting tension fields of raw unstructured information form ever greater complexities of entangled information, eventually giving structure to the ubiquitous bodily exo-extensions to human life.

One might argue that after another billion years, the universe will eventually turn into a super-designed, which is highly unlikely, ultra-low entropic information structure. Perhaps, all tangible matter will evolve into immaterial virtual information structures. Even humans might advance to become fully immaterial informational structures themselves. No more ashes to ashes as in the material world, but information to information as in a possible future immaterial world. Eventually, the material world will dissolve into a super-aware immaterial world. In our era, we are already witnessing a progressive fusion of the real and the virtual. It seems plausible to me that the virtual will take over more and more, and eventually be all there is. Virtual reality is purely an informational construct, bits and bytes being its building blocks. Virtual reality is more real than our commonplace notion of reality, as I wrote back in 2001 in my inaugural speech titled E-motive Architecture at the TU Delft. Virtual reality is a form of hyper-reality since we know every inch of it, which cannot be said of our past universe and our future quantum omniverse. Looking into the universe is looking into the past while looking into the quantum fields equals designing a possible future, seen from that local and temporal position we call the here and now. We simply do not know what the basic building blocks are of the very large and the very big. And we do not know where these two ends meet and how they might merge into one coherent system. It is my hunch that by building more and more hyper-real constructs, we will in the end replace our daily organic reality with a fully transparent and knowable virtual reality. So, when the hyper-real virtual has taken command, then we might have become fully virtual as well. Eventually, everything and everyone will become pure information. Realizing that we are already sort of cyborgs living inside our progressive evolution from simple conglomerations of cells into something like conscious prosuming entities, it does not seem an insurmountable challenge.

Sound Synthesis

Architecture and music have been close allies since the dawn of human civilization. Many architects play musical instruments reasonably well, they seem to have an ear for the structured and layered components of music, the sounds, the notes, the rhythms, the polyphony, and the melodies. Music comes to our ears in the form of vibrations that are translated by our sense organs into chemical pulses and processed by our brains by a method of wave pattern recognition. Humans developed language to label such structured sounds as music and subsequently started to produce sounds into ever more complex logical structures. Making vocal sounds in a specific pitch, duration, and volume and a specific order has become known as singing, perhaps developed in early civilizations in support of recurring rituals. Later, the bodily extensions known as instruments were invented, finding out that not only the human voice could make sound but also banging on skin, wood, iron, and pulling strings of different lengths would be effective to create a specific atmosphere. Instrumental music is an emergent phenomenon of material manipulations, by designing new combinations of matter and information. Together, the basic components of music, the pitch, chord, rhythm, tempo, dynamics, accents, melody, harmony, and counterpoint are composed into a precious layered structure. Composers are the designers of sound assemblies. The notion of the composer comes close to the notion of the component. Composers create sounds and structures until then unheard, sometimes groundbreaking to form fertile ground for further developments. The basic constituting component of all possible sounds is the pitch, the height of the single tone, measured in Hz. The pitch in music is what the letter is in the alphabet. The chromatic scale has become the standard alphabet of sounds. But, in principle, all possible pitches between the notes of the chromatic scale are equally valid. Arabic music also uses half-tones and thus has many more tones than Western music. In Western music, the billions of combinations of the 88 keys on the grand piano can create almost any kind of musical piece. More pitches combined at the same time form harmonies or dissonants. Pitches in a recognizable sequence form the melodies. Every combination of tones in harmony or disharmony is in principle possible; there are no strict rules. A specific set of preferred combinations establish a musical style, which parallels a specific language in linguistics or design language in architecture. In musical scores, especially in the more complex structures like symphonies, the components are mutually synchronized to form the complex being that unfolds in real time when executed.

In buildings, such mutual synchronization of components also is the norm; structural components are usually clad with exterior components and interior components. In my work, I aimed at a full synchronization of structure, interior skin, and exterior skin into one single complex layer, comparable to the principles of minimal music. In minimal music, simple rules create complex patterns by gradual transformations of otherwise simple themes, which is comparable with the gradual transformations we achieved using associative design methods. The optimal integration of structure and skin implies the optimal integration of strength and insulation and the optimal integration of esthetics and maintenance. Integrating multiple functions into one complex component means having a better-informed component. The ultimate solution is to have one project-specific highly informed parametric detail for each specific building, whereas one single hybrid adaptive component integrates the functions of strength, transparency, climate control, insulation, esthetics, and maintenance. The complex components are put together using dry assembly methods to form the larger whole. In that sense, my work is minimalist, although

179

I would rather refer to it as a form of <u>simplexity</u>, simple rules that create a coherent form of complexity. It is complex, not complicated.

Parametric design compares to music that is based on simple rules that create non-repetitive complexity. Not surprisingly, my favorite composers are Brian Eno, Terry Riley, Philip Glass, <u>Simeon ten Holt</u>, and Ryuichi Sakamoto. In one of our earlier exhibitions in Galerie Westersingel 8 back in 1990, we invited composer and synth designer <u>Rob Hordijk</u> to 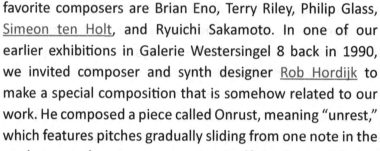 make a special composition that is somehow related to our work. He composed a piece called Onrust, meaning "unrest," which features pitches gradually sliding from one note in the chromatic scale to another, never repeating itself, in always different order. Hordijk describes unrest as a dynamic sound process, whereby harmony functions as a resting point between periods of continuous unrest. The result is mesmerizing and certainly an enrichment to modern music. Hordijk installed his Onrust piece on the first floor, from where the sounds were gently flowing like a viscous liquid down to the main exhibition space.

Around the same time that we designed the Saltwater Pavilion Ilona, I proposed an interactive outdoor installation for R96 Festivals in Rotterdam. Despite a tight budget, we wanted to have something as big as possible, a sculpture of which the interior would be accessible to the public. We had just finished our bespoke international Sculpture City event that took place in the RAM Gallery in Rotterdam in 1994. The leading theme of Sculpture City: "A sculpture can be a building; a building can be a sculpture." Now we got the opportunity to test the idea, although on a limited scale. We proposed a smoothly shaped 20-m-long volume called <u>paraSITE</u>, with the head and the tail parts lifted from the ground, and a voluptuous body part between head and tail. The interior functions as a web lounge, accessible for the public to interact with the language of ParaSITE. The interaction entails absorbing sounds from the immediate environment and processing these sound samples in a sound synthesis program. We invited composer <u>Richard Tolenaar</u> to write the code, in the then-just-released synthesis program <u>Supercollider</u>. ParaSITE language composer Richard Tolenaar crossbred in real time street culture sounds with high classical culture sounds. We recorded the real-time sessions during the day, and ParaSITE performed the pre-recorded sessions during the night hours. The basic components of ParaSITE's music are the sound samples; ParaSITE combined traditional classical samples with a new alien component, namely sound samples from the streetscape. Like Edgar Varèse and John Cage pointed out, before us, any type of sound is a possible material to make music. A sound sample has a very complex structure, not just a single clear Hz pitch.

We developed ParaSITE as an instrument for the public to play with. The public enters sideways via a gill in the silvery fabric of the inflatable. The inflatable is kept under constant pressure; people entering the sculpture cause a temporary slight deflation of the volume, which makes the body of paraSITE feel alive. Inside, we installed a web lounge, the first of its kind in Rotterdam. We placed some of our desktop computers inside the ground; so the public could navigate the World Wide Web, which back then was a novelty. Every 15 minutes, paraSITE would play aloud the output of the synthesizing program, always building upon previous runs of the program. The context-aware paraSITE learns from its environment. The paraSITE sculpture is designed as an instrument, capturing data from its immediate environment, and processing the raw data by using the captured sounds to mix with existing classical musical scores. The processed music was overwhelmingly beautiful, especially at night. At night, the interior lights were controlled by the synthesizing process as well. ParaSITE lived its week-long performance in Rotterdam as a dynamic input-processing-output vehicle. After R96 Festival, we sent paraSITE on European Tour; it landed in Graz in Austria, in Dunaújváros in Hungary, and in Helsinki in Finland, and returned to be installed at the large atrium of the City Hall in The Hague, designed by Richard Meyer, to present what it had learned during the European Tour, using all the sounds from the visited places for its final performance, this time in collaboration with composer <u>Johan van Kreij</u>.

Real-time Behavior

    In the same year, i.e., 1996, we worked with composers Edwin van der Heide and Victor Wentink, interaction designer Bert Bongers, and computer programmer Arjen van der Schoot, to compile the interactive soundscape and interactive light environment of the Saltwater Pavilion. The interior design concept for the Saltwater Pavilion is to build an environment of spatial movement, spatial light, and spatial sound to form the base layer of atmospheric information representing the water cycle. More layers of information would have followed over time, but unfortunately within half a year after the opening in 1997, a new management was installed, and the artistic direction was completely reversed from our abstract informative concept into a blunt exposure of realistic items like skeletons of whales. The new management meant effectively "game over" for the interactive interior of the Saltwater Pavilion. What we managed to realize was groundbreaking though. The Saltwater program captures raw data from a weather station that is positioned on a buoy on the North Sea and converts the data from wind speed, saltwater content, and wavelength into Midi signals driving the fiber optic lights and the ambient sound system. Thus, the data coming from an existing weather station at the North Sea form the base components for the behavior of the interior of the Saltwater Pavilion. We worked closely together with a team of composers, interaction designers, computer programmers, visual artists, and steel manufacturers to merge their various disciplines into a comprehensive work of art, a Gesamtkunstwerk as the Germans put it. Playing with one of the two sensor boards, the light and sound conditions change in real time. With the other sensor board, one navigates in the projected virtual environment, which covers the complete opaque polycarbonate interior skin, like a virtual sky above the undulating surface of the ocean. We designed the Saltwater Pavilion as an instrument to play with, a spatial environment to interact with. The public negotiates with a base layer of information to manipulate its visual appearance and its soundscape. We created an environment that proactively proposes moods, and which is at the same time open for interference by the users with their human moods. The designers of the behavior of the Saltwater Pavilion were not confined to narrowly defined boundaries of their disciplines, but I stimulated them to feel free to cross over to other realms. Thanks to digital programming languages, we could process combinations of digital components and modules, from whatever source these data are coming from or are sent to. We spoke a common language, i.e., the digital language. The graph that shows the input-process-output data flow becomes an intrinsic part of the design concept. More clearly than before, we discovered that design is, in its essence, a set of rules and formulas that can be precisely described in a programming language.

Scripting

Since the Saltwater Pavilion, scripting has become an inevitable basic skill for our design teams. Like natural languages, programming languages have a clear distinct structure, based on their constituting components. Components in scripting are programming components, like predefined lines of code to describe a certain action or complex commands. Together, the integrated parts form the whole of the script that describes the design in painstaking detail. Scripting is, in its essence, a form of writing, writing code, line by line, and module by module, to describe the actions to be taken by the computer. As in mathematics and natural physics, the more compact the script or the formula, the more elementary and hence elegant the internal structure. A convincing example taken from our design practice is the script we wrote for the Acoustic Barrier (2005). The file-to-factory script describes exactly the positions of the reference points, assigns a unique number to each connection between the reference points, describes the geometry of the constituting components, their distances, and the angles relative to each other, and the positions and dimensions of bolts needed to assemble the parts into the structure as a whole. Relations managed by the script are typically relations between neighboring components. In the Acoustic Barrier project and other ONL projects, we primarily manage the relations between the neighboring components, as in the swarm of birds. Driven by the elastic lines that inform the basic 3D model of the Acoustic Barrier and that of the A2 Cockpit, each of the thousands of components is different but only slightly different. From one steel profile to the next, the length varies from just a few millimeters to some decimeters. The same applies from one piece of glass to another and from the rubber profile to the next rubber profile. In the optimization process, traditional engineers suggested arranging the differences into groups with a certain bandwidth, assuming that it would be easier and less expensive to build. The opposite is true. It quickly became very clear to all involved that such a script would need extra lines of code, hence becoming more complicated instead of simpler. Arranging the components into groups and categories means adding exceptions to the rule. Having one inclusive rule without exceptions is always the most efficient way to go. This simple procedural innovation of the assembly of a well-integrated series of uniquely shaped components soon became the public secret to

 our radical approach to architecture: no exceptions – the exception is the rule, as I summarized in my paper Game Changers (2014). My designs are based on simple rules creating complexity, not complicatedness.

The traditional approach in architecture is to select components from a given library of products and put them together in a meaningful way. The cold connections of half-products from a catalog do not allow for effective scripting, while almost every detail would need many exceptions to fix the different elements together. For the Acoustic Barrier, we designed the library of components ourselves, based on available basic raw material, and merged these few different components into one parametric detail that we mapped onto the thousands of reference points. Along the full length of the 1.6-km-long barrier, we did not need any exceptions but just the rule. The Acoustic Barrier is a dry assembly of 40,000 pieces of steel, all of them with a unique number, all of them different, even when they virtually are the same. Even the two rounded endings that are flattened like the beak of a platypus form an integral part of the script. The unique identification numbers are stamped into the steel components during the CNC-driven manufacturing process. The programming strategy that we have developed is to find those kinds of rules that need no exceptions, and those kinds of rules that create a visually rich and simply complex architecture. One building, one detail, as is ONL's sublime motto. The complex yet simple detail comprises all constituting components: a steel profile, a plate of glass, a rubber profile, and bolts, described in their mutual relationships.

Action Painting

It may sound unlikely to even think of components in the visual arts, but there are many parallels to be found between the role of the component in art and architecture. Musical scores are usually well structured, chopped into measures, keys, beats, melody lines, choruses, and counterpoints. In expressionist art, we may look into the character of the brush strokes to find the component, rather than into the storytelling through naturalist figures. In abstract art, the story is no longer leading, and the strokes and the mutual relationships between the brush strokes have become the main components. Components in the visual arts can be recognized in well-structured art forms like abstract procedural art. I should rather say "concrete" procedural art since the works are not abstractions from reality but constructs of a new reality. I will discuss two works of art, seen from the point of view of interacting components, put on the canvases using a procedural way of working, the analog form of running a computer program. The first one is a well-known drip painting Autumn Rhythm by Jackson Pollock, and the other is the abstract robotic painting Machining Emotion by Ilona Lénárd and her multidisciplinary team, in which I was involved as her technical consultant. I know the creative process of intuitive gestural paintings from the inside out, from Ilona's initial digital sketches from the early eighties to the recent, robotically painted canvases.

In the case of Jackson Pollock's action paintings, which are extensively analyzed, I will interweave the popular interpretations with my findings. Pollock's drip paintings represent only a relatively short period, which spans basically from 1946 to 1951, but especially the drippers have been crucial to establishing his fame. "Take some paint. Do something with it. Take some more paint. Do something else with it." This quote is attributed to art critic Harald Rosenberg and is most likely a variation on a notebook entry by Jasper Johns; the emphasis is put on the action. It was Rosenberg who coined the term Action Painting in the December edition of Art News 1952 after Pollock had already abandoned the gestural drip painting method. The term Action Painting refers to Pollock's technique of dripping paint onto a canvas. Instead of using the traditional easel, he placed his canvases on the floor and dripped, splattered, and poured paint onto them from a can, using sticks, trowels, or knives. There is enamel paint, and there is an actor who does something with the enamel paint. The tools for his paintings are by and large the same as used by any other painter: canvas, paint, and brushes. There is paint and there is the painter, who acts upon the paint and runs a procedure on the paint to compose the constituent components for the art piece. What makes the components of the abstract painting specific is the intention of the actor-painter, who paints instances from his/her artist's universe. Pollock uses brushes, sticks, trowels, or knives to act upon the paint on the canvases that are spread out on the floor, as with his work Autumn Rhythm. Like writers and composers, painters customize otherwise generic elements to become the components to play their tune. Basically, "makers" process matter; they transform base materials into something highly complex, reversing the second law of thermodynamics and reducing the entropy level in the closed system of the process of painting. The making of a painting is in evolutionary terms a highly unlike event. Pollock reaches the canvas that is put on the ground from the sides, working all around the canvas to avoid a preferred viewpoint, thus evoking the experience of a universe that has no bottom, top, left, or right. Occasionally, he would step on the canvas to reach the inner parts of larger canvases. According to Pollock, he works on the floor of his barn and works around to get the feeling as to be *in* the painting, inside that parallel universe that is created against all thermodynamic odds. Like a choreography or a musical score, Pollock's paintings are procedural, a series of quasi-repetitive strokes, as if following a program. The procedural logic of the painting takes command over the artist's mind and body, and the artist becomes the executor of the program.

Let us set this groundbreaking work of Pollock against a recent robotic painting by visual artist Ilona Lénárd. Ilona has always worked systematically and intuitively; as she stated in her book Powerlines: "You have to train your intuition to steer your logic." Training intuition is a form of de-learning; she describes it as becoming a sort of "idiot savant" as in the movie Rain Man, a person with special abilities – and disabilities, but that is not relevant in this context – who has direct access to the brain's database. Intuition is a form of not letting yourself be ruled by ample considerations, but rather by direct intuitive actions, based on deep understanding, knowledge, and experience. Knowing the procedure of Ilona's painting from nearby, she typically chooses a basic gesture with a specific character from a pool of predefined types of strokes. The characters of the strokes are well defined but not the exact execution of the strokes. Each stroke with the acrylic marker is different each time she strikes, as in varying the parameters in a parametric design. Lénárd covers the entire canvas with intuitively generated variations of that basic gesture. Like Pollock, she puts the canvas on the ground, as to be able to work all around, and even steps upon the canvas to add strokes with the acrylic markers from that inside perspective, from the middle of the canvas. Layer by layer, she builds up her universe of analog parametric gestures. The gestures can be smooth and circular, like in her earlier Flow series. Flow sketches form the gene pool of sketches, as in our joint Sculpture City project (1994), in which sketches are translated into sculpture buildings. Even for herself, it is unpredictable in which direction exactly the fierce fast strokes would go, as first experimented within the Tangle series, which followed the Flow series. The built-up energy is released onto the canvas. Characterizing the paintings, the notion of abstract calligraphy immediately comes to mind. The abstract calligraphic strokes are devoid of any literal or metaphorical meaning. The characters are bodily gestures, an abstract choreography of muscular movements, without a predefined meaning. She creates the base material for a new personal language that materializes through energetic strokes. The important distinction to traditional calligraphy though is that her gestures are extremely fast, not deliberate at all, allowing her intuition to unleash. The similarity with the principles of parametric rule-based design is striking. Like in parametric design, she invents the rules of the game and then plays her game. Intentionally, the rules of the game are defined to form a framework to self-surprise. The end product is not foreseen but only anticipated in generic terms. While the character of the individual strokes is defined, the intention of the painting as a whole only develops along the process. The process of making an abstract painting is an intense feedback process. After each layer of gestures, the artist responds to what is on the canvas first and then decides how to move on. It can go to any direction during the process of forcing the acrylic paint on the canvas while keeping faithful to the basic character of the gesture and to the rules of play. Similar to parametric design, the diversity and richness of the painting are based on a set of simple rules. Ilona uses large-tip acrylic markers as the brush. The acrylic markers allow for fast and furious movements, which are necessary to act intuitively, to avoid super-conscious considerations. In the later

Pattern series, she used patterned paint rollers, imposing her gestures onto the roller, making similar quirky turns as with the markers. Layer by layer, the ubiquitous gestural components entangle themselves to form another material manifestation of her evolving personal universe. After the Tangle series, Ilona produced the deeply universal Omniverse paintings in her studio at the Van Nelle Factory in Rotterdam, the Netherlands, and the Q series and the Pattern series in her Doha Fire Station studio and exhibited at the Doha Fire Station Workshop 4 and in the old summer palace of the Faisal Bin Qassim Museum during our two-year stay in Doha, Qatar. Each new series of paintings and each painting from that series bears a specific character, based on special painting-specific gestures. The pictorial components are what the words and the sentences are in a language. The abstract painting is like a well-structured minimal music improvisation, while the constituent components do not repeat themselves exactly.

Upstream

Taking industrial household products apart gives a great insight into how the constituent components are related to each other. At Qatar University, I asked my first-year students to take apart simple household products, as artist Todd McLellan raised taking things apart into art in his series Things Come Apart. My students brought with them: a placeholder for a mobile cellphone, a robotic vacuum cleaner, a toy car, a computer mouse, and other simple daily things. I asked them to disassemble them down to their smallest parts, to the screws and bolts. Even the smallest household product turns out to be a fascinating object, full of ingenuity and complex relations between the adjacent components. Each component is aware of its neighboring component because they just need to fit together, properly, exactly, and seamlessly. Each component only has a limited room to move about in the available working space. Sometimes components need to transfer forces, sometimes they transfer electric current, sometimes they need to receive a screw, sometimes they are shaped for ergonomics, and sometimes they are shaped just for esthetics and styling. Most of the components are customized especially for that particular position with those particular neighboring parts for that particular unique product. The smallest components like batteries, bolts, and screws are usually generic elements, exchangeable with similar elements from another brand. It quickly became clear to the students that each component has a specific function, made in a specific material, defined to have a specific performance. In our global economy, the various constituting parts usually have a different origin of production; some parts may come from China, some may be made in Germany, and others may come from the USA. The designers have worked together over long distances to synchronize their designs, as to guarantee that the parts will properly fit together. Although usually just taken for granted, even simple objects like household products are very special in terms of evolution. Designing and producing a simple product requires collaboration between thousands of people from different cultures, countries with different social regimes, global transport facilities, and packaging methods, different design cultures, delivery systems, sales points, and money transfer methods. Before a product enters your household ecology of products, it has seen the world. Almost everything that makes a society work is condensed into one simple household product. That is what I made students aware of when taking things apart.

Opposite to taking things apart, the other way round, the synthesizing of parts is the essence of synthetic architecture, usually in a combination of generic and unique customized parts. As is the case with molecular life forms, the design of a product represents a highly unlikely upstream event in the bigger downstream picture of the universe. Components, not elements, form the ubiquitous constituent parts of the larger objects and built constructs. Everything that has been shaped in the past by nature, and that is shaped in the here and now by nature 2.0 in action, is going upstream against the second law of thermodynamics. Designing household products and larger buildings alike means synthesizing thought components and physical components in a meaningful way, built up from within, based on the informed relationships between the constituting and neighboring components. The brand typically is the last chain in a vast production chain from raw material to half-product. In principle, even for the smallest products that are produced in quantities, the whole world comes into action to make that specific product. Raw materials are mined in one place, processed into usable basic materials in another place, designed in yet another place, produced in its constituting components wherever in the world, transported all over the globe, and eventually assembled where it is probably cheapest to do so, and then brought to market by the brand. The brand puts its name on it and is responsible for the marketing and the product's evolutionary success.

The principles of parametric design to robotic production will cause a radical change to the world of global brands and the global trade of products. The production according to parametric design to robotic production uses wherever possible local materials, produces as close as possible to where the product is consumed, and harvests energy and water as close as possible to where it is produced, where, when, and as needed. Only the design part will remain super-global since the digital workflow on the internet effortlessly connects designers and engineers from all over the world. "Physically local, virtually global" would be a nice slogan for my Another Normal. Thanks to the digital workflow, the super-local is intimately connected to the super-global, as is the individual to the swarm. The digital part of the production process is more effective when operating worldwide. In one day, three teams of designers can work on the same project, working in three shifts of eight hours a day, while each shift takes place in their continent, subsequently in Asia, Europe, and USA. As the world turns, they effectively work continuously as a non-stop design team.

Building Relations

Writing my article Vectorial Bodies for the Dutch magazine Archis, also included in my book Architecture Goes Wild, I was especially triggered by the way car designers are embedding performative components and styling components in the larger whole of the car body. Car designers have taken a big leap from a collage of mass-produced elements toward assembling unique customized components to form the larger whole. In the first decades of car design, a headlight was just another lamp. The German Bosch headlight would equally fit on a German car or an American car. In those early days, a headlight used to be a generic product, standardized to be mounted onto any car body. Same with doorknobs, speedometers, tail-lights, bumpers, windows, seats, wheels, mirrors, etc., you name it. Headlights were mandatory, but instead of styling them with the car, they were initially simple additions to the body. In general, the cars of the twenties and thirties of the last century were still based on the conventional approach of stacking generic elements on top of each other – elements, not components. The car used to be a compilation of a chassis plus a carriage plus additional paraphernalia added for the sake of communication with the driver and with the traffic. Today, cars are designed completely differently. Headlights are integrated components that uphold a dynamic relationship with their adjacent components. Headlights, and all constituting parts, are customized in shape and performance for one specific car. A typical headlight of today is fully immersed in the body as a whole. That specific headlight fits only there, then, and therefore, and where, when needed, and as needed.

Modern international agreements for standardization require a guaranteed level of performance, demanding safety, and circularity, but not demanding the same constituent components to comply with the performance requirements. Modern standardization is rule-based, not solution-based, intentionally stimulating rule-based open design methods. Car designers translate the rules into minimum respectively maximum values for the parameters in their digital designs and finite element calculations. More and more, cars are equipped with sensors and processors, in the motoric parts, in the mirrors, at the front, the sides, and the back, adding complexity and real-time awareness to the car body. These are elements that are not custom designed but like the screws and bolts in-built constructs, mass-produced parts that are mounted into the body parts. Designers are fighting the current car's obesity, especially now that SUVs have become the dominant type of car, by minimizing the weight of the constituting components. The new international standards only partly affect the shape of the constituting components, while new CNC-driven production methods give the designers new freedom to design to produce unique components quickly and cheaply. There is no need for a large series of the same cars anymore. Soon cars may have become one-offs completely, tuned, and customized for one unique buyer. Furthermore, when 3D printing of bigger components has become a cost-effective and esthetically pleasing option, unique car bodies will be printed for unique customers.

When the parts are informed to form a bigger whole, we speak of components. An emergent property of the technological progress to build cars as unibodies is the phenomenon of the styling of the parts. While the components are physically connected to form the whole of the body, styling connects the parts in an immaterial way. Feature lines of modern cars continue all the way from the backlights via the side fenders and the side doors to the headlights into the front part. The continuation of feature lines across the different constituting parts is highly interesting for my take on architecture. The concept of folding lines and feature lines shows the way forward for design industries other than car design, like architecture. I took on the challenge and successfully applied the design concept of character-building feature lines on the larger scale of buildings. The A2 Cockpit and the Acoustic Barrier, in which the Cockpit is embedded, are subjected to such styling components. The A2 Cockpit, commissioned by the private client Hessing Holding, features the styling component of folding lines that gradually appear and gradually fade out. The Acoustic Barrier, commissioned by the Projectbureau Leidsche Rijn of the City of Utrecht, features a continuation of that same styling component, in combination with another styling component that gradually transforms concavity into convexity and vice versa. Stretched over the full length of the project, the A2 Cockpit is fully embedded in the Acoustic Barrier, as an integrated, inflated part, locally pumping up the volume, exactly like a pilot's cockpit that is embedded in the body of a Lockheed Starfighter jet plane. While the Acoustic Barrier has no other function than blocking the noise from the A2 Highway for the housing areas behind the barrier, the A2 Cockpit "en passant" functions as a barrier but is a showroom for luxury cars. By the invention of fusing the A2 Cockpit with the Acoustic Barrier, we created a commercially valuable property that otherwise would not even exist. The fusion brings in revenues for both the private client and the public authority, aka a win–win situation.

Style

Components are subject to styling, and feature lines are passed on from one component to the other, to its immediate neighbors. Zooming in on car design, basically on any car, one can see that feature lines aka the folding lines are passed on from the side to the front and from the side to the back. Folding lines continue regardless of whether there is a door or a fender. Doors in cars are almost invisible and unobstructed cut-outs from the mass of the body. Traditional elements such as doors and windows have disappeared in favor of merging with the body as a whole, becoming components. Explicit functional elements disappear, and the components are fully subordinate to the whole, as in the swarm. We applied the principle of disappearing elements in many of our designs, from our patio housing project in Dedemsvaartweg in the Hague in 1991, the body of the Saltwater Pavilion in 1997, the Web of North-Holland in 2002, to the design of the A2 Cockpit in 2005, basically in all of our art and architecture projects. Feature lines are passed on, not only within the same material but more interestingly also jumping from one material to the other. The folding lines of the Acoustic Barrier are passed on to the Cockpit, which has a different skin structure and a different detail; the folding lines act upon two different structures. The Cockpit passes the feature lines again at the other side, securing a continuous experience of transformation when driving by on the A2 highway, enjoying 60 seconds of architecture.

Structural forces and styling features are transferred from one component to the other. The visual dynamic in car styling is passed on from one component with one specific function, for example, the side fender that functions as a dirt cover to the wheel, to an adjacent component, for example, the headlight in the front that functions as a torch to illuminate the street. A stylistic touch binds the two entirely differently functioning components together. Headlights are no longer focused straight ahead as before when headlights were mere standard lamps, today's headlights are rounding the corner, such that they are seen from the side and the front, while in modern cars like the Citroen C5, the headlights turn the corner with the wheels. The automotive styling of cars and hence of buildings connect the back to the side to the front. Car designers have long abandoned the static representation of the volume, which means drawing the side, the front, the back, the top, and the bottom, separately from each other, as is unfortunately still common practice in the design of buildings. Today, car designers look at the body of the car as a whole, as a complex three-

 dimensional body that is in motion, even when standing still radiating the dynamism of a body in speed. Compare the fourth generation Ford Lincoln Continental from 1961 with the <u>Bentley Continental GT</u> from 2003. There is 40 years of styling development between one and the other. The styling language has a similar majestic approach to it; both cars look like credit cards on wheels, thanks to their overall simple basic shape, the straight contour lines that connect back and front, and the rounded corners. The major shift in general car design practice is that the rounded corners in the Bentley are truly three-dimensional, while the rounded corners of the Lincoln are designed from a side view only. Even in majestic luxury cars, the dynamic design paradigm of the three-dimensional body has become the norm. Bodies have become objects that are designed to be seen and experienced as dynamic bodies moving in a 3D space, acknowledging that car bodies are objects with an internal drive and a desire to move. Cars want to demonstrate their ability to go fast, even when not moving. The new dynamic styling fully reveals itself when slowly circling around and alongside the unibodies. The experience of speed and dynamism is catered by passing on the folding lines from a component to the adjacent component. The product components are the members of a swarm of components that act in sync to create that dynamic spatial experience. A collage of standard elements would not be able to convey such an experience. I have been keen on applying a similar dynamic as car styling to non-moving building bodies. Applying dynamic styling to otherwise static bodies gives them the illusion of vehicles that want to go places and gives the viewer the experience to go along with them, to go where they have never been before. We are still in the beginning phase of adopting such a three-dimensional dynamic styling paradigm. Not surprisingly, there is, at first sight, less urgency to do so since the built constructs do not move as cars do. But then again, for cars that in reality do not drive much faster than 100 km/h, there is not a dire need for dynamic styling either. The importance of dynamic styling lies not so much in the functional need for it but also in the desire to go places, to see the world, and to do that fast and in a unique way. The current trend is to add more and sharper folds to the car bodies, to suggest a more aggressive form of dynamic, coming closer and closer to the esthetics of combat machines, like the <u>Transformers</u>-style <u>Lamborghini Urus</u>. Or is it the minimalist combative like the Tesla Cybertruck? Are car designers preparing us for a beautiful doomsday? The dynamic of Another Normal favors a gentler approach, a more relaxed form of the notion of the dynamic. That relaxed form of being dynamic is what I have aimed at when designing the A2 Cockpit, the Bálna Budapest, and the Liwa Tower. Ultimately, the logical next step in design is to leave the esthetics of a frozen dynamic behind and adopt the principles of real-time dynamic styling.

  Dynamic styling in real-time is the next level. The perfect example of real-time interactive dynamic styling is the visionary concept car GINA, designed by Chris Bangle for BMW just after the millennium-shift. Gina is provided with Festo muscles, pneumatic elements that can be programmed to change their stroke. The pistons are connected to the stretchable fabric of the car's skin. When a Festo muscle tightens itself, the skin of the muscular car body bulges up, as to show its sensuous curves. The real-time dynamic design concept of GINA connects the driver emotionally to the shape of the car; the driver changes the shape by changing parameters, by talking to the car. GINA shows the way for architecture to move toward a more dynamic styling. A dynamic style connects the building body emotionally to its users. And a dynamic style enhanced with interaction connects the built environment with its users in real time. Bangle took the same step with the Gina concept car that I took around that time. I designed the Trans_ Ports project, which is the pavilion that changes shape and content in real-time, one year before Chris Bangle designed his GINA. The other interesting fact is that GINA was first published in 2009, meaning that Bangle and I have lived in parallel universes for a decade, both of us thinking in the same direction of interactive components, taking the same consequences of the need for a dynamic style supported by actual digital-driven technologies of sensors and actuators.

Unique Code

In industrial production facilities, each product gets assigned a unique number. Each car has a unique code, such that it knows itself and is known by others and that it can be traced back to its constituting components. The constituting components have their unique number to trace back to where they are produced, and what the unique characteristics of that component are. Similarly, ONL has applied a system of unequivocal labeling of each uniquely identified component to the design and execution process of the Acoustic Barrier and the A2 Cockpit. In the design process, we assigned a unique code to each component. We are talking about over 40,000 unique pieces of steel and more than 10,000 plates of glass, all with their unique identity, shape, and dimensions. The coding that we applied is not just a random number but a code that can be intuitively understood by the human assembly team. For example, one steel component of the Acoustic Barrier is labeled W18A24, meaning the 24th component in row A of segment 18 in the West sector. Each row of components is produced and preselected in a neat package and transported to the building site. Photos from the building site show the tidiness and the extremely simple organization of the assembly. Assembling the Acoustic Barrier does not require any scaffolding; the structure is designed such that it forms its own scaffolding. The steel manufacturer and the main contractor Meijers Staalbouw managed to complete a 30 m' stretch of the Acoustic Barrier in one day, with a team of only four people, one of them being the supervisor and the others unskilled laborers. As a do-it-yourself IKEA package, each package contains its own unique set of components. Each package contains a printed instruction sheet on which component to connect to which neighboring component. Due to the unique dimensions and the unique positions of each constituting component, the designer and the manufacturer need to think in advance about how to link the design process to the production process, and the production process to the assembly process, in the most painstakingly precise detail. The design to production to assembly procedure of the Acoustic Barrier shows that nonstandard complexity is perfectly well manageable and easy to realize within agreed budgets, even competitive with traditional design and production techniques. Both the Acoustic Barrier and the A2 Cockpit were realized for less than half of the costs that were estimated by traditional professional parties, who were not familiar with the workings of parametric design to robotic production. If our clients Hessing and Projectbureau would have dropped their ears to the traditional contractors and quantity surveyors, the project would never have gone ahead. Fortunately, our clients allowed ONL to form a design & build consortium with the steel construction company Meijers Staalbouw, who also acted as the main contractor. We managed to build the complete structure within a budget that suits a simple straightforward standard rectangular building. The A2 Cockpit and the Acoustic Barrier were the litmus test for a synthetic architecture that treats a building as a well-informed assembly of unique components. The constituent components are designed, produced, and assembled just as needed and when needed – just that, just there, and just then. The litmus test was overall positive; the designs of the A2 Cockpit and the Acoustic Barrier are well-appreciated by our clients and by the public. The design-to-production method is fully scalable and applicable to all sorts of buildings including social housing. As a small innovative office, we developed this unique method in a close design & build collaboration with the steel construction company, and it is exactly that fact that eventually isolated us from the general architectural scene in the Netherlands. Now, 16 years later, there is not a single building in Holland – or elsewhere – that comes even close to the level of nonstandard complexity, based on the universal principles of the ubiquitous well-informed component. The mechanisms of innovation in the Dutch building industry spin slowly, with lots of talking, not too much substance, small steps at a time, not even coming close to the much-needed Another Normal.

Ubiquitous components are everywhere in nature and nature 2.0. Components are ubiquitous at all levels, from molecules as components for organic life, from DNA structures as components to form new species, from words and sentences as components to build language, from abstract expressionist gestures as the constituent components for the unlikely creation of paintings, from algorithms as the components for real-time behavior of interactive built constructs, and from parametrically designed and robotically produced components for new forms of architecture. Emergence is an upstream process that crystallizes the unlikely, even the alien, out of the abundant presence of its constituent components and their mutual relations. Nonstandard architecture naturally emerges from parametrically designed and robotically produced components. Nonstandard architecture is the embodiment of vast numbers of unique mass-produced components. Interactive and proactive architecture is that new life form that emerges out of the network of real-time relationships between acting components. Although it is difficult to see while we are living and inside the process of emergence and inevitably daily contributing to its evolution, there is nothing more natural than synthetic component-based architecture.

5

Components versus Elements

Not an Element

Alarmed by the Venice Architecture Biennale in 2014, curated by Rem Koolhaas (RK), titled Fundamentals, a clear distinction must be made between what an element is and what a component is. While a component is a constituent part of a bigger whole, an element exists as an entity in itself, without having defined a relationship with its immediate neighbors. Components maintain a deep relationship with adjacent components, while elements have no such relationship at all. A brick is just a brick, a floor plate is just a floor plate, and a window frame is just that. Elements are those products, usually mass-produced, that have not been positioned in a broader context yet. Once designed to be positioned among other elements, a weak yet violent relationship is born, after they have been cemented, welded, or hammered into place. Other than elements, components have a preconceived relationship, not as an afterthought, components are informed on their future relationship. I consider only those parts to be components that have been assigned a unique identity in the design process. Only those parts that fit in one specific position in the larger whole are considered components. Only those parts are addressed as such from the earliest stages of design to the production and assembly. In the up-and-coming economy of Another Normal, components are put together in a dry assembly using bolts and screws or fit tightly together without mass-produced connecting elements like bolts and screws at all. My work at ONL and Hyperbody aims at establishing an

201

architecture that is based on the dry assembly of components, for the following many reasons:

1) possible reuse of the structure as a whole,
2) possible reuse in a different context of its parts,
3) easy replacement of broken components,
4) quality and precision of the constituent parts,
5) accuracy of the project planning,
6) verifiability of the design to production to the assembly process,
7) optimization of the performance of each identified component,
8) accountability for clients, designers, manufacturers, and building contractors,
9) digital identification of individual parts throughout the process,
10) absence of waste on the building site,
11) absence of scaffolding, and
12) absence of molds.

The downside of using molds is clear: a high-quality steel mold can be used for a maximum of 10 iterations of a concrete element; a wooden mold can only be used once. Experimental robotic developments offer the possibility to quickly produce unique molds for unique components, namely 3D printing of wax molds that can be melted and the wax reused. Temporary wax molds may offer a sneaky way out of the downsides of making molds, but that technique is far from commercially feasible, and one still has to make a mold first before making the component. In addition to my explicit choice for dry assembly of components, I am a strong advocate of avoiding molds. Using molds means building twice, which is a waste of energy, time, and money. In traditional building methods, a building is built three times. First, as a scaffolding, second, as a mold, and third, by pouring concrete and/or welding steel. In our work, the idea of the component is taken to its logical core value, meaning building only once, avoiding scaffolding, and banning the use of molds. On top of that, in my work, the idea of the component is taken to the point where it can be produced robotically, that is by CNC machines and by robot arms. To produce robotically, the designer must embark on a 100% digital design process; to design for diversity and complexity requires a parametric design to robotic production process combined with a dry assembly of unique components.

Elements and components are different species of building blocks. The idea of what is fundamental about a building block has gone through a substantial evolutionary change in thinking and making, as the architects who are operating inside the digital revolution convincingly have demonstrated during the last three or four decades. Elements are no longer the fundamental building blocks of architecture, which is a tragically outdated point of view. Stubbornly ignoring the IT Revolution, RK categorized the fundamental elements of architecture as autonomous categories, a wall, a floor, a stair, and a ceiling. Indeed, these elements are categorized as such in, for example, the Dutch Elements Method that architects typically use to describe the parts of their buildings. Our nonstandard architecture challenges the Elements Method; there is only one category where our nonstandard design would fit, and that is the category of the "building frames that are constructed from wall, floors, and roofs as a single box construction." As a consequence, in the Elements Method, a nonstandard building, for example, our Web of North-Holland would be described as one single special element, and hence is completely inaccurate to describe its constituent components. The Elements Method may be fit for a building that is a collage of a series of the same mass-produced elements but certainly not for a nonstandard building that is an assembly of a series of unique mass-customized components. Throughout my nonstandard designs, we typically build a project-specific catalog of unique project-specific components. Considering the current state-of-the-art of architecture, labeling constituent parts as elements is misleadingly clinging to outdated ways of designing and building. The constituent components of a nonstandard building are connected to their nearest neighbors, while the mutual connections are an integral part of the design of the component. Isolating parts of a building into mass-producible categories fit the agenda of those architects, whose esthetics build by and large on a collage of traditional elements, a stacking of simple platonic volumes, post-connected in a cold clash of materials and building systems.

Architecture is the result of the connections between the components, the interweaving of components, diversity, and variety, the flocking of components that are unique in size, dimensions, and performance, about establishing bidirectional relationships between people and things, operating in real time. Parametrically driven designs are using the currently available digital toolbox to code and script the connections, as opposed to 3D modeling of a series of autonomous elements. An element is not a component. Elements are dead and unrelated, and components are alive, and connected. A stair is not an autonomous element, a stair is about connecting levels;the stair is an invention to extend life beyond the ground floor level, an invention to go one level up from Flatland up to Spaceland. On the ground floor level, people can only pass each by avoiding bumping into each other. With the introduction of a second level, people can pass over each other. The stair is the connection between the levels. The stair is an uplifting idea, not a static element. The idea of the stair facilitates the action of going up.

By the end of August 2014, I went to visit the Architecture Biennale in Venice. I took with me my prejudice, as I was beforehand skeptical about the theme of fundamentals. The actual visit to RK's Biennale left me even more critical. I saw a cynical and deliberately planned denial of the last 30 years of architectural innovations as if nothing substantial happened after his years at the Architectural Association.

First of all, we need to understand that the elements of architecture that RK identifies are not elementary at all but are derivatives of a higher dimension. Just like molecules or atoms *de facto* do not exist as separate categories, which are only functional as a static outsider view of an otherwise dynamic system, one should avoid putting parts of a building into fixed categories like stairs and floors and consider these categories elementary. Labeling atoms and molecules as the elementary building blocks of all matter would seriously throw science back some hundred years. Quantum theory has taken over for a century, demonstrating the dynamic character of the omnipresent force fields, leaving us today in a pleasant uncertainty about what these fundamental forces are. Static images of atoms and molecules are fundamentally misleading. Atoms and molecules should be presented in the form of living diagrams to illustrate that they are local and temporal instances of a dynamic system. Similarly, downgrading constituting parts of architecture into fixed categories, where they never belonged, seriously compromises today's architectural achievements. According to current architectural theories and practices, doors, windows, facades, columns, beams, ceilings, etc., are nothing more than temporary and local instances of continuous variation. We should not think of doors, windows, facades, columns, beams, and ceilings as separate categories at all. All such subjective elements are in a state of continuous transition from one instance to the other. Doors and windows are specifications of a larger system called the skin. The skin is a performative organ interacting with the larger system called the body, implying that such specific products as doors cannot be treated as separate categories in an architectural framework. And when doors, windows, floors, and walls are mass-produced as generic items, they are best described as industrial products or half-products, not elements of architecture but items from a catalog.

According to Wikidiff.com, the difference between a component and an element is:

> "...the difference between a component and an element is that the component is a smaller, self-contained part of a larger entity which often refers to a manufactured object that is part of a larger device while the element is one of the simplest or essential parts or principles of which anything consists...."

Following this definition, the fundamentals of RK are not what he claims they are. Perhaps, in traditional architecture, mud bricks or wooden logs might comply with the notion of fundamentals, being the simplest or most essential constituting parts of a traditional low-rise building. Walls and floors are not the simplest parts, nor the most essential parts. Walls and floors are per definition hybrids of different elementary materials put together, either *in situ* composed of concrete with rebar, bricks, and mortar, or prefabricated into mass-produced parts. Handcrafted or industrial hybrid products come closest to what RK defines as elements. RK's elements are typically products that

205

can be ordered directly from the building market, whether mass-produced or handcrafted. A component, on the other hand, cannot be ordered as such from the market; components are always the result of an integrated design-to-production process, produced when and as needed. Components are unique parts that together constitute a larger entity, parts that are intimately connected to their nearest neighbors. Elements are like products – not fundamental, and not elementary. By defining elements as fundamental, one effectively denies 30 years of progressive developments in architecture, which is the paradigm shift from mass production to mass customization. Except for the corridor, which merely indicates a function, and cannot be purchased from the building systems market, RK's other elements are product categories. The Fundamentals Biennale is therefore not about architecture but represents an outdated view of the traditional building industry that relies on the mass production of a series of the same products. Intentionally, Fundamentals is not about the state-of-the-art of architecture.

Not a Floor

In the Fundamentals show, the listed "elements of architecture" are the floor, the wall, the ceiling, the roof, the door, the window, the facade, the balcony, the corridor, the fireplace, the toilet, the stair, the escalator, the elevator, and the ramp. Is the floor an element? Or a component? To answer this awkward question, we need to figure out how a floor performs. What does it do, and how does it act in its context? What does it do for people and things? A floor on the ground floor level typically covers the ground with a clean, smoothened, and durable crust. A floor on higher levels spans typically between load-bearing walls and is designed to bear the weight of people and furniture, machines, and interior walls without sagging, keeping them safe and comfortable. A floor element as an industrial product will certainly meet the required performance criteria for its fitness to bridge the gap between the walls. However, it is the connection to the walls that defines their identity. It makes a difference whether the connection is

1) a mere stacking of loose elements,
2) connected by an *in situ* executed wet connection, as is welding, or injecting with concrete, or
3) designed with an integrated connection detail, ready to be assembled, and eventually to be taken apart.

The third option marks the difference between an element and a component. There are many hybrid forms between an element and a component. For example, prestressed hollow core planks are prepared for receiving the *in situ* cast wet connective detail, which is not a part of the element itself.

A component is an element including its connections with its nearest neighbors. I remember RK's statement after the completion of the Kunsthal in Rotterdam: "No money, no detail." That is a *de facto* confirmation of his take on architecture as a cold juxtaposition of elements, where the details are seen as additional to the elements. In my take on design, the detail, which defines how the parts connect, must form an integral part of the component. There is no component without a connection to the adjacent component, regardless of whether this adjacent component is of a different family of components or a parametric variation of the same family. In nonstandard architecture, the constituting components are variable members of the same family and have a quantifiable and qualifiable identity and clearly defined relationship with each other. To be able to run the file-to-factory design to execute an assembly process, each component needs a unique identity and must be fully accountable to perform at that specific position, at that specific moment in time, and for that specific reason.

The Elements of Architecture show ignored the then well-established movement in architecture, originally coined as Liquid Architecture by Marcos Novak, later elaborated into nonstandard production method by Bernard Cache in his book Terre Meuble (1995) and as Architectures nonstandard (1998) by Frédéric Migayrou. The timeline of architectural theories in educational facilities that was shown on one of the exhibition walls in the Italian Pavilion was abruptly terminated exactly in the year of the first publication of Novak's paper on liquid architecture. Liquid architecture, transarchitecture, and nonstandard architecture stand diametrically opposed to generic architecture. In nonstandard architecture, the floor is no longer a separate thing; a floor continues into the wall, and the wall becomes the roof. Floor, wall, and roof are considered one continuous surface, transforming gradually from component to component to perform according to their position in space, functionally, structurally, and ultimately in terms of esthetics. Nonstandard buildings typically have a continuous exterior skin, and a continuous interior skin, separated by the structural system that is fully synchronized with both the interior and exterior skins. This idea of the nonstandard applies to all kinds of spaces, including indoor and outdoor spaces, and smaller and bigger spaces; it applies to pavilions and multistory buildings alike, and urban planning and furniture design as well.

Not a Wall

Likewise, I do not consider the wall an architectural element. In our designs, I am looking at the continuity of the enveloping surface in all its aspects, in all directions, sideways, down, and up. We do not look at the idea of a wall as something separate from its context, something that is imported from a catalog. Instead, I consider a wall a cellular specification that evolved from an initial volume. Architecture is a series of subsequent specifications of an initial single volume that represent the scope of the design task numerically. The wall is a performative predominantly vertically positioned membrane, a section of the continuous skin, which separates and connects. The membrane separates interior and exterior conditions; it separates interior spaces from each other. The membrane also connects; it connects floors to ceilings vertically, and it connects to other membranes horizontally. The membrane is part of a continuous surface that wraps around the volumes as a whole and/ or around particular spaces inside that volume. Neither in traditional architecture nor in nonstandard architecture is the wall isolated as an element; it always is related to neighboring components forming a local and provisional specification of continuity of the built volume. Thinking of a floor and wall as flatland-ish two-dimensional elements, with some

2.5D thickness, leads to a breaking down of space into a collage of flat pieces that are clashed together. A wall as an element is a generic industrial product, a product of the paradigm of mass production of many elements of the same. The Fundamentals exhibition means a retroactive submission to mass production, ignoring the new paradigm of mass customization that is the default production method of the <u>post-Fordist</u> society.

Radically opposite to the post-Fordist attitude, my work strives for a discrete continuity from what is down, up, left, and right, inside, and outside, to form the larger whole. A component is considered a part of the larger whole. That larger whole is what I have called a vectorial building body, a volume that has a motive direction, a vehicle without wheels. We still might use the words floor and walls, not as elementary building components, but just to indicate the relation it has with a person. When we use the word floor, it describes the surface where one sets foot on the ground; a wall is something one can lean against, describing the use of the space but not describing the constituting components of the construct. It may seem overly meticulous to avoid words like floor, wall, and ceiling to describe our architecture technically, but the fact is that these words no longer cover the way we design and construct. They only cover the way we use the space. Typically, we design our buildings as bodies roaming in weightless space, prepared to communicate with possible landing sites that are subject to gravity. We design habitable <u>sculpture buildings</u> that know no up and down, left or right before they merge into their earthbound context.

Not a Ceiling

The impressive centerpiece of the 2014 Architecture Biennale in the Italian Pavilion undoubtedly is the exposed installation space above a standardized false ceiling hovering above the visitors at the main entrance. The public cannot help being impressed by how much installation there is hidden behind the false ceilings in modern office buildings. The commonly used word "false ceiling" is indeed appropriate since the false ceiling does not contribute in any way to the structural assembly of components to form a building. False ceilings are cover-ups for the installations as if it is not OK to expose them. The reason to cover them up is that installations usually do not form an integral part of the design task but are considered purely technical things. Year after year, the budgets for installations in buildings are increasing, especially in special purpose buildings like laboratories and hospitals, but more and more in housing as well, due to ever stricter performative requirements for environmental control.

When applying the paradigm of interactive architecture to building structures, the building eventually becomes a 100% installation. This definitely does not mean that there will be no more design intention left; on the contrary, the design of buildings as installations will inevitably become a collaborative design task, performed by a multidisciplinary team of designers, spatial designers, climate designers, interaction designers, social designers, material designers, and quantity surveying designers. The installation *is* the building, meaning that the development of the installation aka the building is a 100% integrated design task. Still one of the best examples from my practice is the design for the Saltwater Pavilion on the artificial island Neeltje Jans (1997), at least for the part that we designed. We achieved full integration of the installation and interactivity in the spatial design. The installations are the real-time performing components of the building. Water, data, electricity, and indeed goods and people are continuously flowing in, being processed, and coming out. I consider a building to be an acting input > processing > output instrument that is played by its users. While the percentage of installation devices is well above average, in the Saltwater Pavilion, the lighting in the interactive behavior of the building is fully integrated, by synchronizing the interactive fiberglass light lines with the curves defining the interior space. In the Saltwater Pavilion, there is no other light than interactive light. Light is produced by programmable glass fibers and beamers that project virtual reality worlds onto the ice-toned polycarbonate interior skin. The control of air humidity is controlled by the water tunnel. The air-conditioning for the whole volume is integrated into the twisted floor, not as a false ceiling but using the expanded volume of the floor as the spatial duct for the air influx and using the total volume of the Saltwater Pavilion as the exhaust.

The Saltwater Pavilion is still standing out as an unrivaled integrative achievement up to today, an odd 25 years after its completion. The Saltwater Pavilion shows a possible way to regain full control of the integrative design, including the esthetic and performative aspects of the electrical light installations and the climatic installations. In the Saltwater Pavilion, the building *is* the installation, and the installation *is* the architecture. No false ceilings, but a volume and spaces shaped by the integration of the performative components. Not a shocking and cynical display of installation technicians hacking the design process, as the Fundamentals' false ceiling exhibit wants us to believe. Installation consultants and technicians are not enemies; architects must form multidisciplinary design teams that include the technicians, climate designers, structural designers, and any other expert who has in any form influence on the appearance and workings of the spatial experience.

On a smaller scale, architecture can learn from the design of consumer products. When dissecting household products into their constituent parts, one sees that the components, each of them having a specific function and shape, are designed to fit together in one specific configuration. They fit together as the three-dimensional pieces of a complex puzzle, typically according to dry assembly methods using screws and bolts, meaning that they can be disassembled into their self-contained constituent parts when needed to be repaired or recycled. Electrical components are neatly interwoven with ergonomic components. Ergonomic components are designed to fit as a mold to the human body and to be structurally optimized to use as few raw materials as possible. Looking at larger consumer products like cars, we quickly notice that the installations are fully interwoven with the body of the car; there are no false ceilings in the car. There are no discrete elements like a floor, a wall, or a roof; each part transforms into the other. In cars, the exterior skin is synchronized with the structural frame, and the interior skin follows the structural skin from the inside while integrating the lighting from above. A modern car is extensively wired, equipped with 50+ computers. The bigger parts like the motor and the wheels are synchronized with the body designers. The core of the motor block itself may be a given, but everything around the motor block is moldable and finds its place between the motor and the structural skin. This is exactly how architectural constructs should be conceived. Buildings should be designed as building bodies where the function and shape of the constituent components are mutually dependent. Parts of different families and variations of the same are physically and spatially related. Rather than improving on the design of false ceilings, a better strategy is to avoid false ceilings completely and find integrated solutions for the real-time performative components of the built structure. A false ceiling has never been favored by designers but is often seen as an inevitable evil. If only climate designers and spatial designers would join forces, it would not have to be this way.

For virtually all of our realized buildings, we have found integral solutions, for BRN Catering, the Garbage Transfer Station, the Saltwater Pavilion, the iWEB, the A2 Cockpit, the Bálna, and, limited to the shell and core design, for the Liwa Tower, a commercial development where we had no say over the interior design. In earlier days, the Romans integrated central heating installations from under the floors into the hollow bricks of the walls. An excellent example of modern high-rise office buildings is the Nissan building designed by ZZOP in 1991, structurally engineered by John Kraus of D3BN structural engineers. The column-free floors are constructed as castellated beams from side to side, which allow the installation ducts to run through the hexagonal openings to cater to all necessary water, electricity, and air movements. On top of the beams and fastened under the beams, there are big concrete tiles, thus forming a hollow floor deck. The genius of the structural design is that the installations are accessible both from above and from below the deck. The same idea is applied to the outer walls: the steel structure separates the exterior and the interior cladding aka the skin, leaving space for air, water, and electricity risers. This ingenious invention taught me to apply hollow floor decks and hollow walls wherever we have the opportunity. I first used the hollow deck principle in the Saltwater Pavilion and subsequently in all of our designs after the Saltwater Pavilion. There is only one good ceiling, and that is no ceiling.

Not a Roof

In the currently popular academic research into full-scale parametrically designed and robotically produced pavilions, which is the offspring of the liquid architecture movement, there is unfortunately an obsession with roofs. Excellent examples of this research are the ICD/ITKE pavilions (2010–2021), state-of-the-art research directed by Achim Menges and Jan Knippers of the University of Stuttgart, roofs as in dome-like structures that cover an open public space. While their research is fully in sync with my take on the notion of the parametrically designed and robotically produced component, I am critical of their self-chosen restriction to design mainly roof-like structures. Open-air pavilions will not translate into integrative building bodies so easily, if not aimed at from the beginning. Why not focus on complete buildings, and why the obsession with only one functional part of a building? To develop a serious alternative to traditional buildings, one should consider not only roofs but also the spatial continuity from floor to wall to roof, and the connections from floors to walls as well. ICD/ITKE's Maison Fibre pavilion in the Arsenale for the Venice Biennale 2021 is an attempt to build something that looks more like a house, but it fails to match that ambition in many aspects. The carbon fibers that are robotically spun over rectangular steel frames do not come even close to an integrated wall or floor of a house. The fibrous structure remains just a lost formwork, not even properly supporting itself, to which the performative components must be attached. A mature building proposal should have considered the integral performance of any part of the building, including climatic requirements, controlling light conditions, separating, and connecting indoor space from outdoor spaces and other indoor spaces.

A choice for roofs as an isolated element means avoiding dealing with the totality of architecture; it accepts a roof being something in itself, not a constituent component as part of the bigger whole. The prototypical pavilions that every self-respecting university builds these days are at best shelters for informal meetings, follies to be enjoyed in the good weather. The reality of designing whole well-performing buildings is a much more challenging task. Building just roofs *de facto* means accepting the fact that it is not a serious alternative to the current traditional building industry, not even when applying parametric design to robotic production methods. It implies accepting a roof as an element. Sure, the main performance of that element must be to protect the spaces below from getting wet or to provide shade. How can we avoid labeling that part of a building body that is facing the outdoor climatic elements as a roof? Why is it important to discredit roofs as an element and why do we need to change the vocabulary? In our work, we design discrete components that together constitute the total envelope, including the roof-like parts, the wall-like parts, and the floor-like parts; we simply call this envelope the skin. The skin is a membrane that transforms continuously from position to position, in discrete steps as all natural phenomena, from one constituent component to the other. As for the Web of North-Holland/ the iWEB, it is impossible to determine where the floor ends, where the wall begins, and from which point the wall becomes a roof. Floor, wall, and roof form part of the same system. In nonstandard architecture, we are not referring to the constituent components as floor, wall, or roof components. We refer to the components that make up the entire envelope, as a single skin organ that is wrapped all around the body. The components constituting the skin are in terms of their dimensions and the frequency of their tessellation synchronized with the load-bearing structure.

 The Web of North-Holland is a self-contained spaceship that makes a soft landing on Earth. The Web of N-H originates in weightless digital design space, gets contextual data from the building site and its climatic conditions during the design process, chooses a position to land, calculates the necessary structural strength of the constituent parts for that specific position, just there and just that, and finally softly embeds itself in the local site. My buildings are not made up of elements that are stacked from the foundations upwards; they are three-dimensionally assembled with high precision as objects that have strong structural integrity in themselves. The diagrid structures of our buildings are so strong that indeed one Godzilla monster could pick it up without the structure falling into pieces; it will hold perfectly together. Eventually, another monster, the Raptor of a building destruction company, came to the abandoned iWEB to destroy it brutally. The Raptor grabbed the constituent components of the iWEB with its gripper, one by one ripping the components out of the triangulated structural net by brutal force. Like in the tragic downfall of the World Trade Towers, it was the bolts that proved to be the weakest link. The structure of the iWEB proved to be so strong that grabbing one component that is bolted to its immediate neighbors did not affect the stability of the rest of the structure. The Web of N-H does not have a roof that is just loosely stacked upon the walls and which just manages to stay there because of its dead weight. In the Web of N-H, the components are ubiquitously structurally connected, distributing the torsion and gravity forces that work upon them. Forget about a roof as an element but think of the upper area of the body as an integral part of its synchronized structure and skin instead.

Not a Door

Doors have somehow always intrigued me. Not as an element but for what they do. Doors are switches, letting stuff and people in and out. Doors are componential parts of a predominantly vertical membrane that can be switched on and off. When switched on, the door is open, and when switched off, the door is closed. Doors are connectors and separators alike. Doors can be visually hidden as in palaces, like a cut-out of the wall paneling and the wall tapestries, giving access to the hidden system of corridors, stairs, and working spaces for the service personnel, usually leading to poorly lit areas in the basement. Such disguised openable parts are not doors from a catalog, not a product, but a cut-out of something bigger. With the design of the Saltwater Pavilion and the Web of North-Holland, I took another radical step. I decided that the door had to be integrated as a further specified part of the envelope. The two entrances aka exit doors of the Saltwater Pavilion are only visible from the outside as a cut-out of the exterior skin. The

215

same applies to the interior skin. The door is a cut-out of the structure, the exterior, and the interior skin. The part that was once known as the door is now a cut-out from the structural skin, hung on hinges from the top, as the doors of the iconic <u>Mercedes 300 SL</u> (1954) with its flying gullwing doors. As a general feature, doors in cars are always cut-outs of the body, securing the continuity of the skin of the body. There is a stack of reasons why doors in cars are what they are and do what they do. First, there is the need to streamline the body to avoid drag and noise around the car. Second, it is a styling issue; the continuity of immaterial stylistic gestures is passed on by the openable parts. Third, cars are entered sideways, not from the front or the back as in traditional buildings. From the world of automotive design, we adopted these three design components, with similar, though different, good reasons:

1) the continuity of the web of structural components guarantees structural integrity,
2) the streamlining of the body avoids unnecessary heat loss by compacting the enveloping surface area, and
3) styling components are just as important in building design as they are in automotive design.

In terms of styling, we typically prioritize the continuity of the interior and the exterior skin above the traditional emphasis on the entrance. The Saltwater Pavilion is entered sideways; also the emergency exit leads the public sideways out of the body. The same not-a-door strategy applies to the Web of North-Holland; one enters and leaves sideways, thus prioritizing the body as a whole, as a material consistent sculpture, above a compound of elements as in a traditional building.

We name the openable door-switch a passage, or a gate, to avoid labeling the door as a separate element. The position of the gate as a switch for people coming in and out does not need to be emphasized when closed; it only shows itself when in action. The design question I posed myself is: "Why would there be a door when no one is entering the building?" Automatic doors behave similarly; they form an unnoted part of a larger glazed surface when closed but appear to be a possible entrance when you come closer. The performance of the passage is what counts, not its physical expression. A passage opens, closes, or holds somewhere in between, and a passage behaves, either on manual operation, automatic or proactively. Doors may open on demand, unlocked by a key or a code. The keykeeper is the one who is authorized to unlock and lock the gate. Modern key keepers show their irises or fingerprints to get access. Doors are not elements but access points in space. Doors are extremely discriminatory in letting people or goods in or out; they are drilled to maintain a very selective admission policy. Behind each passage, there lies another world. Think of front doors facing the streets. No one knows what happens behind those doors, and you do not want to know either. Doors are gates to the unknown, at times to the absurd and the unreal. Behind each of those hundreds of thousands of closed doors, thousands of reality shows are unfolding. It is the real-life performance of the passage that is the intriguing aspect of the door-switch, not the door as an element, not as a nondescript mass-produced item, but as a performative switch between unique worlds of their own.

Not a Window

Recently, there were five standard roof windows placed on the roof of our Danube house in Hungary. The Velux windows are bought from the construction market store as a generic element. Would I still argue that these windows are rather components as seen in their relationship with the roof and the environments at large? Once placed, they are considered components only in their immaterial performance of offering a view onto the majestic Danube river and the Felegvár ruin at the top of a 300-m-high mountain at the opposite side of the Danube. They bring in the much-needed light into the art studio in the attic. Windows typically are opening to another world, to the as-of-yet unknown. Not accidentally, Bill Gates named his operation software Windows. Microsoft Windows authorizes you to have access to the other, to the alien, to the unknown, to the parallel worlds of the virtual. Think of virtual extensions to our daily reality, which is what television and computer screens do as well; they are extensions to our eyes. Our mundane windows also secure ventilation for the 180-m² loft space. In other words, the windows that are elements in themselves; once placed on the roof, they perform as performative components in the context. And that is what they are designed to do as a product, and that is what the product designers had in mind. The standard windows only cover one aspect of what defines a component, namely the performative aspect, not the geometric context. The five windows that now have become part of our existing farmer's baroque home, built in the tradition of the thirties of the last century, are at best considered as a post-informed element, rather than an integrated part of a building envelope system. The wooden roof with the ceramic roof tiles and the standard windows remains exponents from two different systems that somehow must find a way to post-connect. Once we start a new design and synchronize the exterior skin with the main structural system, we pre-connect the parts to become integrated components. Then the window becomes a specification of an otherwise continuously varying envelope system, and the component formerly known as the window becomes an actor in the design-to-production process. In uncompromising nonstandard architecture, windows and closed parts are parametric variations of the base component. In our architecture, a window is a further specification of the base component. While the base component may be visually closed, the window-ish component offers views to the outside; it may be manually or automatically openable. But by all means, the window-ish components have the same type of connections with their immediate neighbors as other roof-ish components. Window components and roof components are members of the same family, talking the same connecting language, like the containers on a supertanker, like the birds in the swarm, and like the cars in traffic. Components are all different; they have their unique identity, yet members of the same system.

The envelope as a whole may be designed as a performative set of windows or as a windowless surface. Arrays of adjacent components may get assigned a gradually changing materialization and transform from translucent vision glass panels to opaque spandrel panels, all of them integral parts of the larger skin. The idea of windows as autonomous uninformed elements has lost its meaning. Transparency, insulation, vision, accessibility, view, smell, ventilation, and light are the transformative and performative aspects of the components in the complexity of the adaptive design game, based on simple design rules. Dimensions and shape constitute the stereometric aspects of the window-component, while the materials, the locking system and embedded climate control mechanisms are its physical subcomponents. The geometric, performative, and physical aspects are integrated into the building component. Every component performs according to the parameters that are assigned to them in the design script, and every component comes with a physical appearance that is designed to fit exactly to its neighboring components, like a three-dimensional puzzle.

Not a Facade

Having dealt with windows that are no longer windows, doors that are no longer doors, roofs that are no longer roofs, and walls that are no longer just walls, I have pretty much summed up what my take is on the nature of an envelope of a building body. Facade systems that are available on the building market typically are composite products that consist of arrays of elements of the same. A special version of facades is a system that is parametrically designed and features arrays of elements that differ in size and shape. Typically, a facade system has severe limitations in the angle of sloping surfaces it can be applied, while the detailing usually is meant for vertical applications only.

We have witnessed a huge mistake by the general contractor of the Bálna Budapest, who insisted on applying a standard two-dimensional Schüco curtain wall system to the double curved surface of the Bálna. Naturally, we did what we could to demonstrate their severe misconception of the Bálna envelope, but they stubbornly refused to understand the crucial difference between a two-dimensional and three-dimensional system. We were vigorously opposed to the contractor's choice, which was by and large based on his ignorance, and fed by greed for maximum profit. Eventually, we had no other choice than to step out of the execution process. Only the visual appearance of the Bálna was secured following the building approval documents, which the project developer, i.e., the contractor, was obliged to respect but unfortunately not the much-needed technical details. As predicted by us, the three-dimensional joints started to leak almost immediately, which was provisionally dealt with by injecting tons of silicon, which might do the job in the short term but certainly not in the long term; the skin of the Bálna will need an extreme technical make-over. After years of political games, we were, some months before the official opening, asked to do a proposal to redesign the skin, but it was rejected by the local politicians whose short-term thinking seems to match well with that of the main contractors. Ten years after the Bálna's completion, the same contractor [!] who made the mistakes was asked to redesign the technical details of the skin, of which I was not informed.

When an exterior performative skin is wrapped all around a double-curved building body, like the skin of mammals, to which group of vertebrate animals we humans belong, we can no longer speak of a facade. In nonstandard architecture, the continuous surface is a skin that transforms from the feet to the crown, from left to right, and from inside to outside. The word facade simply does not cover the complex transformative performance of the skin, which forms an integral component of the body. Just like mammalian skin is, in medical terms, one single organ, the skin of a building body is one coherent system, spatially synchronized with the structural system. Synchronization of structure and skin is one of many specific characteristics of my designs. The A2 Cockpit showroom for luxury cars in Utrecht along the A2 Highway and the Liwa Tower in Abu Dhabi are examples of deep synchronization. The dimensions of the glass plates, the sandwich panels, and the length of the structural rods are aligned in such a way that there is no longer a secondary structure needed to bridge the dimensional gap between skin and structure. In traditional column and beam structural systems, the spacing between columns is much larger than the maximum dimension of a glass plate or a sandwich panel. We refined the grid of our triangular structural network size to match the maximum dimensions of the glass plates. That synchronization allows us to abandon a secondary structure between the main load-bearing structure and the skin, which reduces the costs of the structural skin substantially.

Moreover, according to the conventions of the traditional building technology, the skin should not be directly connected to the main structure. The traditional theory is that a secondary structure is needed to mediate between the movements of the skin caused by the weather and the tensions in the main structure mainly caused by the wind forces. With diagrid structures, this is no longer true. Diagrid structures are based on a triangular web of components that hardly move under rotation forces that are caused by strong winds. Zero tolerance is the norm for triangulated structures, and diagrid structures are rigid as a crystal. Diagrid structures are scripted, not modeled, to define the precise dimensions and the unique shapes of the constituent components. There is exactly one way the components fit together. The beauty of scripting is that designers can and must take responsibility for the exactness of the dimensions of the components. Scripting architecture means that either all components are wrong, or they are all perfectly correct, there is nothing in between, and there are no gray zones for the contractor to exploit.

The Liwa Tower is known for being the sturdiest steel structure in the Middle East, and maybe in the world, due to its uncompromising bottom-to-top enveloping diagrid steel structure. There is hardly any rotational or bending movement due to wind forces. And that is why we can refrain from an intermediate secondary structure. Leaving out one complete layer in the build-up of a structural skin has serious budgetary benefits. The saved money can be spent to develop more interesting or better-performing nonstandard, mass-customized components of the complex geometry. Something that can never be achieved using traditional building methods like *in situ* cast concrete columns in combination with mass-produced curtain walls.

Not a Balcony

Balconies are mostly designed as attachments to the main volume of the building. Balconies do not have our special interest, since I have always been opposed to expressing functional parts like elevator shafts, stairwells, engine rooms, canopies, and indeed balconies as additional masses to the main volume. I have always strived for full integration into the main mass of the building body. The concept of the spatial arrangement of functional blocks notably comes from the Bauhaus design philosophy, later known as modern or modernist, whereas each distinct programmatic function gets its separate architectural expression. Form follows function, as goes architecture's most misused expression, but not before the Bauhäusler isolated the different functions from each other! Before the Bauhaus, stairs, installations, and also balconies were usually well integrated into the main mass in virtually all buildings, whether palaces or housing blocks. There has not been that kind of logic to express the stairs, installations, and balconies as add-ons.

Add-ons are not functional with respect to energy consumption and the quantity of enveloping surface area. Adding separate smaller volumes to the main volume means increasing the surface area of the envelope, thus increasing building costs and increasing heat loss. Form follows function may be functional in one aspect but dysfunctional in many other aspects. Likewise, the extensive use of large areas of glass wrapped around add-ons is not exactly functional when taking all performative factors into account. In modernism, the extra costs for the envelope are compensated by stripping the buildings from their decorations and using a minimum of material for the walls and the floors. Thus, modernism had no other choice but to become minimalist. The slogan form follows function is, to a large extent, an esthetic choice, and not a functional choice, or at best one-eyed functional.

Moreover, the form-follows-function way of thinking leads to the separation of constituent parts into isolated elements that have not much other relationship than being put next to each other. If I would have practiced architecture in the Bauhaus period, my preference would have gone to the likes of Gerrit Rietveld and Piet Mondrian in the arts. They listened to the science and internalized new findings from natural physics, and they have drawn their consequences to create design universes of their own, which are not functionalist and separatist but synthetic in heart and soul. They focused on building up a new language, instead of stripping and taking apart existing languages. Although functionalism was leading to some strong architectural statements in the beginning, it soon watered down into boring boxes with poorly designed installation huts on top of the block, open-air emergency staircases, and narrow balconies only to take a bit of fresh air, solely for reasons of cost-efficiency. Since my studies at TU Delft, I decided to further explore the synthetic road, choosing for the robustness of one single open volume to contain a variety of functions in one sculpted body, absorbing (semi) outdoor spaces as well by its spatial porosity. Thanks to a deep fusion of art and architecture in our practice, we developed back in the early nineties the concept of sculpture buildings, which we thought of as inhabitable sculptures. Our motto for the Sculpture City event (1994) is that a sculpture can be a building and that a building can be a sculpture. We were anticipating what later became popular as sculptural architecture. But I certainly do not like the word "sculptural" because that means the structure only looks like a sculpture, but it is not. We wanted the built structures to *be* a sculpture and a building at the same time. A sculpture validated by the visual artist, validated as a meaningful sculpture in the portfolio of the artist, and at the same token, a building guaranteed by the architect to perform structurally, spatially, and socially, i.e., to be architecture.

We can clarify the difference between a sculpture and a sculptural building with three distinct examples. An example of a building that I consider a sculpture is the Rietveld house in Utrecht. Walls, floors, windows, and doors are equally free-floating elements in the assembly of the whole. The door is expressed in the same consistent language as the other elements of this stunningly new design universe. Rietveld composed a highly personal library of elements, abstractions from functional elements, not elements from a catalog, but still elements while clearly expressed as such. The consistency of the whole makes it into a sculpture that is a building or a building that also is a sculpture. Example 2: the entrance of the Guggenheim Museum in Bilbao by Gehry looks like a traditional shop front, not related to the otherwise sculptural expression performed by the cladding. As in many of Gehry's projects, the doors are just doors, elements from the catalog. Gehry willfully combines the banality of functional elements with an exuberant decoration of the shed. Gehry's design attitude leaves the building a building, and Gehry's buildings cannot be considered sculptures; they can at best be regarded as sculptural. Example 3: the Saltwater Pavilion has no visible doors at all; its "doors" are openable parts that are winged cut-outs from the skin. When the gull-wings are closed, there is not an explicit sign that it is a building; it can be a sculpture just as well. When the hinged parts of the skin are in their open state, it is communicated that the volume is accessible, and thus must be appreciated as a building. Depending on the position of the door as the on–off switch, the Saltwater Pavilion is at one time a sculpture and at another time a building. The same applies to the Web of North-Holland and, to some extent, to the larger buildings we have realized as well.

In its appearance, and different from Rietveld's design universe, my design universe aims at the fusing of art with architecture. I do not assemble free-floating elements into a certain configuration as Rietveld does; instead, I aim to build unibodies with a continuous structure and skin. The Waste Transfer Station (1994) is the first building that I realized deserves the label "building body" and a big first step toward becoming a sculpture building. A building body is the architectural equivalent of a unibody in product design. There are no clear signs of doors or windows, and there are no interior columns, only free-span components as integrated parts of the envelope; there is just one streamlined body. I deliberately chose the exact opposite of a typical functionalist arrangement. The brief, like typically any program of requirements, asked for some functional volumes with different sizes for the different functions such as office, storage space, main processing hall, and water treatment installation. We decided to bring the different functions together in one smoothly shaped volume that varies in height and width, thus matching the differences in height and width with the required functionality of the parts. With the design of the Waste Transfer Station, the idea of the building body was born. The Waste Transfer Station is a sculpture building indeed, internationally acknowledged, and honored by the prestigious Business Week/Architectural Record award and the OCÉ/BNA award for Industrial Architecture (1996).

The Saltwater Pavilion (1997) was the second award-winning building that is a sculpture we managed to get built, receiving worldwide acknowledgment for its unique achievement, catapulting ONL into the international architectural arena. The Saltwater Pavilion represents a movement that is intentionally not dealt with in the Fundamentals Biennale show. A blind spot for the steadily growing movement of parametrically driven nonstandard architecture? I know first-hand that RK was well aware of our Sculpture City event, to which he chuckling referred when I – a bit teasingly – praised OMA's partly "sculptural" design for the <u>New Luxor Theater</u> on the Wilhelminapier in Rotterdam. Did the important <u>Digital Real - Blobmeister</u> show (2001) curated by Peter Cachola Schmal get unnoticed? What about the <u>Transarchitectures</u> shows (2000, 2001), curated by the design journalist Odile Fillion, then Jean Nouvel's spouse, and Michel Vienne? Just as important for the development of architectural theory as the CIAM conferences were for modernism, the <u>Archilab conferences</u> (1999, 2000) curated by Frédéric Migayrou and Zeynep Mennan, were not to be missed. These events did not fit in Fundamentals' generic elements agenda. The nonstandard architecture movement meant a threat to the concept of the Elements of Architecture that needed to be ignored and at times ridiculed. Besides us, my pioneering peers like Marcos Novak, NOX, Bernard Cache, Greg Lynn, Roche&Sie, and many others developed an architecture of continuity without floors, walls, false ceilings, doors, windows, facades, and balconies, an architecture that, at the Fundamentals Biennale, has intentionally not been acknowledged as a major movement in the theory and praxis of architecture.

Definitely Not a Corridor

Having defined floors, walls, and false ceilings as elements, also the corridor has been declared to be an element of architecture. Like false ceilings that contain the space reserved for technical installations, corridors contain the space for people's communication between designated functional rooms. The corridor is thus seen as an immaterial spatial element, connecting or disconnecting rooms via its stretched spinal shape. Similar to my objection to considering walls and floors as elementary categories, categorizing a corridor as an element does not work for me either, probably because I do not have an obsession with voluntary prisoners. According to the Oxford dictionary, the etymology of the word corridor dates back to the late 16th century as a military term denoting a strip of land along the outer edge of a ditch. Corridors are not an invention to connect people but rather to separate different classes of people. The extreme version of separation by corridors is the double corridor system in all courts worldwide, where the evil people are separated from the good guys. Another perfect example of the cultural importance of corridors is the hidden corridor system in the palaces of the sun kings, separating the servants from the elite. Corridors are a modern invention; in earlier days, neither houses nor official buildings featured corridors. Most of the rooms were connected "en suite." The corridor became a method of separating functional elements. Corridors are the spatial translation of oppression and selection, designed to separate the good from the bad, the bosses from the employees, the private from the public, and the servants from those being served. Especially when corridors are minimal in size and do not allow for sitting and waiting, their narrow functional interpretation becomes awkward.

I have no affection for corridors at all. Typically, I avoid corridors, as they take up space that is useless for a building's key activities, bringing down the gross floor area (GFA) percentage by some 10%–15%, and hence the cost-effective performance of the building. Corridors are only functional in one aspect of a building's spatial performance but counterproductive in many other ways. Corridors are dead additions leading to separation and increased inefficiency, and certainly not a constructive component of a building. I prefer continuous transformations from open to closed spaces, as in the open office floor plan of BRN Catering. The connectivity between the components is as important as their separation. I prefer spaces where one can navigate around freely, where one can freely choose a direction and meander from one atmosphere to the other, instead of being tunneled by straight jacket-type situations where there is only one way to go. None of my buildings feature narrow corridors, with the exception perhaps of the commercial Liwa office tower, of which the interior is, as noted before, not designed by us. Yet, by its compact layout, the length of corridors is limited, realizing a high GFA value, over 90% effective, whereas 85% is generally seen as the optimal value for commercial office buildings.

Not a Fireplace

When the corridor is the non-materialized element in RK's supposedly fundamental categories, the fireplace is both a material object and a social category. The fireplace used to be a wood-burning object to warm the house. There is a direct relation between the place where the wood is burnt and where the heat is consumed. While no longer burning wood but using electricity-driven heat pumps instead, there are new opportunities to renew the idea of distributed climate design, when, where, and as needed. The fascination with swarms and networks inevitably leads to a fundamental revolution in design thinking. Distributed systems are favored over centralized systems. In terms of climatic performance, we must leave behind the separatist of a central engine that is pumping hot water, conditioned air, electricity, or data into the building body. Central engines are paradoxically mostly placed off-center, usually in a cheap industrial shed on top of the building. Central engines require long trajectories for the distribution of heat, water, and electricity to the places where people live and work. The idea of decentralization is simple: why not generate the desired climatic conditions exactly at the spot where it is needed, and where the demand is generated? And why not personalize the climatic conditions instead of only offering a standardized equalized climate for all? Why not radically distribute the climatic system to be at the service of each individual? Instead of one central engine, one would have dozens of small engines in a house, hundreds in a school, or thousands of smaller engines in an office building, generating the desired performative qualities locally. Distributed climate systems generate fresh air, conditioned air, customized temperature, airflow, humidity, electricity,

227

data, and drinking water where needed, when needed, and as needed. Distributed climate systems could work like this:

1) hot and cool air are generated by small heat exchangers integrated into the skin of the building, embedded between the exterior skin and the interior skin, and networked with other small heat pumps,
2) electricity is generated by solar cells and/or small wind turbines, locally embedded as close as possible to the place where the power is needed,
3) difficult to generate super-locally, yet in principle possible, is to extract water from the surrounding air directly, especially in climates where there is a relatively high percentage of relative humidity,
4) data can be transferred wirelessly to its nearest neighbors to form a robust network, in a dynamic equilibrium, instead of having a centralized data management system; small climatic components would communicate with their nearest neighbors to create optimal situations for as many as possible.

The optimal granularity of distributed climatic systems may vary from design to design, from site to site. Ideally, it should be taken to the extreme, to empower every single spatial unit, and even cater to every single individual inside one spatial unit. What are the possible benefits of distributed climatic systems aka the modern equivalent of the fireplace? First of all, in distributed climatic systems there is no need for large lengthy ducts and long open gutters, and therewith with all possible failures and spatial clashes with other building components in the often-complex trajectories of the ducts; distributed climate systems make false ceilings completely redundant; and a great relief for clash controllers. Second, distributed climatic systems allow for intense interaction with the users of that particular spatial unit, using networked sensors that talk to each other and the users of the space. Where there are multiple users, and that is mostly the case, individual users set their preferences for temperature, humidity, and light, and communicate their preferences in real time via the sensors with the super-local small climatic devices. The individually controlled climate travels with the user through the building. Users thus leave climatic traces throughout the structure, interweaving with numerous other traces from other users, whereby the conditions are in real time adjusting to new usage patterns. Third, the skin of the building body turns into an active adaptive organ, an adaptive membrane negotiating between outdoor weather conditions and indoor user preferences. Fourth, on top of the above technical aspects of individualized comfort, the distributed approach has a substantial social implication, while social behavior will take on a new form. Individuals will not just compete with each other to generate their preferred individual climate and their social distance; they will also need to negotiate with fellow users on how to create favorable atmospheres for groups of people. Think of activating LED lights that are turned on only where there are people present in the space. When someone moves into a larger open space, the illumination would dim where there are empty seats, and light up where more people are gathering together. Now apply the same concept to heat and humidity control. Distributed climatic systems support fine-grained informative environments and stimulate social dynamics. Being informed and informing your fellow users of the space, communicating in traditional dialogue augmented via informed devices, creates the basis for enhanced social dynamics.

Compare environments plus their users, who are equipped with ubiquitous sensors and actuators, with self-driving cars in a distributed infrastructure system. Self-driving cars are well-informed about the specifics of the road and their fellow road users; they keep an eye on each other, and they are programmed to keep a certain distance to and from their immediate neighbors. Since the real-time running programs are, in essence, parametric, that is, open to streaming refreshing variables, the distance is adjusted in real time for any possible reason. The distance may be adjusted top-down as in crowd control, and bottom-up according to changing local circumstances and changing whimsical behavior of individual users. Comfortable unobtrusive social distancing requires wearable sensors and senders that communicate with your nearest neighbor. The wearables communicate with the people and the things in their immediate environment. The system would work simultaneously bottom-up and top-down, for example, to constrain the outbreak of an infectious viral disease. Persons will bottom-up inform each other about their health status and are top-down informed on the risks of the virus. Top-down: medical authorities update a new set of scientifically proven values, adapting in real time the appropriate safe distance between people, in a range from zero to some meters, and automatically adapting per person for outdoor or indoor situations, size of the group of people gathered. Bottom-up: your personalized app measures the distance to your nearest neighbors, in front of you and behind you, and sends signals to both the invader, informing that s/he is coming too close. As with an outbreak of the flu, as with driving self-driving cars, people are in it together, and people act and interact to avoid bumping into each other literally or invading the personal sphere of the other. Smart environments are the fireplaces of smart homes, bringing people together, or distancing people at will. A ubiquitous soft parametrization of our daily lives lies ahead of us, deeply affecting our social behavior. Imagine that every staff member and boss, every student and teacher, and every salesperson and customer will eventually organize their behavior prompted by a "fireplace" app, perhaps enhanced with an unobtrusive form of augmented reality vision that visualizes the potential danger, and assume that their preferences are set adequately and that they act unobtrusively accordingly. It would work discreetly and unobtrusively, with no disturbing beeps or alarms, just silently informing each other, not to force people into an all too strict regimented behavior.

At the time of writing this book, after years of living with Corona, the world is nowhere close to constraining the further spread of the COVID-19 virus. If we would have had a ubiquitous social distancing app installed, it would be much easier to control. Adaptive social distancing is bound to become a measurable, manageable, and well-accepted normal behavior, facilitated by the use of personalized apps and smart wearables. Environments that are abundant with sensors connected to people and things alike inform the users how much distance to take and send silent warning signals when they come too close to a person that carries a contagious virus, similarly to how my car warns me unobtrusively for deviating from the lane or coming too close to the car in front of me. People carrying the virus thus visualize an otherwise invisible field around them, protecting others, and informing them, as a piece of advice only, not to come too close. Of course, the proper working of the app requires knowing the medical condition of each inhabitant; it needs recurrent ubiquitous self-testing. Testing should become as generic as checking blood pressure and measuring one's weight, one's heart rate, and one's oxygen level. Social distancing could be part of a general health app. Smart wearables are bound to become the digital equivalent of wearing masks. Distributed networks that give shape to our informed and customized behavior are the inevitable future. We are all in it together; our brains and bodies are wirelessly connected to form one super-organism. Real-life actors and groups of actors, which are the acting processors of information, communicate with their nearest neighbors to form a robust resilient network. Nearest neighbors talk to other nearest neighbors and spread information with the speed of light. Ubiquitously using social media during pandemic outbreaks, we simultaneously counterattack with a ubiquitous info-demic, whereby information spreads across all connected platforms and persons. It is no longer the central fireplace that draws people together for comfort nor the modern replacements like central heating or other comfort-enhancing installations. A fully decentralized and personalized orchestration of people's free movements in space will lead to new patterns of social gathering in Another Normal. The behavior of people must no longer be manipulated by central devices. As of today, in real time, informed people are moving about in a real-time informed space, in continuously changing social patterns, and in all possible gradations from extreme density to extreme seclusion.

Not a Toilet

"There is only a barrier when you need one" is the motto for one of my proposals for interactive structures, the Dynamic Sound Barrier (2010). Based on the principles of the Interactive Wall project (2009), we applied the knowledge acquired to actual design challenges related to the societal need for acoustic barriers. No one likes the boring look of the acoustic fences along highways and train tracks. Typically, they block the view of the landscape, as seen from the point of view of the noise-producing vehicle, and the other way around, barriers form a visual obstruction to the open landscape or the city. The design challenge I posed was: "Why should there be a barrier when there is not a vehicle that produces the noise?" This simple question naturally leads to the concept of adaptive barriers that are down when there are no noise-producing trains or cars, only to rise and envelop the source of noise when the train is approaching. This logical but alarmingly disruptive concept applies to almost any aspect of spatial design. In general, one could state the following: "Why should there be a <whatever> when there is no need for it at that particular moment and in that particular place?" Why should there be light if there is no one in the room? Why should there be a door when there is no one entering the building? Why should there be a canopy when there is no rain or sun to protect the public when entering the building? Why should there be a room or a building at all when there is no one to use it at that moment? And, in the context of the Fundamentals exhibition, why should there be a toilet taking up space when there is no one using it? The design question that I have put forward is whether we can design a toilet, i.e., a bathroom, in such a way that it pops up only when

 one needs one. One decade ago, I proposed exactly this as one of the features of our Pop-Up Loft concept. We applied the concept of adaptive furniture not only for toilets that are popping up on demand but also for all possible functions of a typical apartment. We imagine the Pop-Up Loft to be a 50-m² space, when entered devoid of all furniture, like an empty classroom with a generous height. Yet, the room secretly offers the luxury of a larger apartment, simply by hiding the functional components in the hollow floor and ceiling when not in use and having them pop up when and as needed. The toilet, bathtub, bed, kitchen, eating table, lounge chairs, closets, home cinema screen, and working place, all common functional components pop-up on command. Imagine the luxury of having a 50-m² toilet all for yourself. That is not just a toilet; it is pure luxury. Enjoying a bathroom of 50 m², a sleeping room of 50 m², a kitchen, a dining room, a lounge corner, and a home cinema, each evoked activity would give you that spacious loft feeling. Naturally, any combination of functional components popping up is possible. The programmable Pop-Up Loft is operated by an app running on your mobile phone where one requests the desired configurations of the apartment in advance.

In the all-interactive concept of the Pop-Up Loft, the toilet is no longer a serial product from the mass production industry but a customized component, an integrated and personalized part of a larger whole. Technically, the components that pop up are dormant in a generous 1.20-m-deep space between flooring and ceiling. I already mentioned the type of spatial castellated steel structure and the structural placeholder between the upper and lower interior skin of the room. The in-between space houses the ducting for the installations and the programmable furniture components. The open steel structure of the floor leaves space between the trusses for the toilet, bathtub, bed, kitchen equipment, tables, chairs, and sofas, while the open space inside the vertical skin components of the pop-up unit leaves space for bookshelves, wardrobes, and home cinema equipment. Some components, notably light fixtures, and an air exhaust system for the kitchen, will come out of the ceiling rather than rise out of the floor. Based on the above description, now imagine again this 50-m² loft. When not activated, it is an empty Zen-like space with a pleasant, wooden finish of the enveloping surfaces, the bottom, the sides, and the top – a multimodal space with apparently nothing in it, a space that is patiently waiting to be activated and to be used for any possible social activity.

Having developed this idea for the programmable Pop-Up Loft unit, we went one step further in the MANIC research (2018), executed at Qatar University. MANIC stands for Multimodal Accommodations for the Nomadic International
Citizen. The basic programmable unit of 50 m² forms the basis for a ubiquitous booking system, an integral combination of Booking.com, Airbnb, Trivago, Zomato, and similar booking sites. MANIC is the home of hundreds of fully programmable units. MANIC is the prototype for a new form of flexible homing. Homing, not housing, because the 50-m² unit will ultimately feel like your home, whilst set to match your preferences, including all sorts of digital and analog paraphernalia

that people like to surround themselves with at home. The ubiquitous booking app allows you to book the Pop-Up Loft for any activity and for any period. The explicit purpose of the ubiquitous booking system MANIC is that the building that contains the programmable loft spaces becomes a colorful social mix of short-term and long-term users, booked by different target groups, for a variety of activities. Activities would range from working, sleeping, meeting, dining, to fitness classes, and yoga workshops, at any time of the day, 24/7. The MANIC building would become a natural habitat for 24-hour round-the-clock usage, a commercially highly attractive enterprise. MANIC could have a more than 100% occupation rate, while different forms of occupation may be booked during one day for the same unit. The multimodal accommodation is the home for the new international citizen, made their own only locally and temporarily. What started as a rethinking of an acoustic barrier developed into a radical reformulation of what a home is.

Not a Stair

Stairs are connectors, connecting one-floor level to another, facilitating upward and downward movements from flatland to Spaceland and back. Stairs create tension between the dimensions, between 2D and 3D. Stairs may structurally span between floors, hang from the upper floor, or jut out from the lower floor. A stair is not a stand-alone element, not without a context. I am not looking at stairs as autonomous generic mass-produced elements from a catalog, but at stairs in their connected state, embedded in a context. Stairs are components that build a relationship to components of a different category. What if the stair would be an acting switch, like the door that can open and close, like the window that lets the fresh air in or keeps the climatized air in? In such a programmable state, the stair becomes, just like the door, a connecting or separating device. What if a multitude of stairs would be programmable to switch from one floor to another, connecting to the upper floor one day, and to the lower floor the other day? That would create an adventurous always changing spatial layout of the building. The building would never be the same;

it would never have a dull moment. The three-dimensional layout of the building would change along with different requirements set by its users, for different reasons. That building would become a real-life connection machine, the building scale equivalent of the difference engine of <u>Charles Babbage</u>, and a three-dimensional construct that can have an endless number of different connections. It is said that his difference engine is inspired by the Jacquard weaving machines. A building designed as a three-dimensional connection machine is capable of proactively establishing connections as well as separations. A performing connection machine is also an exclusion machine. When something can connect, it can also disconnect, as does an electricity switch, as does a door. How such a hypothetical interactive environment may be used is up to its base algorithm in real-time negotiation with bottom-up users.

A frozen version of the building as a giant connection machine is our design for the <u>Dutch Pavilion for the Dubai Expo</u>. The 3000 m² pavilion is an assembly of hundreds of stairs and landings connecting up and down, left, and right. All stairs include slopes to accommodate small electric vehicles for the physically challenged. Thousands of different trajectories to navigate the building are possible, as in a three-dimensional urban grid. From each point in the by-and-large open pavilion, the public can choose to go forward and go up, turn left or right, then turn left down or right up, etc. The stairs and landings enclose the nine open voids that offer the public a strong spatial experience and a good sense of orientation, informing them where they are in the complex three-dimensional array of stair-like components. From each point in the structure, people have the choice to continue in two or three different directions, which turns the structure into a social condenser. You might come across people you have met before and eventually start a social interaction. I identified three predefined themed trajectories, the water path, the energy path, and the food path, crossing each other at crucial points. The public is free to deviate from the pre-programmed themed paths and choose their trajectory through the three-dimensional maze of connections. The spatial sloping connectors are designed as somewhat steep ramps that can be effortlessly climbed by small electric vehicles.

One side of the ramp is used as a stair and the other half as the slope. Thirty years ago, we designed the slope that goes up to the panoramic view in the nose of the Saltwater Pavilion in a similar way. We designed the slope as partly ramp and partly stair. The flights of the stair part gradually arise from the ramp and disappear again at the other end. This stylistic design technique gives the impression of continuity from floor to floor rather than emphasizing the stair or the ramp as a discrete element. Stairs and ramps are specifications of walkable surfaces; they are not autonomous elements, which is similar to how human hair is a specific local fold of the skin, forming the hair follicle from which the hair grows. Likewise, the stairs aka the ramps of the Saltwater Pavilion and of the Expo 2020 Dutch Pavilion are local folds of the walkable surfaces, and thus embedded specifications of the variable three-dimensional components.

Not an Escalator

Recently, I had the pleasure to take the longest possible escalator, given the technology of today, with a stroke of a whopping 125 m, connecting the underground Admiralteyskaya metro station in St. Petersburg to the street level. Since they had to go even deeper than the 125 m that the escalator allows, it was decided to build one long escalator of 125 m up to an intermediate level. From this level, a shorter escalator – still another 25 m – eventually leads to the street entrance. The total depth of the station is 86 m, which makes it the deepest metro station in St. Petersburg. I took a video of the 125-m part and measured exactly 2.5 minutes before we reached the end. The video shows that people do not hurry anymore; they take their time. No one rushes down as in shorter escalators; they have time to check their mobile phones and even read books. For commuters and tourists alike, the slow downward dive means a 2.5-minute experience. At the next station in the NW direction, Sportivnaya, we enjoyed the even longer travelators to pass under the Neva river. Standing on these long travelators grants you one of those precious moments of self-reflection, especially since there are no advertisements along the way, which is one of the superior qualities of the St. Petersburg metro system. The Admiralteyskaya escalator offers a time-based spatial experience. The dynamic time-based experience forms an *escalating* design component, rather than a static element. The word escalator in itself indicates that it is something that is acting; it performs the act of escalating, expanding, and stretching time.

Likewise, I conceived the Acoustic Barrier as a 60-second experience of architecture, stretching along the A2 highway near Utrecht, with an embedded showroom for our client, the Rolls Royce dealer Frits Hessing. We wanted the structure to transform smoothly along its course when passed by on the then just opened 14-lane highway. Because of the speed limit of 100 km/hour, the driving experience lasts exactly 60 seconds. In the opening days of the A2 Cockpit, there were often traffic jams, due to the combination of the unique glazed diagrid structure and the unashamed exposure of Rolls Royces, Bentleys, Lamborghinis, and Maseratis. When not in a traffic jam, the smooth 1.6-km drive is a pleasantly slow transformative experience. Concave sections gradually transform into a folded convexness, while the linear yet elastic upper contour line is lifted halfway to accommodate the volume of the showroom. The folding line fades out toward the middle of the pumped-up volume, while at the other end of the showroom, another fold fades in and continues up to the end, where the vertical barrier flattens into a horizontal platypus-style beak. The barrier is designed as a volume rather than as a fence. A volume that is squeezed vertically along most of its course, flattened horizontally at both endings, and pumped up in the middle to form the Cockpit. The elasticity of the lines, the gradual transformations, the different reflections by the fold lines, and the sudden expansion in the middle part contribute to the 60 seconds of architectural experience. The experience of driving along the Acoustic Barrier is similar to being carried by a long escalator, offering an expanded, enlarged, extended, amplified, visual, and emotional experience.

Not an Elevator

When I was co-designing the prematurely demolished Zwolsche Algemeene Headquarters in Nieuwegein back in 1982 together with lead architect <u>Peter Gerssen</u>, the client was very explicit that they wanted as few elevators as possible. They came from a slim high-rise office building in Utrecht of only 400 m² per floor, where they suffered from poor internal communication between the different floors of the building. The stairs were hidden and were not designed for communication between the floors, and the users complained about long waiting times before the lifts. Long waiting times are a common complaint in high-rise office buildings, especially during rush hours. As an insurance company, they were in dire need of better communication between their insurance sectors. We proposed a four-story building with four long trapezium-shaped wings rotating around a square central atrium containing the connective components, thereby stimulating the internal communication and active movements of people. The lift we designed as a panorama elevator, and we gave the staircase a generous width allowing for three people to pass each other, and positioned it centrally in the dark mirrored open atrium, effectively inviting the ZA personnel to use the stairs instead of taking the elevator. We made sure that the passages to the wings are wide enough to avoid the corridor effect and to stimulate social encounters. We celebrated the panoramic elevator as an experiential communicative component rather than as a closed, blind, and purely functional element, since I applied the open communicative core concept in virtually all of my designs, in the BRN Catering building (1987), the A2 Cockpit (2005), the Bálna Budapest (2012), and in numerous non-realized designs.

A unique approach toward the elevator that is not just an elevator is the elevator office at the corner of the 17-story <u>Bata</u> <u>shoe company headquarters</u> in Zlin in the Czech Republic, designed by Vladimir Karfik. Karfik had previously worked with Frank Lloyd Wright and LeCorbusier. This elevator is not just connecting floors; it is designed as a mobile office, a 6 × 6 m² air-conditioned room complete with an office desk and a phone. The vision of the architect and the client was to stay close to the supervisors and workers, to shorten communication lines. Moving from floor to floor, the boss would temporarily form part of a specific production team, starting from the 8th management floor, to inform and be informed as directly as possible by reducing one degree of separation. Can we radicalize the Bata HQ concept and build a multi-story building without elevators as an add-on, but with moving platforms all over, making new connections when and where needed?

Not a Ramp

Can we design a building that needs no elevating platforms at all but that consists of sloping floor parts that gradually merge into each other? Can we build an architecture of the oblique, as was propagated by Parent and Virilio back in the late sixties?

The Dutch Pavilion for the World Expo 2020 is designed such that there are no elevators needed at all. Instead of having lifts, I proposed to use small automated electric vehicles, quietly crawling up and down the ramps, programmed to follow a predefined path or, in their freestyle mode, free to move in any direction. The ramp-stairs are wide enough to let the pedestrians and the electric vehicles pass each other. The idea to use distributed small electric vehicles instead of a central elevator solves many problems for logistics. The vehicles blend seamlessly with the flow of visitors, while a central elevator would need a central core-oriented layout. The ramp-stairs concept allows us to have a fully decentralized three-dimensional building layout. The Dutch pavilion is designed for a specific world expo experience and at the same time a generic model for a distributed form of internal communication.

Designing according to the principles of swarm logic, distributed systems, parametric design to robotic production, and, more than anything else, the radical fusion of art and architecture, not a single traditional architectural element can be taken for granted anymore. Taking the elevator for granted as an element of architecture is not a good starting point for any design. Thinking of relational components rather than mere elements opens up a radical rewriting of the theory of architecture. Framing the rather arbitrary selection of 15 categories of elements as in the Fundamentals biennale is intended to demystify the modern but ended up as a retroactive chronicle.

In the Fundamentals show, some clumsy ramps were put together,
as an homage to the late French architect <u>Claude Parent</u> and
Paul Virilio, the inventors of the "Architecture de l'Oblique,"
the architecture of the sloping surface, stimulating a more
dynamic use of the floors. Parent was already well known
for his bold brutalist designs for nuclear facilities. Having designed and
built several massive nuclear plants, using tons of concrete to protect
the environment from radiation had become a new normal for Claude
Parent. Not surprisingly, he became obsessed with the solidity of the
desolate North-Atlantic bunkers, and after a short intense period of
collaboration with the French philosopher Paul Virilio, Parent started to
question the dominance of horizontal floors and became interested in
the oblique, in tilted walls that become inhabitable, as Parent describes
his vision. He wanted to activate people's behavior and demonstrated
the idea by staging dance performances on sloping surfaces. Parent's
slopes were meant to be slanting walls rather than ramps. Being
inclined, the wall would become an augmented experience of dealing
with gravity, as were the cityscapes imagined by the French
architects of <u>Architecture Principe</u> (1963–1968). The theory
of the oblique deals with how a body physically experiences
a space. The slope implies an extra energy boost when
moving up and a balancing braking posture when coming down. The
oblique energizes people who are moving in space. Parent challenges
the dominant static verticality and introduces the idea of equilibrium
and motor instability, to be used as the winds in the sails of a ship,
as he states in the introduction to his book Architecture Principe. He
breaks the barriers of the vertical separating walls and replaces them
with accessible ramps that increase the useful surfaces of the dwelling
and stimulate social connectivity. Certainly, Claude Parent wanted to
enhance grace in gravity, while the reference to Parent's
oblique is sorely missed in Lars Spuybroek's latest book <u>Grace
and Gravity</u>, which so eloquently deals with the inherent
instability of the space between the vertical (hand space)
and the horizontal (foot space). Spuybroek builds upon the thoughts
of philosophers rather than colleague architects and yet comes to
similar insights as Parent. Ilona and I went to see Claude Parent in his

studio during our one-year stay in the <u>Van Doesburg artist</u>
<u>house</u>, Rue Charles Infroit 29, Meudon, Paris (1988–1989).
He was the only person present in his office, a soft-spoken,
gracefully mustached gentleman in his late sixties. When
Parent showed us his brutalist sketches for cities of the oblique, he
acted somewhat disoriented, out of pace with the world. Back in
the sixties, he was a celebrity, mainly in France, adopting the image
of a dandy. He provocatively drove a Rolls Royce, although he could
hardly afford it. At the time of our visit, he had gone out of fashion,
and only, recently, his star is rising again, posthumously, thanks to the
appreciation of the likes of Zaha Hadid and Jean Nouvel, and above all
through the visionary efforts of the Centre FRAC in Orléans to safeguard
his legacy, drawings, and models. In Parent's Cities of the Oblique,
the ramp constitutes the very structure of a building; the ramp *is* the
building, and the building *is* the ramp. For Claude Parent, the ramp is
not an add-on; it is an integrated, crucial part of the structure of the
building. Parent's sketches of autonomous cities in the open landscape
feature fluid contours and modular systems arranged on gentle slopes
like giant horizontally stretched beehives. Parent's cities are a bold
move away from straight-jacked modernism, toward an urban scale
continuity of the floor and the wall.

Ten years after our visit to Parent, the continuity of floor-to-wall to
roof became a leading theme for my architecture. I had the pleasure to
meet Claude Parent again at an exhibition at the Centre Frac in Orléans
in 1999 where a black 1:20 scale model of our Saltwater Pavilion was
on show. His apparent appraisal could not have pleased me more. The
Saltwater Pavilion is *de facto* one continuous ramp, folded back into a
three-dimensional lemniscate. Coming from the Freshwater Pavilion,
the loop takes the public down into the belly part named the Wetlab,
sloping up toward the nose part, offering a dramatic panoramic view
of the sea through the only window in the whole building. Then the
internal lemniscate route makes a U-turn and brings the public up to
the twisted floor of the Sensorium, only to sneak out of the body via
the side door, which is the before-described skin-structure-skin cut-
out of the body. There is no horizontal comfort surface in the Saltwater
Pavilion; all of the foot space is gently sloping. Seen from the top and

seen from the sides, the shape of the trajectory traces the number 8, a three-dimensional figure of endlessness. Again, the building *is* the ramp, and the ramp *is* the building. There is nothing else defining the walkable surface of the Saltwater Pavilion than the smooth lemniscate-shaped surface, formerly known as a ramp. When asked five years later to design the North-Holland Pavilion for the World Flower Expo 2002 in the Haarlemmermeer, my first concept, the North-Holland Pavilion version 1.0, was to have an upward and downward spiraling walk, a 300-m-long augmented reality experience, compactly nested in an aluminum envelope that includes the floors, walls, and ceilings, to shape the folded volume. The pavilion would be structurally composed of one material only; 2-cm-thick aluminum plates would be bolted together as in a (space)ship. The preformed components were to be pressed into shape by a shipbuilding company. The 300-m-long trajectory passes through eight immersive interactive sectors and folds back into itself. Information pulsates in real time like a hot data stream through the sectors, interacting with the passing visitors, immersing the visitors in a mesmerizing experience. The continuous motion of the building body of the North-Holland Pavilion version 1.0 has no beginning and no end. Space has turned fluid; the folded long stretched sloping volume of the one-way path attracts, absorbs, enfolds, excites, transforms, and informs the public, only after an intense experience to eventually loosen its grip and let the public go. The first version of the North-Holland Pavilion is synonymous with a ramp; the ramp *is* the pavilion, and there is nothing else than the ramp, but it is *not* a ramp, and certainly not a ramp as an element, but a long-stretched interactive experience, shaped by streaming information and curved

steel plates. A much earlier example from our joint art and architecture work is the concept for the TORS sculpture building (1990). The towering softly twisted elliptical sculpture consists of one long spiraling sloping surface from the bottom to the top, corridor-less connecting the subsequent levels. The endless slope revolves around a central atrium, which changes shape from level to level. The most radical aspect of TORS, however, is the design of the structural skin, anticipating our later building method of a steel structure with an interior and exterior skin, as we first realized in the Saltwater Pavilion. Furthermore, the TORS project established an early statement on the integral fusion of art and architecture, which we elaborated upon with the sculpture buildings for the Sculpture City event in 1994, eventually realizing the Saltwater Pavilion in 1997 in the unruly reality of the building industry. Thanks to the support of master planner Ashok Bhalotra and the clients, we were given the opportunity to demonstrate that a ramp is not an element but a connective component as acting part of the balance-seeking experience of the walkable lemniscate loop. The ramp is a vehicle, a device that educates you to continuously find your balance, a vectorial body that invites you to go places, to gradually change your points of view, and to invite you to curiously explore the environment, like hikers in the mountains.

My Fundamentals

Together with many colleagues, I was annoyed and irritated by the so-called fundamentals of the 2014 Biennale. Patrik Schumacher was irritated for similar reasons and wrote in an article in Architese – in the German language – that the Biennale should belong to practicing architects, not retroactive theories. He felt that his efforts to promote parametricism, backed by numerous supporters, had been wrongly dismissed. He rightfully believed that a false picture was painted of what is relevant today in the architectural discourse. In our shared view, floors, walls, ceilings, roofs, doors, windows, facades, balconies, corridors, fireplaces, toilets, stairs, escalators, elevators, and ramps are not standardized elements but customized parts of a larger whole. Where my worldview differs from ZHA is that we are not happy with traditional techniques like welding and plastering to realize fluid geometry. Since the Saltwater Pavilion, I strongly have denounced any wet building technology; I rather build up fluidity from a dry assembly of constituent unique components. Full-fledged component-based architecture is only incidentally adopted by ZHA, mainly in exterior cladding tiling, while in ONL's work, it forms the firm basis of a structural synchronization of the robotically produced components of the main structure and the

243

skin, both interior and exterior skin. ZHA's realized buildings are heavily relying on traditional *in situ* pouring of concrete, onsite welding of steel structures that are following a different system than the (too) many layers of the exterior skin, and traditional plastering of the interior skin to achieve that feeling of continuity and fluidity. While many architects accept traditional building technology to shape their otherwise fluid designs, robotic production and dry assembly are essential to achieve the desired component-based fluid continuity. This means many "no's" to achieve the "yes" to component-based architecture:

1) no to mass-produced standardized elements,
2) no to clashes of generic half products from a catalog,
3) no to molds,
4) no to the onsite pouring of concrete,
5) no to structural wet joints,
6) no to onsite welding,
7) no to scaffolding,
8) no to waste on the building site,
9) no to the separation of structure and skin,
10) no to many different details,
11) no to exceptions to the rule,
12) no to representational 3D models but a yes to scripting,
13) no to a linear building chain but yes to a decentralized network of experts,
14) no to deconstructivism, and, for sure,
15) no to parametricism as a style.

A multidisciplinary design team with designers and fabricators, one building, one detail, a parametric design process that synchronizes structure and skin, locally specified to adapt to different local conditions and climatic requirements. These are my fundamentals for the new synthetic dimension of architecture.

The Synthetic Dimension

 In 1991, I wrote a manifesto on the occasion of the exhibition The Synthetic Dimension that Ilona and I organized at

Museum De Zonnehof in Amersfoort, a cute little exhibition
pavilion designed by Gerrit Rietveld in 1958, adjacent to
the large Levob building complex designed by my late
father J.H.Oosterhuis (Joop), who was one of the leading
local architects in Amersfoort. The Levob has recently been
listed as a modern monument and is fortunately saved from
premature destruction, as was the unfortunate fate of his
Dudok-style PGEM office building in Arnhem. The exhibition
The Synthetic Dimension was hosted by curator Paul Coumans, who
gave us carte blanche to use the whole pavilion. In the Synthetic
Dimension manifesto, I questioned LeCorbusier's five points for modern
architecture and demonstrated that his five design commands are
outdated, mainly due to the effects of the emerging digital revolution,
the evolution from mass production to mass customization, and the
cross-disciplinary fusion of art and architecture.

1) I reject the structural skeleton (pilotis) in favor of the notion of
   the self-supporting building body. Learning from current car
   design, which abandoned the outdated chassis plus carriage
   paradigm for the structural unibody concept already back in
   the fifties of last century, I realized that structure and skin must
   become synchronized in every inch of its complex geometry. Six
   years after my manifesto, the concept of a geometrically complex
   sculpture building was realized in the Saltwater Pavilion project.
   The Saltwater Pavilion has no internal columns, and no pilotis,
   but is an object that features a structural unibody, a monocoque
   structure like a ship, an airplane, or a car. LeCorbusier flirted
   with cars, airplanes, and ocean steamers, none of which feature
   internal columns but failed to take the radical consequences of
   how to translate such vehicles into architecture. The structural
   skeleton of the Saltwater Pavilion coincides with the envelope.
   The integrated structure of the Saltwater Pavilion features neither
   columns and beams nor an exoskeleton, but a skin-integrated
   skeleton instead, as in shipbuilding. The only compromise we had
   to make was to connect the integrated steel and skin structure of
   the body to a whopping 1-m thick concrete base, to prevent the

sea-facing building sculpture from being lifted by rising sea levels. The alternative that we proposed, to let the sea water into the belly, was not accepted.

2)   The idea of the roof garden, although popular now, since many cities have ordained that the flat roofs of midrise and high-rise buildings must be greenified, does not fit in my concept of building bodies. I do support living plants to be interwoven in the metabolism of the building body, to form an active part of the climatic installations as in our City Fruitful project, as in our BULB office and Expo 2020 designs, but certainly not confined to the flat roof with a garden on top. Virtually, none of my buildings have flat roofs. The only good reason I can see for having a flat roof would be to collect water for internal usage of flushing toilets, but that is a bit far-fetched since there are better techniques to squeeze water out of the air.

3)   The idea of the open plan I reject and substitute for the concept of the open volume, internally and externally celebrating fluidity and continuity. For a long time, I do not do plans and sections anymore; they are replaced by the design of three-dimensional bodies in digital space and orchestrated interactive spatial experiences. Plans and sections only represent poorly informed sections out of a three-dimensional volume. Flatland cannot be anything more than a poorly informed slice of Spaceland. Space cannot be experienced from the point of view of the plane. The practice of representing a design by plans, whether open or traditionally arranged, and sections has become obsolete. Once one has seen the potential of the oblique and takes advantage of the spatial challenges of complex three-dimensional geometry, the open plan has lost its meaning. Since the introduction of 3D modeling software, our designs are pre-conceived as three-dimensional volumes in weightless 3D space. We name our designs open volumes rather than open plans, as they are parametrically moldable, scalable, and adjustable three-dimensional monocoque structures. After pre-conception, the weightless virtual objects are communicated with the local conditions of the site and eventually adjusted to build the physical connection.

4) The horizontal window is, for many reasons, no longer a valuable design guideline, while windows are seen as a specification of a continuous exterior membrane populated with transformative components from open and permeable to closed and impenetrable. I have radically abandoned the dominance of the strictly horizontal and the strictly vertical. As Zaha Hadid used to say, there are many more angles than the 90º angle. And back in the sixties, Claude Parent declared that there is a third important axis, besides the vertical and the horizontal, namely that of the oblique. The Saltwater Pavilion represents the architecture of the body, of continuity, of the oblique, of the interactive, in a more radical form than Parent could have dreamt of.

5) The back then liberating idea of the free facade I replace by the notion of the free volume, the open volume, moldable, scalable, and adaptive. Skin and structure are synchronized to adapt to changing exterior and interior circumstances. I continued my manifesto stating that the newly defined open volume inevitably would become sensorial, leading to a synthetic architecture that is thriving in an artificial ecology.

I wrote my manifesto back in 1991, well before we designed and built the Saltwater Pavilion. We knew back then where we were heading, and even now, 30 years later, the provocative text has not lost much of its meaning. The five points of LeCorbusier are by and large still common modernist practice, often literally promoted as such by advisors and city councils when looking for modern architecture or a mandatory style as in the current Bauhaus-style BMW brand guidelines for their showrooms. Examples of radical synthetic architecture still can be counted on the fingers of one hand. The world is changing fast but, unfortunately, much slower than needed architecture is only taking small steps at a time to cope with the rapid changes. Architecture is in a permanent crisis since the emergence of computation, whereas we experience every possible architectural movement active at the same time: modernism, regionalism, historicism, high-tech, postmodernism, deconstructivism, parametricism, eco-fundamentalism, DIY, nonstandard, robotic, 3D printing, blob-ism, and eclecticism, anything goes. Many of these styles come together in the work of Wolfgang Prix of Coop Himmelblau, who notoriously stated, much like a spoiled child, that thanks to digitalization, everything can be designed and made. Koolhaas made a similar statement concerning building the CCTV. He stated that engineers just will do anything, regardless of the technological challenge, thanks to digital technology. Koolhaas mischievously mocked them for it, cleaning his conscience. The Fundamentals biennale is a blunt denial of the raging IT revolution, while the 15 outdated pre-digital elements of architecture support the five outdated principles as set by LeCorbusier, one by one. As we are living here and now, I am keen to distill value from the current crisis in architecture. My 1991 rethinking of LeCorbusier's five points back then forms the basis for my step-by-step evolution into a component-based synthetic dimension.

6

The Component

From Element to Component

Considering that a building is a well-informed assembly of unique components, the new kind of architecture is, in its essence, synthetic. An architecture in which the unique components are parametrically designed, robotically produced, and assembled, just as needed and when needed and just that, just there, and just then. While elements are generic products that have no context (yet), or at best a hypothetical generic context, components are unique performative parts that only fit in one specific context. In the car industry, previous standard generic parts have been developed over the years into components that are designed to fit only in one particular context, in a stylistic relationship to their neighboring components. Now, it is architecture's turn to follow the example of automotive design development. This is a choice that I have already made back in the early nineties of the last century. No longer are we satisfied with the radical prefabrication of many of the same elements, as we applied in the making of the minimalist BRN Catering (1987). Informed by developments in the arts, notably by the interactive spatio-dynamic works of Nicolas Schöffer and the colorful spatial works by Frank Stella, I realized that there is another way and that there is life after minimalist architecture. After we started our joint innovation studio ONL (Oosterhuis_Lénárd), which we positioned as "the fusion of art and architecture on a digital platform," we incorporated the most recent technologies of computation and CNC production into our workflow. While the acceptance of mass production of elements of the same inevitably leads toward visual minimalism, we discovered another road that naturally embraces a maximum of diversity, notably mass customization of a series of unique components. Our early bird belief that mass-customization will replace mass-production methods came with the emergence of parametric design methods in combination with the availability of ubiquitous computer numerically controlled (CNC) production machines. The game-changing promise of ubiquitous CNC production became obvious by the end of the eighties and early nineties. The inconvenient truth is that architects typically are late at adopting new technologies, while our profession is deeply rooted in tradition, while voluntarily imprisoned by the traditional building industry. The building industry is like a supertanker that is slowly moving forward in a viscous substance, unable to change its course quickly. There are so many stakeholders in the building industry who rather stay put in known inefficient building methods than be willing to adopt new more efficient and liberating ones. Yet, one only had to open one's eyes and be willing to see the beneficial consequences it would have for one's profession. Nonstandard design opens up an unheard level of diversity and to user-oriented multimodality. Unique nonstandard components live their performative life next to their nearest neighbors only, not as a generic element that would fit anywhere. The cultural game-changing promise was, and still is, that it is no longer advantageous to produce a series of the same, but a series of unique components for the same price or cheaper.

From Generic to Specific

The open attitude toward technical innovation meant a real paradigm shift in our work and a full revolution of how buildings are designed and built and how they appear visually. We wanted buildings to be big inhabitable sculptures in their own right. We turned upside down the notion of what is beautiful in architecture and what is not, as is the title of my TEDx Delft lecture that I delivered in 2011: "We are changing your view on what is beautiful and what's not." Repetition is making place for rule-based diversity. Repetition is no longer beautiful. Now, 30 years after switching to the nonstandard, complexity is unfortunately hijacked by the deconstructivist elite, and by and large limited to expensive prestigious buildings like pavilions, museums, and opera houses. A successful application of the new rules for the new economy on housing projects is seldom seen. What is direly missed is the fulfillment of the promise of parametric design to robotic production, the promise being the accomplishment of a cost-effective form of diversity. Our social housing project Fside in Amsterdam probably comes closest to that ideal.

Our buildings stand in many aspects in stark contrast with ZHA's, Coop Himmelb(l)au's, and Gehry's featured buildings. One of the main differences is the building costs. ONL's realized nonstandard buildings are definitely cost-effective, and theirs are typically five to six times more expensive per square meter. Our nonstandard achievements are positioned at the lower end of the spectrum of building costs, while their deconstructivist designs hover at the higher end, which does not provide a fair basis for comparison. Typically, there is a factor 5 or 6 difference in price tag and a similar substantial difference in the level of the incorporation of parametric design methods. Only when going full throttle on parametric design by scripting to robotic fabrication of the unique components will the cost-effectiveness of the nonstandard play out. As a compact office, we were able to realize the cultural center Bálna Budapest for a very competitive price tag, while ZHA, with their rejected Szervita Square project, but also Foster and Partners, and Asymptote, failed to realize their projects in Budapest. ZHA and their project development partner ORCO obviously did not wish to comply with the locally feasible price tags for commercial buildings. Despite being developed as a cultural center, the Bálna is a purely commercial project, realized without support from the European Union, which had to deal with a very strict budget.

The way we handled working within the typical Hungarian context was to start a Budapest-based office ONL Hungary kft., operating as an independent entity cooperating with ONL in Rotterdam. We worked in Revit working groups to stitch the models of ONL and ONL Hungary in real time together. Eventually, ONL Hungary got awarded the Revit BIM Experience Award for its innovative efforts. ONL Rotterdam developed the nonstandard steel structure, while the more conventional concrete parts like the underground parking garage and the renovation of the existing buildings were developed in Hungary. We did not treat parametric design to robotic production as a fashionable and elitist showcase. We treat parametric design as a driving force from within an all-inclusive building system, based on an empathic understanding of how to link the digital design to the robotic execution and the dry assembly of unique parts of the structure and the skin. It is hard to underestimate the level of radical empathy for the new rules of the new economy that one needs to incorporate in the design-to-production process before one can fulfill the promise of the cost-efficiency of nonstandard architecture. While Gehry *cum suis* combines the generic and the specific into inefficient complicated hybrid structures, I have chosen to go all the way to an all-inclusive building system. The theory and practice of "One building, one detail" allows one to regain control of the complete design-to-build process, working in design & build teams with the manufacturers.

Looking at the detailing of the entrance of Gehry's Stata Center in Boston USA, one cannot be but disappointed at the banality of the architecture; there is nothing parametric about it, and nothing is CNC produced but just a random stack of elements from a standard catalog clashing at different angles into each other. Gehry's deconstructivist attitude stands in stark contrast with the inclusive synthetic architecture of the Web of North-Holland (Web of N-H). The entrance of the Web of N-H is conceived as a further specification of the skin, like a hair on the human skin is a specification of that skin, formed by folded pockets out of which the hair grows – and not a detail that would clash with that skin. No, the entrance door of the Web of N-H grows out of the parametrically synchronized structure-to-skin detail. Basically, for that specific location of the door of the Web of North-Holland, just some additional lines of code were written to distinguish the door from the other areas of the skin and the structure. The peripheral steel parts are doubled, and hinges are implemented at the upper side as needed; electronic pistons are added to open the part, but, otherwise, the same parametric detail is mapped to this specific openable component as in the non-openable areas of the skin. As the huge door hinges up to form a canopy, it also fulfills the other radical paradigm of "when needed." There is only a canopy when there is someone who is entering the pavilion and needs to be protected from the sun or the rain. When there is no one inside and no one entering the Web of N-H, the huge, hinged door remains closed, and the pavilion returns to its role as an autonomous work of art. When the door is closed, the openable part is hardly distinctive from the rest of the skin; it has visually merged into the volume as a whole. When the door is open, it communicates that it is accessible to the public and thus functions as a building, a work of architecture. The entrance door thus has become a switch from architecture to art and back from art to architecture.

Cost-effective Diversity

Why do I value the implementation of an uncompromising parametric design to robotic production procedure so much? One of the main reasons is that if one is not ready to go all the way, one will experience conflicts between the mass-produced generic elements and the customized nonstandard components. Only when going all the way can the benefits be harvested. Among the benefits is the cost-effective diversity, which is beautiful by nature. The synthetic evolves from nature itself, effectively shaping a new actual form of nature. When one does not go all the way, putting together a hybrid of the traditional and the nonstandard as the likes of Gehry typically do, the result might be impressive but comes at a huge cost. Gehry is complicated, not complex. And because complicatedness comes at high costs, it will not redefine the daily practice of architecture and will not have the power to change the bulk of society's building activities. It will not change the course of the building industry supertanker. Particularly therefore, parametricism, a term coined by Patrik Schumacher as late as 2008, will remain a style for the rich and famous, to be consumed in awe by the masses, but not fit to contribute to the housing crisis, and not available to the majority of people to participate in as a level playing field prosumer. The social agenda for parametric design to robotic production aims to offer general accessibility to the nonstandard for all possible clients, individual homeowners, project developers, and governmental bodies alike. The social aspects of the new building digital technologies should not be underestimated either; availability for all is worth fighting for. Parametric design and robotic production will gradually take shape as the long tail of the economy, meaning that the building market will be populated by many more and smaller companies that are digitally operating, thereby breaking down the dominance of the big contractors in the building industry. Parametric design and robotic production will counter-attack the obstruction of the nonstandard by mainstream architects and their clients, often governments, city councils, or established institutions, whether capitalist or socialist. Governments seem to have a little problem with substantial budget overruns, as is typically the case with OMA's, ZHA's, and Gehry's "complicated" projects. Classical tender procedures prevent innovative breakthroughs, while the formation of design & build consortia where the architects form a team with the end-manufacturers from the very first sketch is not allowed, presumably because of the so-called conflict of interest, but in reality, to safeguard the interest of the big players in the field. Accepting this situation means accepting the lack of innovative force from within the traditional building industry.

Inclusive

Mies van der Rohe devoted much of his attention to the perfection of the detail. Taking one step further than Van der Rohe, the idea of the ubiquitous parametric detail comprises the mapping of one self-adjusting parametric detail to a point cloud of reference points that are populating the envelope, from top to bottom, from left to right, and from the inside to the outside. My take on the detail is to condense the constituent components into a generic adaptive system that incorporates structure, skin, and style. To condense all aspects of the design into one single parametric detail that adjusts to all possible specific points on the topography of the design, whether mapped onto flat surfaces, ruled surfaces, or double-curved surfaces. The ubiquitous detail of the synchronized structure and skin of the Liwa tower does just that. The Liwa tower started as a commercial real estate project with a standard price tag to be rented out for profit. After having completed the tower within that standard budget, the client Al Nasser liked the building so much that he decided to move his own Al Nasser Holding offices to the six top floors of the tower. The smoothly upward-moving curves in combination with the color of the skin reminded the client of the gentle wind-shaped curves of the red sand dunes in the Liwa Oasis where he was born. My design choices were based on practical contextual data, and I did not think of sand dunes; the design is not a metaphor. The name Liwa was given afterward, while it resonated so well with the local environment. The budget remained that of a standard commercial office tower, and also the interior fit-out, which was not commissioned to us, remained simple and straightforward. There was one exception – the client Mohamed Al Nasser wanted to be able to step out of his baby-blue Bentley Continental into a lift that brings him from the underground garage directly to his private office floor. To satisfy the client's wish for the direct lift, we had exactly two weeks to change the plans before submitting them for building approval. We had to change everything – the layout of the core and the office area around the core. While everything is connected to almost everything in the parametric system of the Liwa tower, we managed to do this without delay. In a traditional design process, the client's wish would have caused enormous extra work from both the architect and the project developer and would have cost the client (and probably the architect and the developer) tons of dirhams and would have caused a substantial delay.

For the tender procedure, we wrote a book of design guidelines for the parametric relationships between the constituent components: the steel ring beam, the diagrid load-bearing steel structure, the sandwich panels, the oval-shaped windows inside the panels, and the connecting steel parts. The design guidelines of the Liwa tower describe how the components of the complete load-bearing facade structure fit together in the most minute detail. Since we wanted no exceptions to the rule-based synchronization of the structure and the skin that envelopes the whole tower from bottom to top, we did not want to have a corner detail that is different from the field detail. A separate corner detail would have cost us as much design work as the rest of the facade in its totality. We invented a system whereby the connections between the prefabricated X-shaped diagrid steel frames and the triangular sandwich panels are bay-by-bay fanning out from the central symmetry axis toward the corners. Thus, the corner detail is described by the same formula as the field connections between sandwich panels and steel structure. The variable corners of the Liwa tower corner, ninety-plus degrees, are just one of the many possible folds. The corner detail is not an exception to the rule; the universal parametric detail includes any possible angle of the corner detail. The folding lines follow a curved trajectory from bottom to top, and each fold line is different from the other, which gives the Liwa tower its free-flowing character. The layout is mirror-symmetric along the longest axis, perpendicular to the street. Although generally not perceived as such, virtually, all of my buildings are symmetrical, as in car body design and in industrial-designed household products.

The folds in the middle of the fields are a mere two or three degrees. In his time in the heydays of mass production, Van der Rohe went to the extremes of what was technically and esthetically possible; he went for the highest achievable, supreme version of mass production. Van der Rohe uses high-quality materials and well-developed half-products of a high standard, mostly from a building catalog. Van der Rohe needed extra detail for the corners; he needed to bridge the gap between one array of standard elements to the next around the corner. Van der Rohe could only bridge that gap with the introduction of a special detail for the corner. In nonstandard architecture, the corner is no longer an exception to the rule as is in Van der Rohe's designs. One of the slogans I coined in the early zeroes of this century is: "Mies Is Too Much," as a PR teaser for an architecture that is based on one single inclusive parametric detail, inclusive of all possible positions on the envelope. The designer's ultimate goal is to find the supreme form of mass customization. My first realized building, the BRN Catering building, has no corner detail either, due to its flat oval form with rounded caps of the all-enveloping linear structure. In essence, my companion Peter Gerssen took the initial minimalist approach to another extreme, which has proven to be the fertile breeding ground for a future maximalist way of working. The single parametric detail that forms the basis for maximum visual diversity is generated directly from a simple minimalist rule for the detailing. An adequate description of our current approach, which is minimalist and maximalist at the same time, would be to label the Liwa tower as a synthetic form of inclusive design; inclusive, since fusing art and architecture on a digital platform at the grand scale of architecture. Inclusive, since aiming at a minimum effort to have a maximum impact. And universal, because we apply the universal rules of swarm behavior, which forms the basis of all natural processes. Inclusively universal, because we build our design universes from scratch, putting performative components together to synthesize the bigger whole.

An architecture that is based on a simple set of basic building blocks can generate a ubiquitous diversity that we as humans appreciate as natural. Humans naturally appreciate the rule-based diversity of the thousands of different leaves on a tree and the millions of different people on earth. Similarly, humans will appreciate the familiar diversity of thousands of different components of a parametrically designed building, made up of components that are moldable, scalable, and adjustable.

Passive Interacting Components

To illustrate what I consider interacting components, a distinction must be made between two different categories of interacting components. First, there are passive interacting components, as in frozen constellations like buildings, and, second, there are active interacting components, as in interactive installations featuring real-time communication and interaction between the user and the object or environment. We use the term interactive for both categories, while in the design process of synthesizing the swarm, the real-time interaction between components is the default situation, and the frozen instance is the static derivative of the dynamic system.

In any virtual parametric design process, each component is per definition related to its nearest neighbors, whereby the component is tweaked to change shape and content as long as it lives within the software. The real-time factor of active interaction comes to life when the constituent components, equipped with virtual or physical sensors and actuators, can change their performance by responding to streams of incoming data, external to the design software, brought to life by real-time running algorithms. Passive interacting components do not change shape and content; they are passively interacting, after having been saved as a 3D model and executed in the material world. But still, passive interaction involves the transfer of gravity and wind forces and transferring stylistic design features from one component to the other.

Passive interacting components are well explained by a detailed analysis of some of our projects, focusing on the constructive role of the informed component. Not surprisingly, three of these projects are art projects by Ilona Lénárd: the Folded Volume (1991), the Musicsculpture (1998), and the TT Monument (2002). Typically, art projects are the precursors of architectural projects on many levels, probably because of the manageable scale of art projects, whereas one has control over the whole process, including the budget, the design, the engineering, the fabrication, and the responsibility to deliver. The component is characterized by the level of autonomy, the choice and treatment of materials, the elementary structural detail, the spatial qualities of the work, and their esthetics. Spatial qualities in art and architecture are often referred to as sculptural, but as stated before, we are keen to avoid words like architectural and sculptural, while these words describe at best the superficial qualities of the objects but not the inherent material, structural, spatial, and styling qualities. When a piece of architecture is referred to as sculptural, it probably means that it shares some superficial features of a sculpture. When a piece of art is referred to as architectural, it probably indicates a possible functional use of the object. When a building is referred to as sculptural, it probably intends to indicate that it is not just a box or another prismatic volume but is more complex in its appearance. It must be clear though that I do not take the label sculptural as a compliment. What matters are tactile qualities and structural integrity, not superficial resemblances. What matters is how architecture is spatially experienced, how the public and critics experience what it *de facto is,* and how it feels, not what it looks like.

Folded Volume

For the exhibition The Synthetic Dimension in Museum De Zonnehof in Amersfoort (1991) that we organized and managed together on the invitation of conservator Paul Coumans, Ilona made a series of works directly derived from three-dimensional computer sketches. The theme that was developed for The Synthetic Dimension was to bring together new developments in 3D thinking in the fields of art, architecture, music, and industrial design. We exhibited, among other pieces, a majestic wall-mounted piece titled Cytology from the Cones and Pillar series by visual artist Frank Stella, images of Stella's pavilion design for the Groninger Museum, the colorful Tongue chair by designer Pierre Paulin, original sketches of the Cité Cybernétique by visual artist Nicolas Schöffer, the Rubik cube of architect-inventor Ernő Rubik, a one-to-one scale surfboard, my first one-to-one scale realized design the Villa in Lego (1985), the Guitar Mural by artist/mathematician Remko Scha, new shapes of paprika's by visual artist Adri Huisman and bell pepper grower/farmer Jack Alblas, the 36 silkscreens of the Blind Spot series (1991), and a big sculpture titled Folded Volume (1991), both by Ilona Lénárd. The silkscreen series is printed on colorful PVC-coated glass fiber-reinforced canvas, using six different colors for the ink on six corresponding colors for the canvases. The result is that one of the canvases had the same color printed on the colored canvas, thus creating the blind spot, only to reveal itself when passing by and seeing the light reflected from a different angle. The 6 × 6 strategy of the Blind Spot series results in 36 unique paintings, thereby preluding our future interest in parametric design. The image that is printed on the canvas is a flattened projection of an intuitive 3D computer sketch.

Similar computer sketches printed on green glass fiber-reinforced canvas are used for the skin of the large Folded Volume sculpture. We took the spatial coordinates of the 3D sketch and constructed flat surfaces between the data points. Mostly three, and at times four or five data points were aligned in a planar surface, thus forming triangles, quadrangles, and pentagons. Folded Volume is a free-standing complex sculpture – not on a pedestal – that consists of 20 such flat irregularly shaped components, connected to form the closed volume. In the eighties, I started to refer to such a shape as a shaped container. Each component is different in form, varying in size and number of corners, the triangles, quadrangles, and pentagons that populate the envelope. Only after making a concrete solid scale model of Folded Volume, it was decided what would be up and what would be down, while the solid shape could be positioned on the ground in a few different ways. The crucial challenge in the making of Folded Volume was to develop a proper detail for the execution on a larger scale. We decided to have one single, agile detail that would adjust to the different angles, at which the components would meet each other. The main components to constitute the sculpture are the 20 irregularly shaped parts. Each component is constructed following exactly the same procedure. Seen in retrospect, the making of Folded Volume represents in many ways the prototypical production process for the future buildings realized in our joint enterprise ONL (Oosterhuis_Lénárd), which was formally established in 1989. Between 1980 and 1989, we followed separate architecture and art paths, only to merge at the end of the eighties after our one-year stay in the studio Theo van Doesburg in Meudon, Paris. Folded Volume (1991) can be considered as the predecessor of the "One Building, One Detail" design strategy, which we successfully partly applied to the Saltwater Pavilion (1997) and fully applied to the Web of North-Holland (2002) and all of our subsequent projects.

Each component of Folded Volume is an assembly of a wooden frame with the green PVC-coated canvas skin stretched on it, rounding the edges to create an invisible butt joint between the panels, similar to how spandrel panels fit together in common interior and exterior cladding systems. In the process of translating the geometry into real materials, I did not have access to a CNC machine of any kind; everything was done manually in our studios in a former school building in Rotterdam-Overschie. I measured the dimensions and the mutual angles between the components from a scale model, constructed a digital model, and adjusted the positions of the data points to construct planar wooden frames. We did not want twisted surfaces, while the canvas would not stretch perfectly. The 20 odd parts of Folded Volume are bolted together, such that they can be disassembled again and built up at another location. The design concept of a component-based circularity that is prepared for second use started exactly here with the making of Folded Volume. Later, we applied the basic principles of circularity "avant la lettre" in the design of the Web of North-Holland, and its second life as the iWEB in front of the Faculty of Architecture at the TU Delft. Back then, we just felt that it was a natural thing to do, without labeling it as circular. For artists, it is quite normal to think of the work as an assembly of parts, while they often have to assemble, take down, and reassemble their works in different locations. It is part of the autonomy of a piece of art. A piece of art is usually not confined to a specific place unless it is a contextual work that values context over autonomy. Autonomous works of art are worlds in themselves, small and meaningful self-contained universes.

Folded Volume is probably the most innovative project that we have imagined and executed, the prototype for all of our further efforts. During my keynote lecture at the Synthetic Dimension event, at one moment, I pointed at the Folded Volume sculpture and told the audience, without having thought of that before, that one could consider Folded Volume a 1:20 scale model of a 60-m-high building as well, thereby intuitively anticipating on what was to come in the decades ahead of us. The execution of Folded Volume is fully analog; the only digital part is the initial digital 3D sketch. We just did not have the proper technology 30 years ago to go digital all the way; it simply was not there yet, or rather, we were not yet familiar with it. Henk Meijers of Meijers Staalbouw, with whom we built the Saltwater Pavilion, The Web of North-Holland, and the A2 Cockpit in the Acoustic Barrier, once told us that he had installed the CNC machines in his factory already 15 years before we designed and built the Saltwater Pavilion. The same must have been true for operational CNC machines for wood crafting. Although we became aware of the existence of CNC machines well after their introduction in industrial steel and wood workshops, among architects we were the absolute first to take advantage of CNC machines. We prepared ourselves to be ready for the digital design & build era and were keen to pioneer and apply the new working methods to our art and architecture efforts.

One specifically intriguing design detail of Folded Volume is the seemingly out-of-balance equilibrium of the work. A very big portion of the volume is pointing from the highest point down toward the ground, somewhat teasingly just not touching the ground, to challenge gravity. Large cantilevers jutting out of the main volume have since been a hallmark characteristic of our ONL projects. Big cantilevers are an important design feature of

1) the Saltwater Pavilion jutting out over the sea,
2) the small footprint of the Web of North-Holland,
3) the A2 Cockpit where it visually connects to the Acoustic Barrier,
4) the entrances of the Bálna Budapest, and
5) in the TT Monument and the Musicsculpture sculptures in public space.

Musicsculpture

While the Folded Volume is designed as an indoor project to be exhibited in museums, Musicsculpture (1998) is an outdoor sculpture that was commissioned after having completed the Waste Transfer Station (1994) in Zenderen in the far eastern part of the Netherlands. After cultural attaché Rudolf Krudop of the Province of Overijssel had invited us to compete in the design of the Waste Transfer Station, we were selected – after winning an invited competition – to realize a sculpture annex bandstand by the pod in a small romantic park in Oldemarkt, a small town some kilometers north of Zwolle. Designing the sculpture titled Musicsculpture, Ilona started with an abstracted contour of the leaf of an oak tree. The contours of the Musicsculpture were conceived not as a literal copy but through a series of intuitive sketches with that specific leaf in mind. The selected sketch is bent to form a three-dimensional model for the sculpture. In the brief, it says that the sculpture must function as a bandstand as well. The double function of a sculpture as an autonomous work of art and a functional public object meant the first real-world application of the statement that we made for the <u>Sculpture City</u> event (1994) in the Berry Koedam-led RAM Gallery in Rotterdam: "A building can be an autonomous sculpture, the sculpture a functional building."

A bent spinal cord forms the basis of the Musicsculpture, to which the tapered ribs are connected. In one continuous smooth movement, a 3D bent tubular profile connects the ends of the ribs, describing the rounded contours of the abstracted oak leaf. By bending the spinal cord almost 180º, a relatively flat floor is created for the music band, protecting the band from sun and rain by the overhang of the roof-ish part. At the same time, the bent shape fulfills the function of effectively reflecting the sound. The simple act of bending the irregular shape with softly rounded edges ensures a strong spatial presence of the sculpture, experienced when walking around the small lake, on the shore of which the Musicsculpture is positioned. From every angle, the sculpture looks different. The design and making of the Musicsculpture have been instrumental to the development of our concept of buildings as large public sculptures that dynamize public space through their three-dimensional quality. The Musicsculpture is built up from one single material and finished with one single coating – one design, one pavilion, and one detail. Sheets of steel are cut into shape using the exact data from our Autocad 3D model and welded together into three large components that are first bolted and hence welded together, as in shipbuilding. For the green skin of the Musicsculpture, we applied a similar coating technique as for the black skin of the Saltwater Pavilion (1997), realized one year earlier. Sprayed <u>polyurea</u>, a strong durable industrial coating, covers the whole object from bottom to top, from the outside to the inside of the curved shell structure. The result is a visually seamless object, albeit consisting of three distinct complex components.

The concept of assembling the sculpture from three large separate components came from the steel manufacturer, based on the limitations of transport to the site. We have adopted the concept of breaking down a smooth seemingly seamless structure into a series of discrete components ever since. Folded Volume and Musicsculpture are component-based designs that have proven to be an effective strategy both for the production and the assembly respectively disassembly of the constituent parts. The Musicsculpture is a precursor of our future larger structures. Art precedes architecture. When discussing building components, I make a distinction in what the main component is and what the subcomponents are. The main components bear the characteristic style of the project as a whole, while the subcomponents contribute to the build-up of the main component but are lacking character. I consider the three parts as the main components of Musicsculpture, while the steel ribs that are cut out from flat sheets of steel are considered the subcomponents of the three main parts. The subcomponential steel ribs of the Musicsculpture are similar to the subcomponential wooden frames of the Folded Volume. We typically choose the term component for that family of parts that represents the characteristic style of the design as a whole. In the case of the Musicsculpture, it is not the customized steel plates but the three larger more complex components. In the case of the Folded Volume, it is not the parts that make up the wooden frames, nor just the wooden frames that are the essential components, but the characteristic irregularly shaped panels that incorporate the structure and the skin.

TT Monument

The design process of the TT Monument is an effective mix of analog and digital tweaking of physical and virtual material. It shares many of the characteristics of parametric design to robotic production, and, as

such, another precursor of the current maker industry, which forms a growing part of the long tail of the economy, as aptly described by Chris Anderson, editor-in-chief of Wired magazine. As with most of our projects, to design and build the TT Monument, we had to win a competition first to acquire the commission. This time, it was an open competition, whereas the Waste Transfer Station, the Musicsculpture, and the Bálna Budapest were won through invited competitions, and the Web of North-Holland, the Saltwater Pavilion, and the A2 Cockpit based on being shortlisted and selected. In contrast, our housing projects Patio Housing, De Kassen,

Dancing facades, Dijken, Daken, TGV housing, Family, 8Bit, and Fside typically are commissioned directly by project developers, often upon the suggestion of the master planners.

We started the design process of the TT Monument by 3D modeling a racer on top of a MotoGP motorcycle. In Maya 3D modeling software, later acquired by big tech Autodesk, we merged the driver and the motorcycle to form one continuous object. The conceptual idea represents the indivisible unity of the driver and the motorbike while speeding in Motor Grand Prix and Superbike races. Biker and bike are becoming one; together, they live as a fused man–motor entity. The commercial graphics on the bike and those on the drivers' leather suits are so much alike that visually biker and bike seem to fuse into one single entity as well.

After having made the initial 3D model, we carved the model out from hard foam using our compact in-house CNC milling machine and then took a sufficient number of data points with the MicroScribe 3D digitizer arm, transferred the data to the Maya software, constructed one continuous surface, and refined the model by pushing and pulling the surface locally using the sculpt geometry tool in Maya. By swiping movements with the brush, we added more speed to the fused 3D model. To position the 3D model in the context, we rotated the bike upward at the front to evoke the wheelie, the triumphant prancing horse pose the racer makes immediately after winning the race. As a result of the wheelie posture, the TT Monument has only a very narrow 10×10 cm^2 base to fit onto the ground, as in the real-life wheelie that balances on the back tire. The TT Monument wheelie is the real structural challenge. This exposure of the proud wheelie posture is the most audacious cantilever we ever did.

We wanted the TT Monument to be executed as one continuous surface made of cast aluminum. The TT Monument is an assembly of four main components, invisibly welded together in the factory. The mold for each cast aluminum component is first CNC milled with a large five-axis milling robot in an extremely high resolution, to avoid extensive postproduction sanding of the surface. These four foam molds were then used as the counter-mold to form the sand mold that, in turn, functions as the mold to pour the 2-cm-thick hot liquid aluminum into perfectly executed by the Aluminiumgieterij Algietol in Oldenzaal, a company that produces high precision parts for machines. Once released out of the mold, the four parts are polished to create a semi-glossy finish on the surface of the cast AlMg$_3$ aluminum alloy. Each of the four cast components is designed to have a 90º edge all around the contour edge to be able to bolt them together. Then the four bolted parts are welded together to suggest one solid mass, to achieve the intended smoothness of a driver fusing in speed, showing off with a victorious wheelie. Much to our satisfaction, the TT Monument is highly appreciated by the MotoGP drivers themselves; they do feel the excitement of fusion in speed in the posture and the materialization of the sculpture.

Fside

We designed and executed most of the housing projects in the nineties of the last century. Due to the rigid conventions in the building industry for housing, we had little opportunity to test our then-still juvenile component-based design to production method on housing. Instead, we focused on a maximum fusion of art and architecture on the grand scale of architecture. I will just briefly discuss here our early housing projects in greater detail, except for the Fside social housing project in Amsterdam, which took a dramatic turn one week before the submission of the building permit documents. From the beginning of the nineties, the housing projects offered an opportunity to put the fusion of art and architecture into practice. The Patio Housing project at the Dedemsvaartweg in The Hague was originally conceived as an assembly of steel sandwich panels for the floors, walls, and roofs, in an early attempt to build the six patio homes in one building material, preceding the later "one building one detail" paradigm. The exterior of the closed patios was thought of as a sandwich panel with cut-outs letting through daylight during daytime and electric light in the evenings, based on intuitive autonomous sketches. What we wanted was to give the exterior of the six homes the appearance of a work of art. I was already familiar with limiting the number of different materials and details when I was working with Peter Gerssen, especially when making the executing drawings of the steel sandwich version of Gerssen's Stopwoningen. But to find a contractor willing to go in this direction proved harder than I thought. Probably based on a lack of experience and knowledge, the price the contractor offered was way too high; so we had to find an alternative solution. I was not able to break traditional building conventions, at least, not yet. It was decided to build according to a typical low-rise building system, with sandstone blocks for the load-bearing walls in combination with prefab hollow concrete floor and roof elements. The only invention of the Patio housing project I reserved for the exterior, executed as frosted transparent polyester corrugated plates with Ilona's colorful Dazzle Painting, wrapped all around the single-story building block. From the outside, the building appears as one large painting, without the obvious signs of actually being a building. We refrained from any obvious elements like visible doors and visible windows. All traditional architectural elements are hidden to give the maximum possible expression of the painting. Only the house number, the door lock, and the slot of the letterbox are visible signs of habitation.

With the next housing projects, we followed a similar scenario. De Kassen housing project consists of two rows of eight houses, covered with another colorful graphic Dazzle Painting, on the first and second floors. The impressively large dazzled horizontal bar "floats" above a series of eight fragile-looking glasshouses. In the subsequent Dancing facades project, we did something similar. Dazzling paintings determine the very shapes of the first floor, in the form of colorful dancing boxes on top of a long linear brick wall, which is partly open to the intimate patios. Our early housing projects were successful in bringing art and architecture closer together, albeit limited to the surface, not influencing the structure of the project. The crucial innovation though is that art no longer is seen as a decoration, not as an additional brooch to an otherwise untouchable piece of architecture. The early housing projects have been an important testbed for our future sculpture buildings. We were determined that we wanted to go all the way toward an integrated form of art and architecture. In our early housing projects, no opportunity was within grasp to apply the concept of component-based assembly for an integrated structure and skin. It was only when we got to work with the steel structures of the Saltwater Pavilion, the Web of North-Holland, and the A2 Cockpit in the Acoustic Barrier that the component-based approach that we developed in the works of art came within reach.

After having finished the A2 Cockpit in 2005, the 55-home Fside housing project was rebooted after being dormant due to political issues for half a decade. Again, we did not find an opening to go all the way, but we did achieve a truly groundbreaking innovation, which eventually would have a measurable effect on the building industry at large. Initial investigations into 3D milled three-dimensional shaped foam-based facade panels failed due to fire regulations. We could not convince the client that the foam-based panels would be stable enough during a fire to prevent the fire from spreading to the floor above. Our proposal was rejected, and there was just one week left to find an alternative solution; otherwise, we would lose the job. In a crucial meeting, I intuitively proposed a radical alternative solution, of which I had no idea whether it would be feasible or not; I just had to bluff my way through to save the project. Charmed by the natural look of the untreated aluminum of the TT Monument that we completed in 2002, I proposed to apply natural uncoated aluminum panels for the facade, pressed into a three-dimensional shape by a shipbuilder. We turned to shipbuilder CIG-Architecture – then operating under the name Centraalstaal – and in record time developed a design that would take advantage of their 3D pressing possibilities. Our ProEngineer parametric design files were synchronized with their proprietary Nupas software and got their offer, all within one week. The price for the series of around thousand uniquely shaped 6-mm-thick natural aluminum panels fortunately just fitted within the budget, saving the project. The structural basis of the social housing project had to remain traditional, while contractor BAM required us to use their tunnel formwork system, but the skin is all-innovative. It is remarkable how well the untreated $ALMg_3$ alloy that is usually used to build the hull of marine vessels adequately self-protects against the notorious bad Dutch weather. The aluminum panels are the main featured components of the Fside social housing project; the project derives its unique character from these embossed aluminum components. The success story of the Fside skin catapulted Centraalstaal into the building market. Since the Fside project, which was their first architectural project, they have become a major player in the market of three-dimensionally complex shaped structures. They have built the big fat tree-like columns bearing the roof of Isosaki's Conference center in Doha, they developed the components for the connecting bridge of Asymptote's Formula I hotel in Abu Dhabi, they realized the big cantilevering Porsche pavilion in Wolfsburg using pure aluminum parts welded together to form a solid whole, originally designed by Henn in concrete, and they helped Zaha Hadid and Frank Stella out to realize geometrically complex shapes and elaborate works of art. Centraalstaal also delivered the 4-mm-thick triangular sail-shaped aluminum heat shields for the <u>Bálna Budapest</u>.

Climbing Wall

With each project, we improve the principles of the assembly of the main components. Typically, since the Saltwater Pavilion, for each new project, a project-specific design to production system is set up, as for recent projects like the <u>Climbing Wall</u> (2012) in Amsterdam and Dordrecht, for which we won the <u>Wood Innovation Award</u> in 2012, the <u>Body Chair</u> (2016), the <u>Ecoustic</u> parametric acoustic ceiling system for Ecophon (2015), and the <u>Liwa tower</u> (2014). In its most uncompromising form, we developed a radical "one building one detail" <u>protoCELL</u> system for protoSPACE 4.0 (2010) and more recently for our entry for the <u>Dutch Pavilion</u> at the Dubai World Expo 2020/21. The projects have one thing in common: the ruling parametric detail is based on the <u>inclusion of the structure and the skin</u> into one lean and mean component. For the Climbing Wall projects, commissioned by Mountain Network, we developed a parametric triangular component built up from wooden subcomponents that are automatically adjusted per component for their dimensions, chamfering of the sides, hole patterns for the grips, and the holes for the bolts. The all-inclusive system aims to build a climbing wall using one parametric component only, for every possible part of that wall, whether a flat surface, a double curved surface, or an overarching part of the wall. The Climbing Wall supports itself in the indoor space without the necessity of being anchored to the structure of the building. The gravity forces and torsion forces that are caused by the spatial layout and the movements of the climbers are transferred from one component to the other down to the floor.

We opted for a triangulated system for several compelling reasons. First, by using triangular components, every possible topology is effortlessly described, including flat surfaces, ruled surfaces, and double-curved surfaces. Structural triangulation has become a hallmark feature of ONL architecture since the Web of North-Holland (2002), which in turn was based on some earlier studies we made for the daily newspaper Trouw for the programmable, in real-time moldable interior skin for the International Space Station ISS (1999). Second, triangulation has a huge structural advantage over rectangular column and beam structures, while a diagrid structure has inherent structural stability; lessons learned from Buckminster Fuller. A triangle with hinged connections between the edges cannot be deformed, while a hinged rectangle is utterly unstable. Traditional rectangular column and beam structures need to be stabilized by diagonal bracings, therewith effectively forming triangles. The other way to stabilize using traditional construction methods is to have strong moment-proof connections. A construct with triangular components is more efficient in the use of materials; a diagrid structure uses less than 80% of materials as compared to rectangular structures. A diagrid structure is leaner and cheaper than a traditional rectangular grid. Third, in a triangular mesh, one can remove one component without compromising the stability of the structure. The gravity forces will travel around the void at both the left and the right sides and merge again beneath the gap. In a triangular mesh, the triangular components transfer gravity and wind forces both sideways and downward, thus evenly distributing forces over the structure as a whole. This is in contrast with rectangular frames, whereby the down vector takes all the gravity forces, while the unstable structure has to be braced to resolve the wind forces. Taking out one column in a rectangular frame means a total collapse of the whole structure, due to its inherent instability. Taking advantage of the structural resilience and robustness of the freely shaped diagrid system, in the Climbing Wall system, many triangular components can easily be left out without losing structural integrity. In the realized climbing walls, we have left out some triangles to form a square window. We also modeled larger holes and arches at the top of the structure to let the sunshine in, similar to – but not inspired by – the arches in Arches National Park in Utah, USA. The openings are not exceptions to the rule. The parametric design allows for simply leaving some components out of the system, without the need for further adjustments.

Triangular structural systems are inclusive compared to rectangular systems since any double curved surface and any possible straightforward rectangular building can be described by triangulation. The other way round is impossible; using rectangles, one cannot describe double curved surfaces. To describe a complex topography, one needs to break the surfaces down into triangles or irregular quadrilaterals, which are effectively two triangles in one plane. Traditional forms of rectangular grid structures are discriminatory by nature, while triangular structural systems are inclusive systems, opening up the road toward an all-inclusive architecture. While post-production operations should be avoided as much as possible, there is no technical difference between cutting sheets of wood or metal into rectangular shapes or irregular or triangular shapes, provided that the cutting process is data-driven, by the efficient file-to-factory procedure. The amount of waste material in the cutting process of triangular parts is comparable to traditional rectangular designs, while the parts can be nested effectively, and the remaining pieces are mostly recyclable. As long as the cutting takes place in the controlled environment of the factory, the lost material can be effectively repurposed. Based on its inclusiveness, resilience, and efficiency, I choose triangular construction methods more and more as the basis for the designs, to get this head start advantage over traditional rectangular gridded structures. The ubiquitous application possibilities of diagrid systems are the reason why we have been able to deliver complex-looking nonstandard buildings well within standard budgets. The Climbing Wall project is living proof of that, as is the A2 Cockpit, the Bálna Budapest, the Liwa Tower, the Dubai Expo 2020, the Seven Daughters, and counting.

The parametric design to CNC production method of the Climbing Wall we have from A to Z in-house managed and executed, including writing the algorithms to drive the CNC machines and synchronizing with the software of third parties. The design of Climbing Wall is modeled and programmed from scratch in the visual programming interface of Grasshopper to Rhino. Where previously, as in the design for the A2 Cockpit and the Fside housing project, we had to rely on (expensive) software like ProEngineer and Revit, these days, we can do all that is needed for the accuracy of the execution of the projects in the Grasshopper plugin for Rhino. For the design of the Climbing Walls, some upward curves are introduced, driving the geometry of the 12-m-high climbing structure. Between the curves, the double-curved surfaces are populated with a triangular mesh. In this stage of the design process, the frequency of the mesh is optimized to maximize the sizes of the plywood panels, within the standard size of the 122 × 244 cm plates. The 18-mm (Russian) plywood has a high enough quality to secure the stability of the structure and to minimize postproduction treatment of the surfaces. Once the optimal balance is found, the mutual angles are analyzed in Grasshopper, and a red alert is programmed to warn when the angle would become too acute for the ubiquitous connection detail. The edges of the triangular plates are chamfered to accommodate the changing angles of the weighted normal perpendicular to each nodal point. The angle of the chamfer gradually changes from node to node as the weighted normal changes from point to point. The construction of the main component is to have three wooden 6 × 12 cm ribs perpendicularly screwed to the edges of one triangular piece of plywood, but shorter somewhat than the edge, halting well before the ribs would reach the nodal point. No forces are coming together in the nodes of the components of the Climbing Wall, and the gravity and torsion forces are led away from the nodal points toward the edges via the fields of the plywood plates. Plates and ribs are working together structurally, a truly innovative structural aspect of this assembled structure.

In building up the structure, the ribs are bolted together by two people only. Two persons erect the complete climbing wall, one by one mounting the components to form the complex shape of the towering free-form structure. To properly connect the components, the three sides of the plywood plates are CNC milled at gradually changing angles along the length of the edges, adapting to the gradually changing weighted normals of the nodes. The logic is as follows: in a double curved surface, the reference line perpendicular to the surface, the weighted normal, is changing from one point in the point cloud of reference points to the next. The edges of the triangular plywood panels thus need to adjust from one angle to the other. The data generated by the Grasshopper script are directly fed into the five-axis CNC milling router table, resulting in two straight-edged lines that are just not parallel, one at the exterior of the panel and one at the interior side of the panel. The structural beam, which is a non-chamfered rectangular beam that is shortened towards the three points of the triangle to avoid clashing in the nodes, follows the inner edge line, as is the other rectangular beam at the neighboring triangular panel. The structural beams are provided with prefabricated holes in exactly the right place for bolting the components together in the assembly. The plywood panels are provided with a pattern of holes to fix the climbing grips. The distance between the holes depends parametrically on the shapes and dimensions of each panel. The grips may be placed in any of these holes, and their position can be easily changed to change the challenge of the climbing trajectory. The triangular flat panels plus the beams, including the pre-drilled array of holes, form the base component. Each component is different; each component, and also its constituent subcomponents, has a unique identity in the design script and the robotic execution. In the process of building up the complex shape of the Climbing Wall, no *ad hoc* adjustments are needed; it is a simple assembly kit of unique components. The Climbing Wall system is competitive in many ways: competitive in its versatility for the realization of complex challenging shapes, and competitive with its price tag, its painstaking precision, its structural robustness, and its simplicity of assembly. The Climbing Wall is complexity made simple, complexity based on simple rules, without exceptions to the rule.

Body Chair

The next step I took in the development of the component is to open up the parametric design process to end-users. For the Climbing Wall, we have written a user-friendly app, which enables the climbers to design the base geometry themselves. Many smaller climbing walls have been executed using this user application. The app users can tweak the geometry in a way that is easy to understand, pushing and pulling the vertexes of the points cloud, whereby the number of rows and columns can be chosen freely. After having tweaked the 3D model, a mesh is generated. The 3D model is sent back to ONL and imported in a proprietary Grasshopper script, which generates the data that are necessary for the CNC production. This process is to a high degree automated and functions as a prototype for larger structures. More and more, users will become co-designers and use architect-designed applications to configure their own built structures. The Climbing Wall app paves the way for further prosumer-type applications.

A similar procedure is followed for the design and production of the Body Chair. The basic design concept of the Body Chair is to be constructed as one unibody – hence its name – described by 56 triangular components. Usually, chairs are built up from separate elements for the support structure and the seating surfaces. I wanted to have a chair that integrates the support structure and the seating surfaces in a single ubiquitous component-based system. I wanted to prove that what works at the medium scale of a climbing wall also works on the smaller scale of furniture. I wanted to prove that the parametric design to robotic production method is fully scalable. Once that single component with the connections to neighboring components is developed in a parametric model, half of the work is done. The other half is writing the script that exports the production data to the CNC machines. To connect the design script with the execution script, one needs to know exactly what data the CNC machine reads.

In the parametric design model of the Body Chair, the mutual relations between the main component and the subcomponents are dynamically connected. When the shape of the main component changes, the related subcomponents change with it. To avoid clashes in the points where the triangles theoretically would meet, like clashes caused by the thickness of the aluminum, we have rounded off the corners. Rounding off the corners where the corners of triangles would meet means nothing less than an innovation in itself. Avoiding clashes as a rule is always better than solving clashes by exceptions. The radius of the fillet of the rounded-off corners varies with the degree of the angle of the triangular component. The sharper the angle, the smaller the radius. The visual effect of the open nodes leads to irregular star-shaped open corners, which is esthetically pleasing and gives the Body Chair its open transparent character. The structural skin and the connection detail secure the decorative quality of the system. In the design of the Body Chair, the decorative quality is not an added value but an intrinsic one. Two triangles are combined in one component that is folded over the edge between the two triangles. Further folds are applied to the edges of the laser-cut components to facilitate the connections between the components, connected by bolts and rivets. By combining two triangles into one folded one, the number of assembled components and the number of rivets for the connections are reduced from 56 to 28. The 28 folded constituent components are parametrically linked to a central spine. The polyline of the spine is the driver for the shape of the body of the chair. The 3D grasshopper model is programmed to adjust the position of the vertexes on the polyline aka the spine to the dimensions of the legs, the body, and the head of the client; the chair as whole changes with it to build that unique configuration for that unique customer. The components come in a number of pre-set variations: variations in the cut-out patterns and variations in the finishing of the pads. The simplest cut-out pattern is a large opening in the middle of the triangle, with an offset of 3 cm from the edge to form the structural frame. The probably most chosen pattern is the decorative subdivision of the base triangle in three sub-triangles along the lines of the bisector meeting in the center of gravity of the base triangle. A third more elaborate filigree pattern divides the base triangle into 9–18 smaller triangles, whereby the size of the remaining edges inside the main triangle is reduced to 2 cm. The finishing of the seating surfaces of the components that take the shape of the cut-out of the base triangles is also a pre-set variation. The customer can choose between a hard inlay of natural wood, a soft inlay that is clad in leather, or a soft inlay pad finished with a pure wool fabric from Ilona's Jacquard woven Omniverse series of tapestries. The last option combines specifically well with the filigree patterns of the components at the sides, the front, and the back.

A user-friendly Body Chair app lets the customers tune the distance between the data points according to the length of their bodies and limbs. And, by tweaking the trajectory of the spine, the same parametric model transforms into either an active chair position, a relaxing chair that is leaning a bit more backward, or an almost horizontal lounge chair. The trajectory of the spine, the angle of the folds, the holes for the rivets, the pattern in the interior of the triangle, the seating pads, etc., every possible aspect of the design is an acting part of the parametric model, and thus an acting part in defining the main component of the Body Chair. Changing the position of one or more vertexes means that all related data are changing with it. This generic parametric feature forms the basis for the web-based user application that we have developed for the personalized version of the Body Chair, as to fit exactly to one's own body. In the app, the potential buyer enters the dimensions of his/her lower leg, upper leg, torso, and head into the central spine of the Body Chair. Using the sliders, one adjusts the activity factor of the central spine of the chair. The 3D model of the chair in the web-based viewer changes with the chosen values in real time. Further options for the customer are the pattern of the main triangular frame, the pattern of the fabric, and the choice of inlay: wood, leather, or fabric. We gave the web-based venture to bring the Body Chair to market under the name of BYYU, pronounced as "By You," a niche daughter-branch of Hieselaar BV, who developed the file-to-factory system together with us and generously invested in the production of the prototypes. The

Body Chair is the first piece of furniture of its kind, offering an endless series of unique customizable chairs, co-designed by the customer. Not a single chair will be the same; yet, it originates from the same parametric model and follows the same file-to-factory design-to-production procedure. The constituent components inside the Climbing Wall app and the Body Chair app I consider interactive components, components that are open to interaction by third parties, in this case only in the design phase.

Ecoustic

The development of the Body Chair to the Ecoustic acoustic system meant another step forward in the development of parametrically designed and produced products. While the Body Chair is a standalone object, the Ecoustic ceiling adapts to an existing environment. Ecoustic is an all-inclusive prefabricated installation kit of a free-floating swarm of panels; it fits in any environment of whatever geometry. The parametric system adapts to the contours of the space it is installed. The Ecoustic system applies to both ceilings and walls. I have questioned the very existence of the concept of the suspended ceiling, and I must admit that the same critique is to some extent also applicable to the Ecoustic ceiling system. However, the Ecoustic is an open system, with open space between the components, with open corners as in the Body Chair, and integrated the (LED) lights. The Ecoustic design has some innovative aspects that make it generally applicable to any family of building parts that connect to neighboring families of parts of a different nature.

1) Ecoustic spatially adapts to the contours of the space,
2) Ecoustic can be shaped freely in three dimensions apart from the boundaries of the space,
3) Ecoustic applies to both ceilings and walls,
4) the open structure of Ecoustic lets the air circulate freely and reach the concrete main structures to function as a climatic stabilizer and energy saver,
5) Ecoustic is an acoustic ceiling with integrated light fixtures,

6) the Ecoustic system uses the hanging system to fix the positions of the free-floating swarm of panels, simply by connecting the uniquely dimensioned edges of the pyramid to the hinges, thus forming a rigid tetrahedral frame, and

7) Ecoustic synchronizes structure and skin.

Ecoustic is an all-inclusive solution for spaces of any shape with long reverberation times that need acoustic treatment to become livable. The all-inclusiveness of Ecoustic comprises every aspect of design, production, hanging system, packaging, transport, and installation. Together with the client's ceiling expert, we went through every aspect of the design to the production to the assembly, in the most painstaking detail. As we find over and over again when developing a project-specific parametric system, the success of a component-based system lies in the connection between the components. In addition to the mutual connections, also the connections to the concrete load-bearing structure are solved parametrically. Due to the limitations of the available 3D milling machines to cut the acoustic 5-cm-thick panel in shape and the fragility of the base acoustic mineral wool, we could not use the mineral wool itself to form the connections. We had to find a separate connective element that adapts to the changing angles between the rounded-off triangular components. Instead of the unique fixed connectors, we chose to go for hinges, custom designed yet all of them the same. Having hinges between the components, the structure is unstable and needs to be stabilized by a three-dimensional hanging system. The invention specific to Ecoustic is to use the hanging system to fix the positions of the free-floating swarm of panels, simply by connecting the uniquely dimensioned edges of the pyramid to the hinges, thus forming a rigid tetrahedral frame. Again, Ecoustic is an example of synchronization of structure and skin, whereby the acoustic panels form the breathing skin and the pyramidal frames the structure that adapts to the existing environment. For the optimization of packaging and transport, the subcomponents are not pre-assembled as is in the Climbing Wall system. The three-dimensional topology of the swarming acoustic panels self-forms, simply by mounting the subcomponents together to form the featured inclusive components.

Until today, only one single small prototype has been installed, in the Michelin-star-rated restaurant Zarzo in Eindhoven. Although the prototypes and the installation at Zarzo have been successful in technical and esthetic aspects, unfortunately, the new management of the ceiling company has put the development on hold until further notice. The official reasoning is that they will keep focusing on the mass production of standard ceilings, instead of exploring new markets, thereby proving my point concerning the traditional building industry.

Web of North-Holland/iWEB

The characteristic building components of larger projects follow by and large the same parametric design to robotic production logic as the smaller ones. The Web of North-Holland (2002), five years later reassembled into the iWEB in front of the Faculty of Architecture at the TU Delft (2007), hosting Hyperbody's protoSPACE 2.0 lab, is a sculpture and a pavilion in one. Each one of the realized projects has a project-specific definition of the base component, its unique approach to the connections between the components, its choice of materials, and its spatial qualities. Although the design method is in principle fully scalable, the nature of the detail changes with the scale of the project.

After the closure of the Floriade World Flower Expo in the fall of 2002, the steel and skin structure of the Web of North-Holland pavilion was taken apart in its elementary subcomponents; this was anticipated in the design, the connections are bolted (structure) or riveted (skin). The Web of North-Holland is designed as an open summer pavilion, without any form of insulation. I proposed to the then dean of the Faculty of Architecture emeritus professor Hans Beunderman to re-use the Web of N-H and position it in front of the Faculty of Architecture and the Built Environment at the TU Delft to host Hyperbody's interactive architecture lab called protoSPACE 2.0. ProtoSPACE 2.0 is the working space for master design studios to hands-on build interactive prototypes, an interactive virtually augmented environment for immersive design sessions. Beunderman was supportive of the idea and created a budget for repurposing the iWEB, reassembling it, and preparing a budget for state-of-the-art equipment, computers, five beamers, and five double-sided projection screens. The components that were taken apart after the Floriade were stored for years in the backyard of Meijers Staalbouw's factory, awaiting the decision to be redeveloped into its second life as the iWEB protoSPACE 2.0. Eventually, in 2007, the funds were awarded, and the pavilion remastered.

The iWEB is the second-life version of the Web of North-Holland, featuring some improvements in the connections between the structure and the skin. The iWEB had to function all year round, meaning that an insulating skin had to be added, and fixed to the interior side of the steel structure. The interior skin is synchronized with the steel structure. The design challenge was how to insulate a structure that was not designed to be insulated in the first place. We chose to work with Polyned, which company specializes in inflatable tent structures. Taking the data from the 3D model of the existing structure a red-colored PVC skin is inflated toward the inside of the diagrid steel structure. The color red is chosen to evoke a similar red glow seen from the outside as in its first life when illuminated at night. The PVC skin is CNC cut and stitched together, placed inside the steel frame and then inflated. By inflation, the PVC interior skin is pushed onto the triangular grid pattern of the steel structure, which creates the embossed surface with the corresponding triangular pattern. This PVC skin functions as a backing for the polyurethane foam sprayed upon it from the inside, in three layers of approximately 3 cm after expansion of the foam. On top of the polyurethane foam, a strong layer of a bright red polyurea coating is applied to create a robust surface to sit and walk on. The sloping sides are used to watch the projections from the back of the projection screens. The triangular mesh of the steel structure remains visible as an embossed relief in the interior skin. The cantilevers of the former Web of North-Holland are used as an interior seating arrangement without the need to add separate seats for public events in the 150-m^2 flat floor collaborative working space. A series of small ventilators with embedded electric heating is built in for cross-ventilation, with integrated power LEDs for ambient up-lighting of the space.

To become the iWEB, the original red steel structure of the Web of North-Holland was sandblasted and refurbished with a silver-gray coating. The twisted steel plates, the connection plates, and the Hylite aluminum skin panels are re-used; only a few skin panels are reproduced due to some small scratches acquired during the disassembly. Hylite aluminum is an almost magically strong material. The 2-mm composite is built up from a polypropylene (PP) core with ultrathin layers of 0.2 mm aluminum on both sides. Hylite has a strong flexural rigidity against buckling; when you hit it really hard, it flexes back without showing damage from the impact. Hylite combines strength with elasticity. The iWEB is the first application of Hylite on the scale of architecture. Normally, it is used for aluminum covers of reports and manuals. An innovative new detail, suggested by Henk Meijers of Meijers Staalbouw, was developed for the connections between the steel structure and the cold-distorted Hylite skin panels. Short tubular aluminum is mounted on the exterior side of the steel members; its roundness allows it to adjust to the different angles of the skin to the steel structure. To indicate where the tubular subcomponents are positioned, an imaginary circle is projected on the steel structure perpendicular to the weighted normals of the vertexes of the triangulated web. The tubular connectors are chamfered at both ends for both structural and esthetic reasons. The circle of six shiny connectors around each node has a strong decorative appeal. Technically and esthetically quite the iWEB structure to skin detail is an improvement to the initial Web of N-H detail.

What must be considered the main characteristic components of the iWEB by which the body of the iWEB is assembled? The twisted steel plates, the cold-formed Hylite skin panels, or the combination of both including the connective detail? Since the steel-to-skin structure is one well-integrated system, the question is where one component ends and where the next starts. The Hylite aluminum twisted skin plate, the three over their diagonal folded steel plates, and the tubular connectors between the steel and the aluminum represent the subcomponents. One steel subcomponent is shared by two main components. Likewise, each steel plate relates to the two adjacent aluminum panels. The steel plates are twisted over the diagonal to follow the double-curved geometry of the skin. The different angles of the weighted normals determine the twist of the steel subcomponents. The curvature of the surface model informs the curvature of the outer edge of the steel plates, while the interior edges of the steel plates follow a straight line. The steel subcomponents are assembled by straightforward rectangular steel connecting plates, whereas the holes for the bolts are predefined according to the algorithms of the steel manufacturer Meijers Staalbouw. Meijers Staalbouw's algorithms are informed by our output data, according to the direct machine-to-machine file-to-factory process. The algorithms from the side of Meijers Staalbouw are informed by calculations of the main structure by engineer D3BN (now RoyalHaskoningDHV). D3BN calculated the thickness of the steel plates, and the number and size of the bolts, using the then experimental finite element method software Diana. The position of the steel plates and the connectors informs the cold forming of the aluminum skin. As a rule, two adjacent triangular components share the same folded steel plate and tubular connectors. However, in the Climbing Wall system, the connecting wooden beams are split into two half beams, and the connecting steel plate of the iWEB is a shared subcomponent, an integral part of the two neighboring components. The subcomponents of the iWEB are assembled onsite, whereas subcomponents of the Climbing Wall are assembled in the factory to form the integral main component. Organized around the nodes of the mesh, the main component is a virtual component that consists of six one-thirds of the Hylite panels, and six one-halves of the trusses, including the six connectors between the steel and the aluminum.

Burning Down the Faculty

Tragically, protoSPACE 2.0 has functioned for less than a year as our Hyperbody protoSPACE lab before it lost its innocent young second life due to a raging fire in the main faculty building. Out of the blue, back in 2008, the unimaginable happened.

The robust brutalist concrete structure of the <u>Faculty of</u> <u>Architecture</u> designed by Broek and Bakema burnt down. Even while the fire was raging through the building, no one expected it to collapse completely, while the brutalist design supposedly was built strongly. The tragedy that unfolded is a perfect example of chaos theory. A malfunctioning coffee machine triggered the devastating fire. Only a specific rare sequence of events leads to such an escalation of failures. Early in the morning of the day of the fire, a leakage in the sprinkler installation was discovered. To repair the faulty connection, the water supply for the fire risers had been shut off. But, that same leakage in the sprinkler system caused a short circuit in the coffee machine. Normally, after such a minor fire is detected, it is easy to extinguish the source of the fire. As a confluence of circumstances, when the firefighters came, there was no water in the fire ducts, no water came into the fire hoses, and they could not extinguish the fire. Then the fire started to rage across one floor, destroying thousands of valuable books collected by the faculty's staff, notably the precious collection of books of architecture historians. The cabinets in the faculty building are made of wood, which substantially helped the progress of the fire. Eventually, having no water available inside the building, the fire brigade abandoned the building and started to fight the fire from the outside, only to find out that they could not properly reach the fire since the large water basin in front of that side of the faculty building stood in the way. At a loss, the firemen, students, and staff could only stand by and watch the Faculty of Architecture burning down completely, floor by floor, until the concrete floors collapsed. On Dutch television, renowned building experts said, when the fire was raging wild, that the robust brutalist building certainly would withstand the fire, but, just as in 9/11, it was the connections between the floor components and the load-bearing walls that caused the collapse. The steel that holds the parts together melted down, causing the structure to become unstable and eventually collapse.

For the iWEB, which was not directly damaged by the fire, the story did not end there. Despite having survived the firestorm, the iWEB protoSPACE Lab was doomed, marking the most tragic episode of my professional life. The Faculty of Architecture – and therewith the iWEB

– was cut off from electricity, water, and data supply by the fire, causing the double-layered inflated roof of the iWEB to collapse and eventually break under the weight of the collected rain, and causing serious damage to the costly protoSPACE equipment inside. By law, the fire brigade was given control of the site, and they did not allow us to enter the site to place a generator, which needed to deliver the electricity for re-inflating the iWEB roof, thus saving the interior. Awaiting the decision, someone of our Hyperbody staff slipped through the fences to place a generator to save the iWEB and protoSPACE before it was too late, while heavy rains were expected. The authorities did not take this lightly; we were marked as terrorists as the then-new dean Wytze Patijn told me. He told us that he could not do anything to save the iWEB. But at the same time, he, fortunately, convinced the fire brigade to save the valuable chair collection from the basement of the faculty building, which was a much riskier act. I still have mixed feelings when looking at the image of a group of officials at the entrance of the collapsed faculty, including the then Dutch Minister of Education Ronald Plasterk wearing a summer straw hat.

Losing the iWEB feels like revenge from the traditionalists. After the fire, the iWEB was left unused for years. A new destination had to be found for the iWEB, the world's first parametrically designed and robotically produced pavilion. But I did not find an investor or end-user and got no support from the subsequent deans either. No one wanted to burn their fingers to write a recruiting letter of support to save it from destruction. The iWEB was eventually removed from TU Delft's real estate portfolio and doomed for destruction. It would have been possible to turn the iWEB into a self-supporting venue and become part of the Green Campus that was projected on the site of the burnt-down faculty building. But the iWEB and the Green Campus do not speak the same language. In none of the sustainability verification programs, BREEAM or LEED, one gets points for being lean and mean for doing less of the same, not for something radically different, not for the cost-effective and environment-friendly parametric design to robotic production methods. Nonstandard architecture does not fit in the sustainability frame. For example, one gets points for using a more environment-friendly gutter, but no points for leaving out the gutter completely, as we typically do in many designs, including the iWEB. One gets points for optimizing traditional ways of building but not for taking a radical new approach. Eventually, the iWEB was brutally demolished by big raptors taking aggressive bites out of the substance of the iWEB. The only positive takeaway from the brutal destruction is that the structure of the iWEB turned out to be so strong that when the raptor was grabbing one steel plate without removing the bolts, the rest of the structure remained perfectly stable. Again, the bolts of the connecting plates turned out to be the weakest point – the revenge of the diagrid.

ProtoSPACE 4.0

After a failed international competition for the new faculty building, TU Delft Campus and Real Estate decided to upgrade an old, abandoned university building instead, officially to save costs and time. After years of nomadic operation, hosted by IO, the Faculty of Industrial Design, we finally could move the protoSPACE equipment to BK City. Hyperbody was assigned a 150-m^2 rectangular space with four columns in the middle to use as the protoSPACE 3.0 lab, and two small rooms, one for the office and one for the KUKA robot arm. On the floor of protoSPACE, we installed an interactive floor. The interaction with the sensor-equipped floor communicates with the five projectors to build a somewhat toned-down version of iWEB's protoSPACE 2.0 immersive design environment. Hyperbody's high-potential guest researchers Marco Verde and Mark David Hosale, in collaboration with our regular Hyperbody staff, signed for the design of the interactive floor, following the principles of parametric component-based design. The floor panels are equipped with proximity sensors and pressure sensors to detect the presence and weight of people stepping onto the tiles. The tiles communicate with their nearest neighbors to propagate the acquired information, as birds communicate in their swarms. The local tiles are connected to form a global network, which allows for top-down instructions to the tiles and hence to the ambient environment. The data produced by the tiles are collected and used as

285

parameters to inform immersive design concepts. We once worked with the Dutch-Hungarian choreographer <u>Krisztina de Châtel</u>. Dancers are trained to change weight and are thus able to precisely play with the pressure sensors of the tiles. Among ongoing research experiments, we also used the floor tiles to interact with my interactive lectures.

 Supported by a generous budget to produce one-to-one prototypes, we took on the task to design a new <u>protoSPACE 4.0</u> Lab, to be placed in front of BK City. To act more efficiently and result-oriented, we decided to run the protoSPACE 4.0 project in close cooperation between the group of Hyperbody MSc1 master students and the ONL praxis. The MSc2 student group of the following semester assisted in the robotic fabrication of 1:1 components. I asked senior ONL architect Gijs Joosen to join me in supervising the design process and tutoring the students. We wrote a brief that organizes the students into a team of experts, each expert representing a specific task in the design process. The design team of the protoSPACE 4.0 lab consisted of

1) the concept designer and process designer (me),
2) the structure designer,
3) the skin designer,
4) the material designer,
5) the climate designer,
6) the interactivity designer, and
7) the styling designer. In the course of the master projects, I introduced Bodycheck sessions to integrate and update the progress of the expert groups.

Students would swap expertise during the semester to become part of one of the other expert groups, building upon what students before them already developed. The basis for the protoSPACE 4.0 design is the ubiquitous all-inclusive parametric building block <u>protoCELL</u> as suggested in my initial motivational sketch. The floor, the walls, and the roof, integrated into one single parametric component, describe the complete building and integrate the diverse functions that a floor, a wall, and a roof need to have. The initial shape of protoSPACE 4.0 is a symmetrical forward stretched hexagonal volume, leaning down toward one side to form the flat floor working space, while the part that juts upward forms the stepped comfortable seating area, also used as a small auditorium from where to watch performances. Working and observing are the two socially entangled components of interactive learning. Emphasizing the shape of the pavilion as a whole, the basic shape of the main constituent component is the large stretched hexagonal building block protoCELL. The hexagonal mesh is wrapped around the volume, thereby creating one design component that covers the floor, the wall, and the roof. Due to the overall shape of the pavilion, each component is unique in its dimensions and local performance.

The material of the protoCELL building component is polystyrene foam, cut into their unique shapes by hotwire cutting robotic arms. Through the mediation of Hyperbody Ph.D. candidate Jelle Feringa, we acquired in 2010 a second-hand ABB robot arm, and since we had no place for the 2.5-m high and 1000-kg heavy robot at protoSPACE, we made a deal with Heiko Dragstra of the firm Scorpio in Capelle a/d IJssel to temporarily install the robot there. Thanks to our joint Hyperbody-ONL-Scorpio efforts, we produced 20 unique large all-inclusive building components. Even by today's standards, it is probably the most radical, integrative, and innovative design for a parametric building component ever made. The protoSPACE 4.0 component protoCELL intends to build the foundation for a new fully customized building industry. The new parametric design to robotic production factory produces a series of unique components on command; the new building site assembles the prefabricated components by connecting them from the outer shell and the interior divisions, including the floors and the walls. ProtoSPACE 4.0 is Folded Volume reimagined on the scale of architecture. Each component of protoSPACE 4.0 has a different shape but follows without exception the same parametric design-to-CNC-production procedure and is hence not more expensive than a series of the same components would be. Polystyrene is a cheap material for structural mass, in the order of magnitude of 100 euros per m³. Polystyrene is a highly insulating material; the 30-cm thickness of the components ensures the near zero-energy performance of protoSPACE 4.0. The large 2.5-m-long components are coated with a strong gray industrial polyurea coating, both on the inside and the outside, easy to clean, pleasant to touch, and resistant to dents and scratches. The industrial coating upgrades the otherwise inert polystyrene into a strong structurally well-performing composite skin panel. The body of foam takes the compression forces and the moments, while the skin takes the tension forces, as in steel sandwich panels. The foam functions both as an insulator and as a spacer for the exterior and interior skin. The protoSPACE 4.0 stressed-skin building block integrates many different performance requirements, integrating stylistic, climatic, lighting, structural, and ergonomic aspects. As in car styling, the feature lines travel across the different components, fading in and fading out along its trajectory. The curved feature lines are applied to the volume of the educational pavilion as a whole, not confined to one or two individual components.

When having a high insulation index, ventilation becomes an important issue, especially when hosting 50–100 metabolizing people inside. Each individual protoCELL building component is assigned to perform specifically in terms of climate control, level of openness, structural capacity, and level of interactivity. Besides the fully solid components, there are components with a large opening, either for embedding a window, a ventilator module, or a small local heat pump module to control the interior climatic conditions. For the floor, interactive components have been foreseen that form a table, a chair, only to pop up on demand, when needed, and as needed. For the international conference GSM III (2016), the third of a series of four GSM conferences, I asked Hyperbody MSc2 students to revamp the protoSPACE 4.0 mock-up into the GSM III stage. The 20 protoSPACE 4.0 building components were reinstalled, and one of them was transformed into an interactive lectern that would only pop up when a lecturer is coming close to the GSM III podium.

Lars Spuybroek once grumbled that I design buildings like cars without wheels. Although meant as a critique, I take it as a compliment, since I intentionally design buildings with an explicit forward drive, buildings with side entrances like cars have, rather than frontal entrances like palaces. My designs are by and large symmetrical with a dominant vector, I referred to them as vectorial bodies as is the title of my Archis essay of the same name, published in 1999, and republished in my book with collected essays <u>Architecture Goes Wild</u> (2002). Buildings that want to go places, virtually. Buildings where you step sideways into, as in cars, ships, and airplanes. Buildings that have a small footprint and only touch the earth lightly. Buildings that are spaceships that make a soft landing on Earth. Buildings that are designed in weightless digital space, only to process the gravity forces when being calculated to stand on the surface of Earth. Buildings that are sculptures as well as functional bodies, crystallizations from an until-then-unknown personal universe.

The connections between the large yet super lightweight components of protoSPACE 4.0 are spatially anticipated in the shape of the components. A peripheral zone is CNC milled out of the hotwire-shaped PS blocks, all around the component, both at the interior side and at the outside. The recessed edges are reinforced with wooden laths, glued to the polystyrene. The laths are used to host the metal connecting elements. To spread the forces evenly, every 10 cm, there is a hole in the laths to receive the connectors. The building blocks are dry assembled, to be prepared for future disassembly of the components. Just as what happened to the iWEB, the protoSPACE 4.0 prototype has been rebuilt several times, in its last performance functioning as the acoustic lecture stand for the speakers at the GSM III conference in November 2016. The recessed areas are also designed to pull the wires for electricity and the data cables and to offer space for water ducts for the climate control devices. In its performance as the GSM III stage, the recessed areas are used for the programmable lights, interactively responding to the public seating on the MVRDV-designed orange grandstand. The protoSPACE 4.0 building components are prototypes for building complete homes and even for larger multi-story structures. When it comes to multi-story buildings, the PS in the core must be replaced by a more dense and inflammable type of foam to build safely, to get the building approval. In 2007, we proposed a similar component for the facades of the Fside Housing project, as mentioned earlier, but it was rejected by the fire defense experts. Not that the PS would catch fire but the PS would melt down, then lose its structural integrity, and potentially let the fire spread to the floors above. The protoCELLs for protoSPACE 4.0 proudly stand as a prototype for parametrically designed, uniquely shaped components that integrate structural, tectonic, material, stylistic, and climatic aspects into one integrated featured component, interacting with neighboring components in the virtual design phase and well connected to its neighbors during its life in the physical world.

Λ2 Cockpit

Most of the projects have a specific technical and esthetic solution for the main component. Typically, a specific solution is not repeated twice or more. For the A2 Cockpit and the Acoustic Barrier, the main component is centered around the parametric nodal design and includes the subcomponents radiating from that particular node. As for the Cockpit, the node itself *is* the component, orchestrating the double-curved structural skin facing the A2 highway. As for the Acoustic Barrier, the twin node and the tube connecting the front node and the node at the back *is* the main component, organizing the 1.6-km-long structure as a whole. Similar to the nodal points of the iWEB, the reference line perpendicular to the triangulated mesh of the surface, aka the weighted normal of the surface, governs the geometry definition of the subcomponents of the Cockpit. The nodal points are created by populating the double-curved surface with hundreds of vertexes, driven by a master curve that has been tweaked to optimize the distribution of the points. The component design for the Acoustic Barrier follows a slightly different strategy. Almost a thousand parallel lines are projected onto the double-curved surface of the Acoustic Barrier, pinpointing the same amount of reference points. Subsequently, the weighted normals are constructed by scripting from this set of thousands of projected points. In the Cockpit and the Acoustic Barrier, the complexity is compacted in the nodes themselves. Describing the nodes and the relations between the neighboring nodes effectively covers the design. In the nodal design of the Cockpit, the centerline of the 30cm central steel rod coincides with the weighted normal. This solid round iron bar is home to the connection plates that are welded in the exact right positions to receive the ends of the robust tubular steel trusses measuring 33 cm in diameter. Calculations showed that 30.5-cm tubes would have sufficed, but at the request of Meijers Staalbouw, a stack of 33.3-cm tubes could be purchased for a better price. We agreed to the somewhat larger diameter since the overall visual impact was meant to be robust anyway. In contrast to the almost commonly accepted ideology of the modernist, where less is more, I typically choose more over less, more robust over as thin as possible. The practice of building means negotiating between alternatives, and winning budgetary space at one point means that more money can be spent elsewhere.

The point cloud of 300 reference points at the A2 side of the Cockpit is defined by increasing mutual distance toward the middle, positioned along a set of horizontally stretched three-dimensional curves, two of which are direct extensions from the gradually disappearing and reappearing horizontal fold-lines of the Acoustic Barrier. The nodes and the tubular connecting trusses form the physical manifestation of the diagrid network structure of the Cockpit. The dimensions of the skin and those of the load-bearing structure are synchronized to make a secondary supporting structure redundant. The distance from node to node varies from 3 to 6 m; 6 m in the middle part, and 3 m at the edges to emphasize "pumping up the volume." The upper limit of 6 m is dictated by the maximum dimensions of readily available standard glass plates. In the parametric ProEngineer model, the frequencies of the triangulation are optimized, set by the number of points projected from a master curve outside the model onto the curves on the surface of the Cockpit 3D model, to find the most efficient balance between the dimensions of the glass and the frequency of the steel structure in the pumped-up middle section.

ONL was back then and probably still is one of the few architectural practices that use such advanced parametric design software as ProEngineer, a piece of software that is typically used by product designers. It reveals the affinity of the design systemic to that of product design and more specifically car design. Similar arguments are used to come to structural and esthetic solutions. In the automotive industry, manufacturers and designers work much closer together than in the building industry. The synchronization between the structure and the skin of the Cockpit is the result of fruitful cooperation between the architect ONL, the glass producer Pilkington, and the steel manufacturer Meijers Staalbouw. Pilkington did build a 1:1 scale mock-up in their effort to understand the complex 3D logic behind the changing positions of glass in relation to the changing angles of the nodes. As long as the geometry has zero thickness as in the wireframe model, the relations are straightforward, but when the thickness of the subcomponents is taken into consideration, one begins to understand the complexity. The complexity is caused by the economic necessity to use flat building materials to describe double-curved surfaces. The translation of a double-curved complex geometry into physically flat components has been an absolute first for the facade builder Pilkington in Europe and one of the earliest examples worldwide as well. ONL took care of the engineering of the geometry, secured its precision, and took responsibility for the correctness of the data that were sent to the production machines. Executing the principal engineering of the geometry of the characteristic components in-house in the architect's office facilitates maximum optimization.

One needs a perfect understanding of how objects rotate in three-dimensional space to engineer a double-curved structural skin system, which is a difficult thing for the human brain as our orientation organs are designed to balance upright. It is challenging to find the logic of dimensioning the individual glass plates, being all different, albeit only changing some millimeters from plate to plate. With the help of the 3D parametric software, the complexity of the node is modeled, scripted, visualized, and fully understood. The building industry has to come to terms with the nonstandard; the adage of one size fits all is no longer valid, and the new adage is made to measure, meaning that one particular piece will only fit in one specific location, one component out of a massive series of unique components. The dimensions of each glass plate are unique, the length of each steel tube is unique, and all 300 nodes are unique; the components and the subcomponents are parametrically designed and robotically produced. Each subcomponent is assigned a unique number, stamped, or glued on it during the production process. Due to the double curvature of the surface, the three-dimensional positions of the weighted normals slightly change from nodal point to nodal point. This angular change has consequences for the connecting tubes and for the L-profiles to which the glass plates are attached, kept in place from the outside by exposed round stainless-steel placeholders. The gaps between the 34-mm-thick glass plates are filled with silicone, and smoothly attached from the outside, securing the water-tightness without the need for a second level of water defense as in traditional facade systems. This simple and effective detail has been the subject of much debate with the project manager representing the client, since it had not been done before, and hence not considered proven technology. Fortunately, Pilkington was confident to guarantee the solution, and no leaking has ever occurred, since the delivery of the project in 2005. The main component of the structural skin of the A2 Cockpit is the star-shaped node itself, while all complexity is concentrated in the node. One tubular steel truss is shared by two adjacent nodes, as in the iWEB. Each node shares six subcomponents in the field, and four along the edges of the surface. The subcomponents that are semi-automatically welded to the node are CNC cut to the right dimensions.

Special attention has the exact length of the steel tubes. Meijers Staalbouw had them calibrated after welding the connecting end plates because they have to fit exactly in the 3D diagrid. Calibration of the tubes is necessary since a diagrid system does not tolerate any tolerance. Traditionally a steel structure has an allowed tolerance of some centimeters. At the building site, they typically force the elements together, which often causes unwanted yet in practice acceptable tensions in the structure. Building *without* tolerance means a true revolution in the building industry, where the contractor typically states that "equal equals unequal" when the architect desires precision. But the inconvenient truth is that the 3D models of the architects in combination with the chosen methods of execution simply require tolerance. Equal equals unequal does not apply to nonstandard architecture. The running gag is that we have a zero-tolerance policy. Triangulated structures do not accept tolerance, while nothing would fit when it did. The (sub)components fit in only one position, and it has to fit exactly; otherwise, the deviation will proliferate throughout the mesh and compromise the whole system.

While the Climbing Wall's main component is the fully integrated triangular part itself, the Body Chair's two triangles combined in one folded quadrilateral, the main characteristic component of protoSPACE 4.0 pavilion the voluminous all-inclusive building block, the iWEB's and the Cockpit's featured component is the physical node itself with shared subcomponents circling the node. In the design of the iWEB and of the Body Chair, the complexity is taken away from the node. In the iWEB, the intelligence of the system inhabits the folded pieces of steel between the nodes, whereas the cold-formed Hylite aluminum skin panels follow the exterior curves of the steel plates, formed by fixing them to the three shared ones. In the Cockpit, the agency of the component resides in the complex node itself. Although the projects share many characteristics in their appearance, notably because of the apparent triangulation, each project must find its inherent logic. Each project emerges from a different set of design rules, a different set of expectations from the client, a different combination of materials and techniques, and different performative requirements. Each project is positioned in a different context and hence inevitably leads to a unique solution for the featured component. The final design of the leading component inevitably is the result of a project-specific sequence of decisions. Also, the human aspect should not be overlooked. Many projects are developed by different teams of collaborators not only externally but also internally within ONL and Hyperbody; each individual brings slightly different expertise to the table and hence contributes to different design solutions. There is not one single optimal solution for triangulated structures, new component details keep being invented for every new project.

Acoustic Barrier

The genesis of the Acoustic Barrier, in which the Cockpit is embedded, resulted in another solution for the design of the components. The almost thousand reference points are created by bombarding an equal number of parallel lines onto the double-curved and folded surface. Because the surface is not straight but alternately convex, concave, folding, and flattening out, none of the distances between the projected reference points are the same. The point cloud of reference points is the initial condition for scripting the nonstandard detail. Instead of modeling, it showed that scripting is the appropriate way to describe the numerous small differences between the dimensions and shapes of the constituent subcomponents. Scripting the dots to connect (clockwise) to form a mesh self-builds the base wireframe model, which is then used for further specifications of the subcomponents constituting the node detail. Horizontal line segments of varying length, due to the gradually changing geometry along its course, connect the dots on the A2 highway side to the corresponding dots on the other side, the side of the industrial area. The horizontal tubular steel profiles, all parallel to each other, form the basis for the main components of the Acoustic Barrier. Steel plates are welded to the round steel tubes following the angles that are retrieved from the wireframe model. The nodal pairs that are connected by tubular steel profiles are the main components of the Acoustic Barrier, looking like small missiles. These rocket-shaped components are sort of doubling equivalent of the Cockpit nodal component, whereas they cater to both sides of the barrier. The tubular steel profiles are then connected by L-shaped profiles, also diagonally to stabilize the structure. As the steel manufacturer Henk Meijers once mentioned, the acoustic barrier is in its essence a power pole on its side. Once he understood the design in that simplified way, he knew that he could offer the right price. Henk Meijers suddenly realized that it actually is a simple construction and that it is complex but not complicated. The tubular steel profiles, the connecting plates, and the connecting profiles are assigned a unique identity in the script. The same unique number is pressed into the steel during the process of CNC cutting. Each of the roughly 40,000 different steel parts fits only in one place.

The subcomponents that are close to each other, circling one missile component, are packed, transported, and eventually unpacked at the building site. A group of four laborers read the number and put the subcomponents together in the right order, guided by an A4 print-out of the assembly of that specific group of components. The structure self-builds, with no scaffolding needed. Per day, a team of four people puts together a stretch of 30 m' of the steel framework of the 7.20-m-high barrier. The 1.6-km stretch is effectively built in two months. After the assembly of the spatial frame, the glass plates are assembled. Again, each plate of glass is only slightly different, in a gradual transformation caused by the curvatures and folds. The glass plates are framed in rubber profiles and CNC-cut to length. The system with overlapping scales solves the difference in the behavior of the steel structure and the glass plates when heating up and cooling down. The rubber profiles around the glass plates slide over one another when the structure expands and shrinks. The esthetic bonus of the overlapping scales is that coming from one side driving the highway the acoustic wall looks different when coming from the opposite direction. The scales, the folding lines, and the convex sections morphing into the concave sections together contribute to the 60 seconds of architecture, experienced when passing by with the allowed speed of 100 km/hour.

Bálna Budapest

This chapter would not be complete without discussing the two largest realized projects, namely the Bálna Budapest and the Liwa tower in Abu Dhabi. Imre Márton, the project developer and our client for the Bálna Budapest, which was called the CET back then, came to visit our office in a former gym in Rotterdam. Márton traveled together with Iván Bojár, de editor of Octogon magazine, who had interviewed us twice before for the Octogon magazine. They checked out the A2 Cockpit and were impressed by its technical achievement, its iconic presence, and, above all, its cost-efficiency. Márton commissioned us for the open competition and tentatively coupled us with the established local architect Tima Zoltán of Közti, the later designer of the Liszt Ferenc international airport extension. With his own eyes, Márton had seen the advantageousness and beauty of the parametric design to robotic production of the A2 Cockpit and wanted the same system. I did not realize then what I know now that he only wanted the looks but not the integrity of the system. The Bálna project turned out to be a project with as many hurdles and challenges as one can imagine in a design assignment. The local architect presented his fee offer, which would have taken up 90% of the available designer's fee, and I had to split ways with the local architect in a friendly manner and decided to establish an independent ONL Hungary kft. Eventually, we submitted the building permit drawings and the calculations, including the work of our subcontractors, all packed in 40 boxes stuffed with drawings. And surprisingly, not least to the surprise of the project developer, the building permit was submitted within the strict time planning. The building permit, a joint effort of ONL Hungary, ONL in Rotterdam, and 20 or so subcontractors, received unanimous approval without serious amendments. ONL Hungary kft's role as the lead architect meant full responsibility for the work of the structural engineer, the climatic designer, the fire expert, the water defense expert, the construction physicist, the landscape designer, the traffic consultant, the heritage expert, the quantity surveyor, and some others who represented other minor aspects of the building complex. Much to my dismay, the project developer did not allow the prescription of the robotic production of the steel structure in terms of the building permit, although the detailed drawings did not leave any doubt; it was eventually described in more general terms. When it came to the execution phase, it became clear why we were not allowed to describe the efficiency of the parametric design to the production process. The project developer had already in mind whom and how to build the Cockpit system in the context of the Hungarian building industry. He wanted the steel structure to be welded on site, while according to him and his contractor partner, it would be cheaper. He simply did not want to consider the competitive cost-effective offer that was given by Meijers Staalbouw in a consortium with the Central Industries Group for the complete Cockpit skin-structure system. The component-based design thus was put in jeopardy, and as mentioned earlier, it led to a hugely underperforming detail of the nodal component, doomed to be the cause of water defense problems. Their choice for a welded steel structure and the faulty skin detail caused headaches for the Hungarian contracting parties, and predictably in the end more expensive than Meijers' all-inclusive offer. The competitive and performative power of the component and the assembly of components into a structural swarm of components were lost. The components were decomposed. Everyone and everything lost; the building lost the integrity of the design-to-production process, the contractor lost money and missed the opportunity to educate himself, and the project developer lost credibility and eventually sold the project after a series of lawsuits. Until today, the Bálna suffers from poor management by the subsequent owners, alternately the project developer, the city, and now owned by the state. As described in my "What's up Bálna?" blog on my website, only some bars and restaurants have come to life and, to some extent, also the 999-seat flat-floor event space, but the building as a whole is as of yet underused. Despite all of its anomalies, the Bálna has become an icon for the city of Budapest, which is well-appreciated by the citizens. The citizens are not bothered by the underlying problems, and to be fair, from a distance, the Bálna looks just great. Googling Bálna Budapest shows that many pictures are taken, around and inside. People love the iconic structure, especially for its successful marriage of the old and the new. Just recently, I found out that the young Hungarian visual artist Péter Mátyási made a painting featuring the Bálna with Mount Fuji in the background. I could not resist buying the painting.

Liwa Tower

The design in the building culture of the Middle East requires a new approach to the parametric design system and a new look at how to connect the design to the execution. Building in the Emirates is a different ball game. As in many countries, but more explicitly in the Middle East, 99% of all structures are *in situ* cast concrete and hollow concrete building blocks for the non-loadbearing walls. Villas are built like that, apartment buildings, office towers, hardly any exception. Already in the competition phase, we indicated that the Liwa tower is designed as a steel structure, which had not been done before in Abu Dhabi for tall buildings. The client Al Nasser Holdings is active in the offshore business and familiar with building platforms in steel. CEO

Mohamed Al Nasser was charmed by the steel structure that was visible from the inside; he liked the design and put pressure on the project developer Bob Brodie of <u>Northcroft Middle East</u>, who was responsible for managing the project, to choose our design as the winning entry. Brodie came over to visit us in the Netherlands to check out the Cockpit and was immediately convinced of building diagrid structures in steel. It remains a remarkable fact that in the Netherlands, not a single project developer wanted to burn their fingers on our nonstandard parametric to robotic architecture, and in one year, two foreign developers were readily convinced of the feasibility of the systemic approach and by the, at the same time, iconic presence and spatial quality of our designs. Brodie was happy with the visible detailing of the steel nodes; he knew that Al Nasser would like it and was charmed by the simultaneous grand spatial gesture and the all-inclusiveness of the parametric component.

But why were the Dutch project developers not happy with the Cockpit, and what was their problem? It remains a mystery to us why our active role in the Dutch scene came to a halt just after the completion of the Cockpit. Neither the Bálna Budapest nor the Liwa has been published in a Dutch architectural magazine, while they are featured in most international printed and online magazines. Like with the Bálna Budapest, for the realization of the Liwa project, we had to overcome many hurdles too. Brodie had difficulties finding a contractor for the project; the market overheated in 2005, and preferred contractors like the Dubai branch of the Dutch BAM were not interested as they had bigger projects in their portfolio to take care of. Eventually, Brodie found the local <u>Al Rostamani Pegel</u> willing to offer a bid. They found <u>Eversendai Engineering</u> <u>Qatar</u> with Eversandai's factory in Dubai to prefabricate the X-shaped steel diagrid components, spanning two floors at once. Based on ONL's sophisticated parametric guidelines, they built one floor in one week, thus completing the 20-story tower within half a year. The tower stands on top of a two-level podium and a 3.5-level underground parking garage, both of which took considerably longer to build. Originally, we had proposed to continue the steel structure down to the base of the parking garage, but that was a clear no-go from the side of the master planner RMJM, who insisted that the tower should stand on top of the podium. It took us all my persuasiveness to have the tower at least visually touch the ground by the stretched diagrid structure. The top 20 floors with the diagrid structure rest on a 1-m-thick concrete ring beam that transfers the forces to the concrete podium and the subterranean garage. The parametric synchronization between the structure and the skin of the Liwa tower is unique in its kind while skipping the necessity of additional corner details. All details, whether in the field or the fold, all around the tower, follow exactly the same formula. Only the parameters change from the symmetry axis in the middle part toward the corner. The corner is not a corner but one of the many parametric instances of the formula.

Whereas the virtual model identifies the single triangular skin panels and the singular steel tubes as the constituent subcomponents to form the main component, the main component as built slightly changed the theoretical logic. The assembly of parts is based on the larger X steel diagrid part that extends over two floors, while two triangular skin panels are combined into one almost square panel. The pragmatic Arabic logic falls a bit short of the integral Dutch approach but turned out to be considerably smarter than the tradition-driven choice to weld on-site by the Hungarian contractor of the Bálna. I can perfectly well live with the contractor's choice for the bigger X-shaped parts of the diagrid. What stands out is the satisfaction of having built an innovative tower in Abu Dhabi.

A generally accepted rule in the prefab building industry is to have as few possible parts that are as big as possible. This certainly is true for the mass production methods that started to be dominant in the sixties of the last century, but in our era, nonstandard architecture has set new rules for the economy of building. It is no longer bigness that counts but the full integration of functions into one component, which leads to a full synchronization of different types of subcomponents into one coherent building system. The synchronization of skin and structure demands a more fine-grained approach and demands a clear definition of a main component that is all-inclusive. The synchronization of structure and skin has as a consequence that the main structure becomes finer-grained than usual, while the skin panels typically become larger than usual. The two aspects of skin and structure merge into one coherent system in the nonstandard approach. One of the reasons that traditional builders think that our system is more expensive is because of the smaller mesh size of the main load-bearing structure. The advantage of the structure-skin synchronization lies in the fact that a secondary structure between the main structure and the facade is no longer needed. This is unheard of in traditional building methods, while there always is a secondary structure to span between the larger bays of the main structure. It is not difficult to imagine that leaving out one complete layer in the assembly of buildings is guaranteed to be cost-effective. We spend a bit more budget on the main structure but save probably twice that amount by skipping the secondary structure. It is not only the synchronization of structure and skin that makes the Liwa tower a cost-effective one. Also, the diagrid system takes credit for substantial savings. It is stated that the diagrid structure of Norman Foster's Hearst Tower in Manhattan uses only 80% of the amount of steel in comparison with traditional rectangular column and beam systems, which holds as well for our Liwa tower. The Liwa is known for being by far the strongest structure in the region with respect to torsion and gravity forces while using less material. Nothing beats a structural diagrid system, and all the more unlikely that the Liwa tower after a dozen years after its completion still stands out as a unique object in terms of parametric design to robotic production.

Only in those rare circumstances that are off the beaten path a direct link can be established from the main components in the design phase to the same main component in the execution phase. Successful projects in that aspect are the Climbing Walls, the Body Chair, the Ecoustic ceiling, and the protoSPACE 4.0 prototype, which projects share maximum simplicity that leads to maximum complexity, which I refer to as simply complex. The building industry is like a physically inert and mentally stuck supertanker, hard to change course, but not impossible. We have come a long way with the iWEB, the Cockpit, and the later larger buildings, the Bálna Budapest and the Liwa tower, but we are not there yet. The ultimate expression of the parametric to the robotic has yet to be realized. So far, none of the currently successful offices like ZHA, COOP, or Gehry have succeeded either. They typically rely on more traditional building methods. In their projects, the parametric to robotic is at best skin-deep, rarely structural, and certainly not in a full synchronization of structure and skin. The closest to our achievements is the Museum of the Future in Dubai designed by Shaun Killa and engineered by Happold Engineers. I have witnessed the design and building process from nearby. I visited Shaun Killa in his office and discussed the ins and outs of parametric design to robotic production with him. I have offered the selected contractor BAM to share my knowledge of complex nonstandard geometry. What I have seen that is eventually realized is pretty much an optimized steel structure, yet without the much-needed synchronized skin. The synchronization of structure and skin is engineered as a hybrid mix of the triangulation of the donut shape and the horizontal stratification. The skin panels are vertically stretched between the horizontals, which I find illogical since the intimate relation with the steel structure is lost. The conflicts in the system arise where no horizontal floors are possible, that is, at the lower part, the upper part, and the interior part of the donut shape, meaning that exceptions to the rule have been introduced, which would not have been necessary. The choice for the horizontals as the organizing system comprises the logic of the double curvature of the donut shape. Here, a conflict is sneaking into the process between the traditional approach of section and plan and the nonstandard approach that takes the integrity of the three-dimensional shape as a starting point. In the Cockpit and the Bálna, the nonstandard system is dominant; as a result, the horizontal floors meet the structural skin at random points. But do not let me be misunderstood; the Museum of the Future is an almost perfect masterpiece and a great achievement from a cultural point of view. From my personal designer's point of view, one crucial question remains concerning possible conflicts between the logic of a double-curved structure and the logic of a functional interior featuring horizontal floors. In my recent design for the Dutch Pavilion for the Expo World 2020, I responded to that self-imposed question by bringing structure, skin, floors, and walls together into one single synchronized system.

Do by Dutch

Explicitly not related to Qatar University, due to the then ongoing political blockade, I joined the Dutch team as the lead designer, to tender for the design of the Dutch Pavilion for the World Expo 2020 in Dubai. The self-imposed design task for the design of the Dutch Pavilion for the World Expo 2020 in Dubai, which was nicknamed Do By Dutch, pronounced as Dubai Dutch, was manifold:

1) to design a pavilion that is fully self-supporting in terms of food, energy, and water supply,
2) to design an experience for three intertwined trajectories, one showing the new developments in food production, one for energy production, and one for water production and treatment,
3) to design an open breathing structure that would need a minimum of cooling, and
4) to design a structure that is based on one single family of parametric components.

One moldable component for the surfaces where people walk on (formerly known as floors), one parametric variation for the surfaces people are leaning to (formerly known as balustrades), and one for the surfaces separating discrete spaces (formerly known as walls). Being self-supporting does not mean that the building is not connected to a shared network for energy, water, and food supply. It does mean that whatever is consumed and what is produced must be in an equivalent balance with its immediate neighbors. Being self-supportive should never be interpreted as isolating oneself from the world. On the contrary, being self-supportive means connecting oneself to the world in a fair balance with one's immediate neighbors. Isolating oneself would mean that exchange with immediate and faraway neighbors would come to a stalemate, while ubiquitous interconnectedness is key to building a culturally rich environment, the key to world-making. Connecting to immediate neighbors is a process of give and take. It means processing information of different kinds, data, food, energy, water, materials, memes, ideas, and emitting the processed information in a differentiated form, modified by the acting component's preferences.

The three interwoven trajectories have a separate entrance, to distribute people evenly in the pavilion and to avoid long queues. The trajectories lead up to the roof where the trajectories merge in an exemplary well-balanced environment. The three trajectories cross each other at intersection points halfway. The public may follow one particular trajectory or change paths and follow the other; any personal trajectory through the three-dimensional hive is possible. The three-dimensional layout of the pavilion is a complex spatial web of slopes, platforms, voids, connections, and experiences. From each platform, one has the choice to go in two directions (at the periphery) or three directions (in the heart of the building). Each slope is both ramp and stair, suited for small electric vehicles to drive up the elderly and physically challenged, and for pedestrians that have no problem using stairs. The slopes are divided into two zones, one for electric vehicles and one for pedestrians. In the spatial layout, the slopes have open voids to the right side and the left side, which adds to the transparency of the open pavilion, taking credit for surprising interior vistas and an intense spatial experience. Navigating the Do By Dutch pavilion, one enjoys views to the left, to the right, up and down, and diagonally through a series of voids. The three trajectories of food, energy, and water are laid out in such a way that they are passing through all platforms. The trajectories are given their thematic touch and feel, each in their characteristic color. The structure is open from all sides for natural ventilation; it offers a constant refreshing breeze for the visitors, much welcome even in the Dubai winters. The three-dimensional maze stimulates permanent turbulence of the air, as in a termite hill. The freshness of the air will be enhanced by spraying water droplets from the water production path.

The Do By Dutch Pavilion is one giant connection machine. Everything connects to everything through the spatial web of physical connections, and each building component connects to its neighboring components. In analogy, birds connect to neighboring birds in the swarm. The parametric one-building-one-detail design strategy guarantees the integrity of the connections. The parametric design is directly linked to the CNC production of the perforated steel components. Scripting architecture means that either everything fits or nothing fits, what becomes already clear in the early design phases, and maintains its integrity throughout the whole design-to-production process. From the initial parametric sketch, the fundamental components are defined to fit precisely, such that every unique component bearing its unique number is assembled in just that single unique position in that specific order and for that specific reason. The base component of Do by Dutch is a triangular sheet of steel with flanged edges, folded or welded to the triangular base, whichever proves to be the most effective. Simply by bolting the all-inclusive components together, the building will self-erect, and no scaffolding is needed. The Do by Dutch pavilion is a self-building building. In each stage of the assembly, the structure is perfectly rigid and safe to work upon or hang from. The assembly can be done by small teams of relatively low-skilled workers; two persons can assemble a component, one steering a small robotized aerial platform to lift the components into place, and the other to fix the bolts. The whole process of assembly could be robotized completely since the same procedure repeats all over, using varying parameters. Imagine having small mobile crawling robots that take over the work of humans, one to maneuver the component in place and one to fasten the bolts. The "one building one detail" method paves the way for a fully robotic production *and* assembly.

When one would want to robotize a traditional building method, one would need an army of different robots to do the job, while there are so many different elements in a traditional building and so many different ways of connecting the elements. Especially the wet connections that still prevail in traditional building methods are difficult if not impossible to automate. I am seriously looking forward to getting a commission for a building like the Do By Dutch, to deliver proof of the uncompromising parametric design to robotic building to robotic assembly process on a larger scale. Robotic production and assembly guaranteed will pay off in terms of budget, time, robustness, and esthetics. From component to component, transitions are foreseen from maximum open to fully closed. Where there is more material needed to transfer the gravity loads and dynamic wind forces, the components will be more closed and more robust. The material is only there when and as needed. Where more openness is required, typically in the components formerly known as balustrades, larger perforations are opted for. Where integration of exhibition items is required, there are inlays embedded in the components, similar to the soft seating pads in the Body Chair. Where climatic compartmentalization is required, the components are closed or glazed, enclosing a space. There the components are doubled to include the required insulation. The basic component is further specified where and as needed. Where a bigger platform is needed, the parametric model is adjusted to the desired dimensions. Where ducts and risers are needed, they follow the edges of the components, synchronized with the structural components, similar to how in the Cockpit building the sprinkler system follows the horizontally stretched tension profiles of the structural skin. The patterns of the perforations are parametrically linked to the shape of the components, as in the Climbing Wall. The patterns are resonating with but not copying traditional Arabic patterns. The structural performance of the components is calculated using a finite element method (FEM) by CIG Architecture. The outcome of the finite element analysis for the Do By Dutch pavilion underpinned CIG's competitive financial bid, and their price was within budget, which is quite exceptional for such a radical design.

One of the reasons that we did not win the tender that I entered together with event designer Live Legends was that one of the jury members criticized us for copying the French parametric designer architect Marc Fornes of Theverymany, who started business one decade after the completion of the iWEB, which still stands as the pioneering example of parametric design to CNC production. It seems to be the fate of being framed as a pioneer, which means an appraisal and a fatwa at the same time. Fornes is scripting beautiful delicate structures based on his structural stripes, produced from flat material, and assembled to form three-dimensional structures. Fornes is rightfully respected for that. But here is the thing – no one is copying no one here. Fornes' design and construction method and the scale of the projects are entirely different from ours, whereas Fornes generally works with long CNC cut stripe components, which by their connections form complex concave and convex lightweight structures, making up medium-scale indoor and outdoor installations. Unfortunately, that jury member was poorly informed, and the inconvenient outcome was that I could not compete for the realization. I will need to create new opportunities to play out the performative strategy of associative design to robotic building based on one all-inclusive component for larger structures.

The Seven Daughters

The Seven Daughters could be that project. Recently, we presented a 120-m-tall urban sculpture project to celebrate the rise and shine of female power in Qatar to the Supreme Committee and Qatar Museums in Doha. There is an ongoing debate among Qataris on how to evolve from a traditional patriarchal tribal society toward a society that offers equal opportunities for men and women. Having taught female students at Qatar University, I know their ambitions and pushbacks, and their ups and downs. I have witnessed a strong feminist movement in the making. The wife of the former Emir Hamad Bin Khalifa Al Thani, Sheikha Mosa Bint Nasser Al Misned, is a strong advocate of ubiquitous education, equal rights, and opportunities for Arab women, as does Dr. Amal Mohamed Al Malki, professor at the HBKU University in Education City. Qatari women do not repudiate their recent Bedouin past. Qatari women (and men) are keen on a forward-looking fusion of tribal heritage with modern societal insights and technological advancement. Sheikha Mosa is chairperson of the ambitious Qatar Foundation for Education in Education City, where one finds besides the many branches of international universities, Isosaki's Qatar National Convention Center (QNCC), OMA's Qatar Foundation HQ (QF), and the Qatar National Library,

Pelli's Al Sidra Hospital, Pattern's Education City Stadium for the FIFA 2022, and Ali Mangera's Qatar Faculty of Islamic Studies, together with great pieces of public art, among others the controversial Miraculous Journey by Damien Hirst, and a cool (and cooled) urban park Oxygen. Sheikha Mosa's daughter, Sheikha Almayassa, chaired the project for the National Museum of Qatar, designed by Jean Nouvel, a building that I, without hesitation, place in the top 10 of my favorite buildings, if not the number one. The success of these buildings is only made possible because the Qatari royals have deeply involved themselves in the design intentions and actively guided the whole process.

In other fields, female power is on the rise as well, notably in leading positions in state-supported organizations, like the Qatar Foundation, or the new M7 design center. At Qatar University, the study of architecture is only open to female students; the boys are supposed to study abroad, whereas the boys typically study international business law. Female students are culturally not supposed to go and study abroad on their own while traveling abroad is in practice only accepted when accompanied by their immediate relatives, brother, or husband. My students are eager to learn new skills. Many of them see Zaha Hadid as their role model. They think like Zaha, interested in architecture that is emotional on the one hand and based on science on the other hand. We taught them to work with Grasshopper/Rhino and tutored by Ilona Lénárd and we held abstract painting sessions to counterbalance the scholastic lessons from instructors. The students generally have an open mind, which is a great promise for the future of architecture and design in the Middle East.

Back in the late eighties, I had one female Qatari student in my unit at the Architectural Association in London. As of now, she is the only female architect running her own architectural office in Doha; the stubborn reality is that she is still fighting for her independent position as a boss. It was our perception that many Qatari men are, as of yet, not exactly happy to see female colleagues in leading positions. The cultural revolution in Qatar is orchestrated top-down by the liberal royal Al Thani family, not yet whole-heartedly bottom-up supported by the more traditional Qatari tribal culture. It is in this context that the Seven Daughters project is conceived.

For the design of the Seven Daughters sculpture, I worked with Ilona Lénárd and a team of Qatar-based female architects, Yassmin Alkhasawneh originally from Jordan, Hend Gamal from Egypt, and Yamamah Alsalloum from Syria. Hend and Yamamah are former students of mine from Qatar University, Yassmin, an artist in residence at Doha Fire Station I met at an architecture conference in Msheireb's Bin Jelmood House, aka the Museum of Slavery. Together with Yassmin Ilona and I also designed the Monument for the Hundred Heavenly Heroes in Kyiv. Ex-student Nancy Makhoul, also from Syria, later contributed later on to the design process with fine watercolor impressions. Our Arabic team found the Qatar story of the Seven Daughters, a story that is connected to the Zodiac sign of Ursus Major, the Big Bear. The story of the Seven Daughters is the narrative of seven daughters chasing after the brutal murderer of their father. Each of the seven daughters is represented by one of the seven stars of Ursus Major. The "murderer" is the star that only occasionally lights up in the skies, and then disappears again, close to one of the seven permanently visible stars. Traditionally, under the clear desert skies, Qatari tribes orient themselves on the stars at night. Everyone who has been in a desert knows how disorienting an environment the desert can be. As of today, there is GPS everywhere, in the deepest deserts, but using smartphones, one still needs an internet connection. Through the ages, Arabic tribes have been master astronomers exploring every single detail in the skies, which are so clearly visible above their deserts, far from the city's light pollution. We decided to dedicate the 120-m-high sculpture to the rise and shine of emerging feminine power in Qatar.

My initial idea was a structure that has large nodes and connections between, big enough to host visitors and a transport system between the nodes, vaguely similar to André Waterkeyn's Atomium (1958) but in a free arrangement of the nodes, and definitely without additional support structures. After it was decided

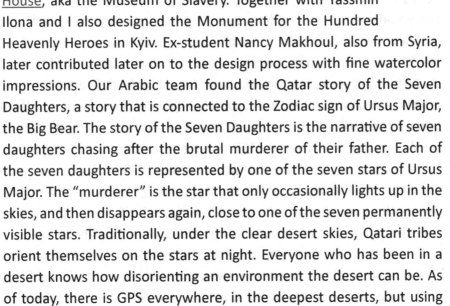

to have the nodes represent the stars of the Ursus Major constellation, the project was almost self-designing. I checked the relative radiant power of the stars and modeled the size of the nodes accordingly. The stronger the radiation of the star, the bigger the node. Then I modeled the positions of the nodes according to their relative distance from Earth, varying between 80 and 120 million light years. The result is a dynamic constellation of seven nodes and their connecting parts, balancing on one of the nodes on the ground. The Seven Daughters sculpture is strategically placed at the end of the pier in the bay of Lusail, the new city in the Northern part of Doha. The sculpture features a hexagonal pattern for the exoskeleton, mapped both on the nodes and the spikes. The design of Seven Daughters closely follows the paradigm of one building, one parametric component, and one moldable detail. The structure is built up from thousands of hexagonal steel frames, all of them unique in size and shape, assembled to give shape to the challenging balancing act of the sculpture. The exoskeleton is structurally connected to a second hexagonal structure, as an offset toward the interior, thus creating a strong spatial structural system. Semi-opaque triple-layered ETFE cushions form the adaptive skin of the hexagons. The cushions are programmed to alternately open and close to block the impact of the sun, by inflating respectively deflating one of the two interior volumes of the three ETFE layers. By closing the silkscreen printed patterns, projections from the inside become visible. Hence, in the Seven Daughters project, the main component is a programmable component, whereby the content of the ETFE subcomponents changes in real time. Images and videos are triggered by the interaction with the public, and at night evoked by arrays of thousands of LEDs. Evocative scenes of the main achievements of female leaders and essential workers bring the seven distinctive categories to life in the seven spheres: social life, education, culture, economy, environment, leadership, and health.

7

Where Are We Now?

UAE

Soon after the commission to design the Bálna, we acquired some challenging commissions in the United Arab Emirates, for master planning projects and international competitions. Our journey to the UAE started in Shanghai at the international conference Contract + Design Asia 2004, where I was approached after my lecture by Yasmine Mahmoudieh, a London-based interior architect from Iran, who invited us to join forces to launch a new hotel concept. It was our first trip to China, and a dozen trips

were to follow – invitations by Tongji University, Tsinghua University in Beijing, Harbin Institute of Technology, Nanjing University, Guangzhou, Chongqing. Connections to Nanjing were mediated by our Hyperbody researcher Xin Xia, and connections to Harbin by Hyperbody Ph.D. candidate Han Feng. Besides lectures, we organized workshops, publications, and exhibitions and were invited to competitions. Unfortunately, none of the competitions in China played out well. After 10 years of coming and going, we were not able to get our hands on a physical building commission. The closest we got was the invitation by Mr. Lu Yun to design a Wedding Chapel in 2007 in the CIPEA district in Nanjing. CIPEA, the China International Practical Exhibition of Architecture, is a cultural building site in the woods near Nanjing, where, among others, Steven Holl has built the Sifang Art Museum. Our design was rejected because Yun did not think that it could be done. At least we did not invest more than we earned, and we ended up with a positive balance.

We were more successful in the United Arab Emirates. The first initiative was the development of a unique concept that was named the Flyotel, a luxurious hotel and conference facility hosting water planes, flying out to see new developments in Dubai like the Palm from above. The Flyotel marks our first step into the Middle East, which I have visited a dozen times since. Mahmoudieh introduced us to Anita and Amir Mehra, sister and brother, also originally from Iran, daughter and son of a respected Iranian medical doctor who founded the highly valued Iranian Hospital in Dubai in the early seventies. The

Mehra's grew up in old Dubai, when Dubai still was mainly a fisherman's village and, at the North side of the Creek, the Gold Souk. The Mehra's witnessed the transformation of a local center into a global metropolis. The only high-rise in their early days was the World Trade Center tower. Amir Mehra was running a small business of food imports from Iran, taking us on trips into the desert with his vintage four-wheel drive Toyota pick-up truck. Anita Mehra is the Vice President of Communications at the Dubai Airport. She eventually received citizen rights in the UAE, which is quite exceptional for inhabitants from outside the UAE. The Mehra's opened doors for us that otherwise would not have remained closed. We were introduced to the project developer, among others, Nakheel, to whom we presented the Flyotel concept. While Mahmoudieh signed for the interior design, we designed the overall concept and made in-house the sculpture model, the exterior renderings, and the videos. Although the design triggered Nakheel's interest, they hesitated to bring the project to life, while it had not been done before. This is how we got acquainted with the Dubai way of working. They want to innovate, yet typically ask where they can see the example of proven technology. They innovate quickly, relative to what they did before, but internationally and technically not ahead of the troops. Innovation in Dubai is by and large a PR issue. The predominantly British and American branding experts have done a great job in Dubai; their on-liners are strong and effective. British and American firms are leading the Dubai building sector and their consultants apply the strictest American requirements for construction and safety. Their contractors apply the simplest building methods, while they rely on the deployment of masses of unskilled poorly paid workers. Only in recent years have some more advanced construction methods been tested but mostly limited to some PR projects only.

Several commissions for urban studies originated from our cooperation with the Mehra's. The first and biggest opportunity that we got was the invitation to tender for the design of a two million square meter development in the heart of Abu Dhabi, on the site of the abandoned palace of the former ruler, Sheikh Zayed. It was Sheik Zayed, notoriously barefoot in sandals, who imagined the Palm by drawing the concept in the sands. Emir Zayed is credited for making Dubai into the thriving city that it is now. When we were visiting Dubai in those early years, the image of Sheikh Zayed was omnipresent along the city-highways and in hotel lobbies. Zayed was the one to bring seven smaller emirates together into what is now the United Arab Emirates. Building upon his already immense popularity, public relations campaigns made him a national symbol. After some years, in Dubai, the image of the current ruler of Dubai, Sheikh Mohammed bin Rashid Al Maktoum, was added to the portrait galleries. His famous sign is the three-finger salute, thumb, index finger, and middle finger, a PR campaign that was started in 2013. The three fingers form the letter W, simply meaning "to win." Officially, it stands for "Win, Victory, and Love." Winning is an obsession for Dubai; every achievement is communicated in superlatives, and all goals are super-ambitious projects of a scale that is not easily found elsewhere. The tallest is the Burj Khalifa, the largest is the Dubai Mall, the biggest is the man-made island, the Palm, the fastest is the Hyperloop, the most innovative is KIlla's 3D-printed office, and the Museum of The Future. Even being the happiest has become an official goal since the installment of a Ministry of State for Happiness and Well-Being. Our involvement in the Middle East marked my fourth major culture shock, after previous travels to the USA (first visit in 1981), to Japan (first visit 1983), and to China (first visit 2005). Except for the bigger part of Africa, I have traveled all continents, most of them by invitation to lecture. Our interest in the Middle East culminated in the realization of the bespoke Liwa Tower in Abu Dhabi and some years later in a full professorship at Qatar University between 2017 and 2019. In Qatar, we became an active part of the Arabic culture in our daily life, and through our professional activities in the context of the role of the component in design and execution.

Big5

In 2006, we participated together with CIG Architecture and Meijers Staalbouw in the Arabic international building and construction show called the Big5. The Big5 is a yearly building construction fair, alternatively running in the five collaborating Arabic countries, Saudi Arabia, Qatar, United Arab Emirates, Kuwait, and Oman. The exhibition at the World Trade Center hosts the main Big5 show, as Dubai is the leading economic and financial hub for the Gulf Region. The presentation of the Triagrid, a joint venture of ONL, CIG, and MS, a component-based parametric to robotic system of synchronized structure and skin, was supported by the Dutch government and presented in a joint exhibition stand with two other Dutch manufacturers who wanted to explore the booming Emirati market. The costs of the prototype itself and our travel and hotel costs were on our company account. The Dutch government took care of the costs of renting the exhibition space. Triagrid is directly based on the structural all-inclusive Cockpit detail. For the Big5, the system was enhanced with insulation sandwich panels. The subcomponents of the 5-m-high prototype were transported to Dubai and assembled at the stand. Triagrid's presence at Big5 had that Calimero feeling, coping with the giants of the building industry. The goal was not to get commissions for the design of roof or facade structures but to present it as a concept for the complete building, such as the Cockpit. That goal turned out to be an ambitious one. The requests for a quote we have got were parts of otherwise fashionable yet traditionally designed buildings, mostly for facades and roofs for atriums like the Fraser suites hotel in Qatar. Typically, Middle Eastern project developers are running confidently into complicated design challenges and exploiting commercial companies to solve their problems, for the lowest possible price. Was this the market we wanted to operate in? In the first place, the potential of the Triagrid system is not taken advantage of when implementing it as an atrium roof and facade. Triagrid system is cost-effective when it forms the main component for a complete structure and skin of a building, and that can only be achieved in the design as an all-inclusive system, not just a non-structural part of it. Our joint effort did not lead to a new project.

In the same period, we pitched our Triagrid system to the municipality of Dubai to build 60 metro stations. The door to discuss with the project manager directly was opened by our sponsor Anita Mehra. Through her position as VP Director of Marketing and Communications at Dubai Airport, she has connections with Emirati decision-makers. We re-engineered the concept design that was delivered by the Dubai-based architectural and engineering firm Aedas into the Triagrid system. The 60 stations had obvious similarities with our earlier designed and internationally published A2 Cockpit. As Aedas' design has the same overall shape, outline, and size, not surprisingly, our system would have fit perfectly. From a distance, our Triagrid metro station proposal looks the same as the Aedes design, yet completely different in the way it is structured, produced, and assembled, and very competitive in price and technology. The Triagrid proposal would require much less steel. We were granted a presentation to the (expat) project leaders but not directly to the highest Emirati decision-makers. It appeared that they were looking for a facade system as the outer layer on top of the already approved steel structure, not as a system that integrates structure and facade. Our redesign would have forced them to go through the building permit process once again; they were not ready to take that risk.

Another one of our efforts had been stranded, for similar reasons as our lack of follow-up appreciation in the Netherlands. The building industry is like a slow supertanker, of which the course cannot be changed by one single smart innovation, and that supertanker sails the world. Deep innovation is only possible when one has all the cards in one's hand. The investor, the developer, the designer, the manufacturer, the contractor, and the operator need to operate as one responsible entity. Which compares to is the position of an autonomous artist, but then with budgets that are a hundred to a thousand times higher. Thanks to our self-confident naivety, I have come a long way, but because of an apparent lack of business-savvy commitment and business-style social intelligence, I have not been able to transform ONL into an all-inclusive enterprise on the grand scale of architecture.

Qatar

Living, working, and teaching in such different places as Rotterdam, Budapest, London, Paris, and Doha, lecturing in over 60 cities worldwide, experiencing deep culture shocks when in the USA, Japan, and Qatar, shaped our worldview, how to proceed from where we are now. Our ambitious goal to initiate projects within the frame of the fusion of art and architecture brought us to a hundred cities in some 50 different countries around the world. When we were invited to lecture, perform workshops, and to design projects, we always took the opportunity to see the city, understand the culture, and navigate the region.

Middle management expats in Qatar, many of them from North-Africa, typically live in gated communities, well segregated from native Qatari and Asian people. One does not get to see how the Asian "workers" live in the barracks. In recent years, social conditions of the lowest income are improving, boosted by the developments in preparation for the FIFA 2022 World Cup. Qatar by and large has adopted the American way of life, indulging in overconsumption of almost anything, driving muscled four-wheel drives but without the inherent violence of the American culture. Doha is ranked among the safe cities in the world. The Qataris typically build boundary walls around their villas and around the gated communities to protect themselves against desert sandstorms in the first place, not in fear of perpetrators. Yet, Qatar is obsessed with security; in every compound, apartment building, shopping center, and office building, there is a large team of security agents, mostly biding their time as the intendants in our museums. Virtually, all security agents are of African origin, while service personnel typically come from the Philippines. In the building industry, it is mostly Pakistanis and Indians, and in the city parks, predominantly Afghanis. Qatar was looking for Nepalese people to fulfill service jobs during the World Cup event. The concept behind targeting selected ethnic groups for specific types of jobs is to secure good communication among them.

In Qatar, we lived for two tropical years in the new city of West Bay. My years at the Department of Architecture as a full professor have educated me as much as I educated female students, while the architecture school is not open to boys. Together with Ilona, I organized workshops with students and installed an interactive sculpture, while Ilona exhibited the work that she made in Qatar at prominent places like the Sheik Faisal Bin Qassim Museum and the Doha Fire Station cultural hub. Working at Qatar University and Doha Fire Station, where Ilona got a studio and mentored the artists in residence, has been more intense than our shorter

confrontations with the Japanese, American, and other cultures. One first gets to know a culture when one has a job, and when one needs to get the necessary permits. Settling down in Qatar means getting a Qatari ID, a residence permit, a Qatari driver's license, a mobile phone contract, renting a car, opening a bank account, renting a home, getting the mandatory healthcare certificates, going shopping for food and household items, and finding an artist studio.

Settling Down in Doha, Qatar

First published on September 8, 2017, on www.oosterhuis.nl

Having assessed our police clearance, diplomas, and marriage certificate by the Dutch authorities, having these assessed again by the Embassy of Qatar, and having them accepted by Qatar University's HR department, we got our one-way tickets to Doha. This may sound easy, but it took us almost three months to get everything done correctly. We sold our beach house in the Netherlands, sold our Citroen C5 limousine, packed our household stuff and stored it in our house in Hungary, and went to Doha to find a place to call our home for the years ahead of us. When we arrived in the middle of the night in the height of the 40º summer at the end of August, we were most courteously welcomed by a hospitality service provider who led us through customs in a most supple way. Then, a driver from Qatar University drove us up to the temporary accommodation they had reserved for us. We were tired after 10 hours of waiting and travel and were hoping for a comfortable suite at the <u>Al Mirqab Compound 2</u>, part of which is hired by Qatar University to house their faculty members, but it turned out to be far from luxurious. The apartment upstairs they prepared for us was empty, except for one double bed, with joint facilities, sharing the kitchen. There was no elevator, and so we had to carry our heavy suitcases to the desolate yet spacious apartment. Not exactly what we expected; so we called it off and asked to drive to the Mövenpick Hotel in West Bay instead. From there, we moved to the massive <u>Ezdan</u> complex in West Bay containing more than 3000 units and an Olympic-sized swimming pool, *ca.* 300 QAR/ night. We found out that many faculty members of Qatar University and other companies live their expat lives there. We stayed for 10 days but wanted something better and more comfortable. We had many viewings in West Bay and the Pearl, determined to find a better option. Finally, we hit home with a furnished one-bedroom apartment in <u>Falcon Tower on Diplomatic Street</u> in West Bay, a premium location for 12,000 QAR per month all-inclusive, slightly above my 10,000 QAR housing allowance, coincidentally managed by the same Al Mirqab company that manages the first compound we found ourselves dropped into. Falcon Tower is known as the pilot tower, whereas Qatar Airways rents 30% of the apartments for their pilots.

In the first two weeks in Doha, we managed to go through all the processes to get the Qatari ID, residence permit, bank account at QNB to do our shopping, the checkbook to pay the rent, and the driving license. Most people were surprised that we managed to get all licenses, home, and car within two weeks; we spoke to several people who said that it took them months. Since the job at Qatar University, and therewith the first payment, first started on the 10th of September, we pre-invested a lot of our private resources for one month of rent, completing the furnishing of the otherwise furnished apartment, and daily living expenses before settling down. Before we got the rental car, we used Uber for almost all trips. Uber service is great in Doha, with usually 3 minutes before they arrive at your exact location, and only 2.50 € for a city ride using the simple UberGo, driving economy cars like Honda City, and a maximum of 10 € to bring us to the outskirts of the city, which we had to do quite frequently to get our licenses. The first thing we had to do was to have a medical examination at the Medical Commission, which is, as we soon got to know, a state-owned facility at the extreme Southern end of Doha, adjacent to a not-yet well-developed part of Doha's lime rock desert. Early morning, we took off on a 30-minute ride with a driver from Qatar University; they have 20 Toyota Avalon luxury cars available for faculty and guest transportation, expected to be delivered at a modern hospital of some sort as we had seen under way, but it turned out to be a rather nondescript low-rise facility, processing all immigrant expats. We realized that we are considered workers, making the Qatar economy work; whether high skilled and well paid, or low-skilled and underpaid, and workers are treated equally. To be processed, one has to know some things in advance, which we had to find out ourselves, and no one instructed us beforehand. First, one has to dress according to formal Qatari rules, meaning covering legs and arms completely, both for men and women, and second, pay by special credit card. Ilona was not allowed to enter as her dress was too short; the kind Sri Lanka driver of QU took us patiently back, and Ilona changed, and we came back one hour later. We could not pay cash as only credit cards were accepted, but my American Express card was not accepted. This was probably anticipated; at a hideaway outside, we could buy credit on a temporary card and went through the blood sample tests and the X-rays, men separately from women. Waiting rooms are separate for men and women too. Passing the test is mandatory to become eligible for getting a residence permit. All in all, it took us a full day, and it gave us a first-hand experience of how the Qatari society works in a combination of rigid bureaucracy and improvisation.

After a medical check, we were sent the other day to The Forensic Laboratory to take fingerprints. They took all our fingers, left and right, and as if that is not enough evidence, they copied our hand palms as well. Give them one finger, they take the whole hand. We are definitely in the governmental system now. As we found out later, one's data are connected to one's Qatari ID. Information on any expat worker is available to the authorities. When entering office towers, both governmental and commercial, one has to leave one's ID card at the reception desk; without a Qatari ID, one does not get anywhere. Before we emigrated, we were told to bring enough portrait photos, which we had made in the Netherlands. Well prepared, we went to the HR Department at Qatar University to apply for the Qatari ID. However, the rules for the formal portrait photos had changed recently, and to get our Qatari ID, they told us that we would need a photo with a clear blue background. They could not tell us where to get those; they suggested one place in a nearby mall, but the driver of Qatar University lost his way, and I was only able through Google Maps to find one not too far from where we got lost in one of the many 2 × 2 km^2 low-rise residential neighborhoods. Luckily, the studio that we found specialized in bridal reports, and they could help us out; one hour later, we got our portraits with the mandatory clear blue background, which by contrast gives more color to one's skin. The photos were accepted, we handed in our passports to HR, and we could fetch our residence permits some days later.

To rent an apartment and to apply for a driving license, one needs a Qatari ID first. We used the days of waiting for our IDs to contact some property agents through propertyfinder.qa and went to see apartments in the Zig-Zag tower at the Lagoona Mall. At first sight from a distance spectacular and tempting to move into, but on closer inspection, the quality of the building was quite poor. The rooms are not spacious; they have great views but have no balconies. But the main problem was the poor state of maintenance, especially concerning the air conditioning. We continued our search, guided by a young female Moroccan broker, who showed us a rather nice furnished apartment in Tower 21 at Porto Arabia at The Pearl. But, with the small balcony, and neighboring tower being too close, we continued our search through several other agents. After having viewed almost ten other apartments in The Pearl, we went to sign the contract for an unfurnished apartment in Tower 3, this one with a very large balcony. The same afternoon that we were on our way to signing the contract prepared by an Algerian agent working for the mediating company Majestic, we browsed propertyfinder.qa for a last check and found an apartment in West Bay that looked sort of attractive. Immediately, we called the representative of the owner and made an appointment for viewing directly with the Al Mirqab management that operates the Falcon Tower. And indeed, this apartment definitely was the right choice, a spacious one-bedroom with a splendid sea view overlooking the whole 7-km-long semi-circular Corniche, a great balcony, a round 25 m' pool, cooling sea breezes, and in very good condition. This apartment in the Falcon Tower, perfectly located between the Hilton and the Four Seasons, along a pier that exclusively leads to the world-famous Nobu restaurant, felt immediately like a home due to its high standard of classic finishing, more modest and intimate dimensions of the entrance hall and corridors, while the finishing of the apartment itself also looked much better than the other apartments in the Pearl towers. Just in time, we hit home. Otherwise, we would have spent our days in the Pearl, which feels like an artificial environment yet without a clear identity as of yet, while West Bay gives that feeling of living in a vibrant metropolitan city. In a fast-developing country like Qatar, rules are changing quickly. Now, having acquired our Qatari ID, our Dutch driving licenses were declared no longer valid. I found this out when signing in for a one-year leased car. They told me to go to Traffic Headquarters, a brand-new classical limestone building along the highway to the Education City, westward of the fast-expanding city of Doha. Not knowing what to expect exactly, we went there hoping to get the license on the spot. Not so, after having paid 90 QAR for the intake document, they told me to go to one of the private driving schools to do the driver's license test. The first driving school we went to turned out to be a private facility run in chaos, probably by Egyptian staff. I was told to come back later as they had no free time slots, but after some persistence, I was told to go through the so-called Signal Test and answer the 20 on-screen multiple-choice questions about specific Qatari traffic signs that I had never seen before. By calculated guesswork, I passed nevertheless without any mistakes. So, I thought that is it and expected to get the license in the neighboring building of this local traffic department in the Khalifa district near the Aspire Dome. Not so, they gave me an appointment for a road test two months later. But I needed my license the next week; so we asked around whether we could speed up. One younger English-speaking staff advised us to go to the United Driving School in another district close to the Villagio shopping mall. The regular staff in these rather shabby sheds told us the same: 24th of October, or otherwise, talk to the captain. That we did, and after some more massage, I was granted to do the road test the next morning. I miraculously passed driving an economy car with automatic gear, which I was not used to but fortunately remembered how to handle the car. I passed by driving two times round the corner in a quiet residential neighborhood. I did not even have to start the car, but as was advised on internet sites, I demonstratively adjusted the mirrors, and I ensured to keep my two hands firmly on the steering wheel and extensively looked into all possible mirrors – left, right, and middle.

The next day, our car, an almost new Honda Civic, was delivered to our residence with the tank only half full. We soon got a feeling of how Qatar is kept going by its 2.2 million expats predominantly from Sri Lanka, Philippines, India, and Pakistan, relative to a mere 300,000 permanent Qatari inhabitants. We quickly switched to an all-wheel drive car to have a higher driving position; in a compact car, it feels like driving in a narrow canyon with the biggest possible SUVs overtaking us from left and right. We are using our new Honda CRV automatic AWD often to do our shopping at Carrefour in the very well-functioning City Center close to our new home in West Bay, the giant Mall of Qatar 20 km west of West Bay, or at the IKEA 20 km north-west from Falcon Tower.

Major ten to fourteen-lane city highways between West Bay, Katara, The Pearl, and Lusail are largely still under construction but almost finished well in time for FIFA 2022, meaning that as of now, each day, the route could be different. The whole country is under intense reconstruction, to be continued until the opening of the FIFA 2022 Football Games, and up to the years of consolidating after the games. Google Maps updates its map only once a month or less and does not give correct info; instead, we use the Waze app, which gives much more accurate traffic information, based on daily feedback information from its many users.

Our longest trip in the first week in Qatar was 30 km westward to the extra-large collection of the Sheikh Faisal Bin Qassim Al Thani Museum, almost in the exact geographical middle of the Qatar peninsula, a large private domain in the desert. It is not unlikely that the now-isolated museum will be surrounded by fast-spreading urbanization within a decade. Sheikh Faisal is assumedly the richest member of the ruling Al Thani family, and assumedly the richest person in Qatar. Sheik Faisal owns many large real estate projects in Doha, both low-rise and high-rise, and owns a variety of companies in construction, hospitality, trading, transportation, entertainment, education, services, and information technology. He is the personification of diversifying the Qatar economy, and in his personal life, he is a vigorous collector of almost everything from cars to carpets, from art objects to a complete and old Syrian house that they took apart stone by stone and resurrected in the museum, and from weapons to symbols of a broad range of religions other than the Islam. His enormous collection, to a large extent undocumented, was shown to us by the then-new Dutch director Kees Wieringa, who is also a professional classical pianist and composer. Kees Wieringa is well known for his recordings of the piano works Proeven van Stijlkunst by the Dutch composer Jacob van Domselaer and the Sostenuto Ostinato by Simeon ten Holt. We felt we had hit home indeed.

The Open Mosque

First published on May 9, 2017, on www.oosterhuis.nl, written during my A380 flight back from Dubai to Amsterdam on the night of May 5, 2017.

I was invited to lecture on May 5, 2017 at the 2nd Mosque Design & Development Conference in Dubai. My participation was kindly sponsored by the Dutch Creative Industry PIB program. It was a small but interesting conference, in many ways worth participating in. It provided ample insight into the social importance and daily use of mosques, substantively deepened by some historical analyses and forward-looking concepts for mosque designs. Several presentations were straightforwardly critical of how mosque design has been trivialized in recent years, especially in the West. Prof Ali Alraouf signaled and evaluated the growing Islamophobia and Mosquephobia in the West and partly accused traditional mosque designers and builders to have triggered the phobia by adopting cheap and banal imagery for the mosque designs. According to Raouf, from the side of the investors, a cool proprietary wind has blown into the social function of mosques, and hence in mosque designs, reducing them to places for efficient 10-minute prayers and the Friday elevation speech, preventing the mosques to function as the active colorful social meeting places that they once were. According to my observations, such downplaying of the social function of the mosque reaches its extreme in the hidden prayer spaces in shopping malls and universities. A researcher at The American University of Sharjah identified the problem and became interested in how to avoid crossing people flow in the dark backspaces of the shopping malls to enhance efficiency and avoid queuing and jamming at the ablution spaces, which I found a very technocratic analysis, without even considering the possible impact of a good design concept, assuming that nothing more could be done than optimizing the flow of people in a black box away from the general public. Therefore, I found that his research only institutionalizes a pitiful situation. The mono-cultural segregation of functions that devastated western urban planning and building concepts has done its destructive work in the Middle East as well. There seems to be a relation between downplayed mosque designs and the blind attraction that some of these places have to more traditional interpretations of the Islamic faith. Poor backward-looking designs by and large go hand in hand with growing narrow-mindedness and therewith give rise to further segregation and polarization. In this context, it is instructive to read Marwa Al Sabouni's book The Battle for Home on the relationship between poor urban planning and building in Homs in Syria, and the heartless fights between the various political factions in recent years, where she finds herself caught in the middle of it. In her view, urban segregation accelerated political polarization and eventually its self-destruction. There was inevitably a strong emphasis by most speakers on sustainability, circular building, energy saving, retrofitting, etc. But again, without many relations to the design language and urban syntax itself, it must be obvious that the very first design idea is the most important to define the possible success of any green building concept. Some presentations rightfully argued for the integration of large surfaces of running water for natural cooling purposes, thereby doing away with the need for air-conditioning, which usually feels either too chilly or not cool enough. The common opinion at the 2nd Mosque Design & Development Conference was that the mosques should become social hubs again, attracting a variety of activities and functions, well integrated into the social fabric, while the design of the mosques should take advantage of the most modern technologies available. A rather nice example of a modern approach to mosque design is the <u>Faculty for Islamic Studies</u> in Qatar Education City, designed by architect Ali Mangera. That impressive building features dynamic sweeps to define the rocketing minarets and the swirling domed interior spaces include everything that defines a social hub: a café, outdoor meeting places, a library, a place for prayer, a place to play around, and a park to stroll around. The modern mosque is an open mosque; every aspect of the design is open for design and open for social interaction. It is a true relief to see such an optimistic forward-looking approach in built form in the heart of a country, which is, in the western world, considered a conservative Islamic society. In reality, Qatar is a country of many speeds, in many ways more advanced than our Western world, and in other ways, notably concerning the rights of workers and minority groups, living in a bygone past. In this sense, it is comparable to China, where we have seen a similar fast development that exists parallel to a living past. Qatar is a country of different velocities, which makes it a prime example of a dynamic society, open to innovation and respectful of tradition at the same time. In my lecture, which to my surprise was voted to be the most interesting contribution, I showed examples of built projects based on parametric design. I explained parametric design as building relations between people and things. Data-driven building techniques like parametric design, file-to-factory processes, design-to-production, mass customization, robotic building, 3D printing, and adaptive environments were generally accepted as the way to go. When asked to design a mosque, I would from the very first conceptual idea establish dynamic links of social, spatial, climatic, and financial parameters to the swarm of constituting building components to be in dynamic control of the design and development of the mosque, as to be prepared to guarantee the desired performance, both socially, technically, and esthetically. After having completed the Liwa Tower some years ago in Abu Dhabi, and after having been commissioned several urban design studies by project developers like Aldar Properties from Dubai, I was and still am looking forward to contributing to further design developments to build in the Gulf Region. A lot of innovations still have not invaded the building industry there yet, but the general mood is to have an open eye for game-changing new developments. At the same time, one should become aware of the severe constraints to innovation by being all too dependent on cheap labor. Robotic techniques will replace the cheap workforce in due time, while prefabrication, including 3D printing of a series of unique building components, will eventually replace the now dominant technique of on-site pouring of concrete. It was a real pleasure to meet like-minded designers and researchers in that fascinating part of the world, which superficially seems to be at odds with the western world. I was strengthened in my belief that the power of a game-changing design proposal can make a substantial difference, to start building a new precious inclusive societal fabric, not only for mosques but basically for every possible building project.

Message in a Bottle

First published on November 13, 2017, on www.oosterhuis.nl

Navigating West Bay in Doha, I got more and more impressed by the towers we passed by when shopping at City Center when coming home from Qatar University or driving the corniche coming back from the airport and Doha City to our home in the Falcon Tower. At first

sight, the <u>West Bay towers</u> look like a randomly placed bunch of autonomous individuals, begging for attention. To realize that most towers in the skyline were built in the last 10–15 years helps to understand the nature of this development.
The Doha towers are *branding* themselves in a brand-new downtown. I will not write a report here on the historic development of Doha; many have done that before me, to learn more about the history of Qatar there are some excellent websites with abundant information and (photo)graphic reports on the new city of West Bay. West Bay is known as Al Dafna in Arabic, meaning reclaimed land. You might want to check John Lockerbie's <u>catnaps</u> website to get detailed information on the development of Doha. Lockerbie is a retired planning consultant who has worked a lifetime in Doha to experience the rapid growth of the metropolis of Doha from the early seventies.

Instead of copying knowledge, I try to understand the city from the perspective of the towers themselves. What do these towers communicate by their existence and, more specifically, by their marketing? Visiting the <u>Radwani House</u> in the new but stylistically retroactive 760,000 m² <u>Msheireb</u> development in Old Doha, one small detail caught my attention: a series of fifties and sixties sanitary bottles on a recessed part of the wall of the bathroom in the house of a well-to-do Qatari family, originally built in the 1920s, unwittingly foreshadowing the explosive growth to come after the discovery of oil and gas. What triggered my interest was the branding of the bottles. These shampoo and perfume bottles are distinguishing themselves from their competitors. It immediately occurred to me that in much the same fashion, the West Bay towers distinguish themselves from their immediate neighbors, by branding themselves, some of them very successfully. Such bottles necessarily share similar characteristics with the real estate towers: base, shaft, distinct capital, and an overall distinct shape. The main difference is the scale of things; the characteristics of the West Bay towers are operational on the urban scale and not on the domestic scale. The shampoo and perfume bottles operate on the scale of the shelves in the supermarket, while the towers operate on

327

the scale of the streets of downtown. The bottles simply seem to have pumped up their volume to become a real-estate-shaped container. The base of the bottle, aka the tower, keeps them from toppling and connects them to the urban context. The shaft holds the substance, whether we are looking at the liquid shampoo or the leasable office space, while the crown functions as the orientation point for the eye of the one navigating the city. The cap on the bottle is the attraction point where the hand goes to open the bottle; similarly, the top of the tower forms the attraction point where the eye of the city navigator is drawn to. Simple, often curved, lines define the shape of the branded containers, whether small as a household product or big as a real estate investment. Both the small bottles and the big towers are containers of strange substances. What substance do they contain? It feels as if the uniqueness of the shape has a positive effect on the nature of the substance. In the public's perception, the substance seems to improve when the branding of the product (bottle, tower) is successful. In the end, the successful branding of individual towers has a positive effect on the downtown area as a whole. Many of the West Bay have no substance at all, while they remain unfinished and underused. Despite an estimated current percentage of 70% vacancy, I believe that in the long run, West Bay will once become a vibrant downtown area, pleasant to work and to live in, a downtown area with substance. Much depends though on the implementation of electric cars, shaded outdoor pedestrian areas, planting of thousands of trees, and drastic narrowing of the 10-lane city highways.

  Once we went to visit the Supreme Committee for Delivery & Legacy to discuss future art projects in preparation for the FIFA championship. We were welcomed on the 33rd floor in one of the most prominent towers, namely the 40-story high Al Bidda Tower. The many offices related to the FIFA 2022 form the solid yet temporary substance for the Al Bidda Tower, not one of the highest towers but certainly one of the most characteristic ones. The sharp-edged and fragmented plectrum-shaped Al Bidda Tower rises in a rotating movement. The semi-structural facade is triangulated to follow the torqued shape. Triangulated facades are of particular interest since we have designed and built the Liwa Tower in Abu Dhabi using a parametric diagrid system for the load-bearing structural facade of the smoothly shaped tower. From the inside of the Al Bidda Tower, one can unriddle how it is built by the Dutch contractor BAM/Higgs & Hill. The first things to notice are the oversized slanting concrete columns, at least 1 m in diameter, even on the 33rd floor! The structural engineers, or was it the architect, have chosen to build the primary load-bearing structure in concrete, with an additional secondary structure using triangulated steel beams and another tertiary heavy layer for the glass facade. In comparison, in our Liwa Tower in Abu Dhabi (2014), we synchronized the dimensions of a lean fine-grained primary diagrid structure with large triangular aluminum sandwich facade components, using less than one-third of the materials as compared to the Al Bidda Tower and saving at least two-thirds of the costs of the structure plus skin.

Unfortunately, branding is not about being smart or lean; it is about being proud of being costly. Branding means boasting, bragging, and bluffing using superlatives; a branding message that tells you about doing more with less does not work, unless some other superlative can be applied. Between the brand and the substance, there is the skin. The skin is the interface between the city and the user, a membrane between an open outdoor and enclosed indoor environment. The quality of the materials that are used for the skin in combination with the level of design thinking tells the story of how a brand, in this case, a governmental one, sends its message to its immediate environment, inside and outside. On the one hand, the brand communicates toward the city and on the other hand toward the users and their guests. The Al Bidda Tower is just one example out of 150 other towers in West Bay in Doha, which all can be analyzed similarly with respect to their branding, substance, and interfacing membrane. The message is in the bottle.

Arch-News Interview

I was interviewed for arch-news.net by Marwa Al Sabouni, author of the best-seller book "The Battle for Home," based in Homs, Syria | published on www.oosterhuis.nl website on June 9, 2017.

MAS: *You identify yourself as "The expert formerly known as the architect," you also give lectures to students to introduce them to the term "information architect"; how has the mission of the architect changed in our modern times, and is technology the only cause for this shift in thought and related terminology regarding the profession of architecture?*

KO: After having written my book *Towards a New Kind of Building* (NAi Publishers, 2011), I asked myself publicly what exactly is the expertise of today's architect? The easy answer is of course that there are many different types of architects, but the question I wish to discuss here is if we should consider ourselves "generalists" or if we perhaps are professionals in a more specific field of expertise. The problem is that at my faculty at the TU Delft, the dominant concept is indeed that of the generalist: hovering over all disciplines, the puppeteer who pulls the strings, the one who makes end-decisions, and the one who is top-down dictating the rules of the game. In our educational system, the students are given the illusion that they will be the ones who decide. But I know from practice that the reality is different; many puppeteers make important decisions covering various fields of expertise, in a most complex way woven together to develop a one-off architectural product. It is my conviction that we all should respect each other's expertise, including that of the climate designer, the structural designer, the quantity surveyor, the user, the client, the interaction designer, and the material designer; I consider them all to be experts in their field. I want to work with them on a level playing field, which is why we at ONL and Hyperbody are developing open collaborative design techniques to work 1:1 with any other expert in any stage of the design to production and the design to operation process. Now, we come to the question I have avoided as of yet: what then is my expertise and that of my team? The not-so-easy answer is: basically, we are experts in the (digital, parametric) programming of the spatial interaction between the constituent components of the construct, be it a chair, an installation, a pavilion, a building, an urban scheme, or a work of art. As an important sub-expertise of ourselves at ONL and at Hyperbody TU Delft, we are experts in the *real-time* behavior of the constituent components, dealing with interactive architecture.

MAS: *You are a well-known professor and professional expert in the field of architecture, teaching at the Technical University of Delft but also practicing in both fields (academic and professional) in the Gulf region; where do those two fields meet in your opinion?*

KO: At the university, I am a professor from practice; I bring in the experience, knowledge, and skills from practice into the university. In practice, I am often considered a scientist, someone who wants to know exactly how things work, and someone who is not satisfied with vague ideas, metaphors, and illusions but wants to see the verifiable proof of the concept. I typically start with coining a hypothesis and then test it on all levels, from design to engineering to production to the actual usage of the building. If someone challenges the hypothesis with good arguments, I listen and adjust the design concept or the design parameters accordingly. I am reaching out for expert opinions from other experts. Likewise, I develop a strong personal opinion, step by step constructing my scheme of things, my personal universe, which is partly emotional and partly rational. In this respect, it helps a lot that my partner in life and business Ilona Lénárd is a visual artist and looks at things from a different perspective. Since 1989, we have decided to join forces and dedicate our studio to the fusion of art and architecture on a digital platform. So it is not only a match between the academic and the practice but very much the cross-disciplinary approach that characterizes our methods of working. Cross-disciplinary design requires empathy with the other, even empathy with the alien, the unknown, allowing the other to bring in their expertise and their emotions, and it requires finding ways and procedures to make the match without ending up with a half-baked compromise. This is the real challenge – how to set the rules of the collaborative design game such that the unfolding of the process generates surprising new insights.

MAS: *The identification of beauty has been an area of philosophical investigation for a long time; today, you are promoting a "new kind of beauty." How do you define beauty today? Namely, what are the main characteristics and values of a beautiful object of the 21st century that are different from an older one?*

KO: In my TEDxDelft lecture, I challenged the audience to understand that we are changing your view on what is beautiful. We also post that statement on our ONL website. The leading theme of my talk was the paradigm shift toward mass customization as the logical further development of mass production. I am not criticizing mass production as such, since it brought our society where we find ourselves right now, but I challenge the production methods to produce a series of unique products, a series of unique building components, potentially leading to an until now unseen natural magnificence of complexity. Complex but not complicated, since complexity is based on a well-balanced set of simple rules. And mind you, such complexity based on mass-customized CNC design to production methods is no more expensive than regular mass-produced buildings. The CNC machines, provided that they are given the proper data, do not care if they produce a series of the same or series of unique elements; for the machine, all data are similar and therefore require no more procedures or production time. Consistent application of mass customization thinking and production, and then I mean not only confined to roofs but the building as a whole, lead to thoroughly different esthetics, where repetition of any kind will no longer be considered as beautiful or even functional, but rather as a remnant of past eras. The century that lies ahead of us will be dedicated to the beauty of continuous variation. The New Kind of Beauty is a natural beauty, revealing its inner beauty without having to compensate with layers of superficial makeup.

MAS: *Many of the modern works of architecture today adopt what has become known as "Parametricism." Do you think of parametric design as a distinctive architectural style or as a method of production? And how so?*

KO: In fact, the number of realized works based on parametric design methods is still very scarce. I believe our LIWA Tower is the only one now in the Gulf Region. Although it tries to adopt the parametric look and feel, Tower 0-14, nicknamed the Swiss Cheese Tower in Dubai by Reiser Umemoto is not parametric. And when parametric design methods are used, they usually are confined to parts of a building, not to the whole, not in any of ZHA's buildings. Parametric design is only taken to the full extent in some experimental pavilion designs, but even then, usually confined to a roofscape only. I must confess that I do not appreciate the term parametricism at all. First of all, since parametric design has existed since the early eighties of last century, and second because I am against all -isms, I am always looking for the real thing, and not for its derivatives. Maybe this is my Dutch background to look for a deeper meaning, similar to what the painter Piet Mondrian was striving for in his quest for the universal. And third because I think the term parametricism narrows down the subject too much to a technique, to an external appearance. The New Beauty encompasses much more than that; the idea of continuous variation by building dynamic relations between all constituent components must settle firmly between the ears of the designer in all stages of the design until completion, in all conceptual assumptions, and all procedural programming for an open design environment.

MAS: *As an architect, where do you draw your inspiration from?*

KO: My favorite reading is the magazine Scientific American. I draw more inspiration from science and art exhibitions than from architectural magazines. The inspiration I draw from architecture is by going there and seeing the buildings themselves, not so much from looking at published images. The only way to fully understand a building is to go and see it in real life, to feel the spatial layout, to touch the materials, and often to experience how much traditional effort it took to produce the photogenic image. This is particularly true for the interiors of ZHA and Gehry, whose designs create illusions of continuity, which can only be achieved by many layers of traditional plasterwork. This is for me fully contradictory to the essence of design to production. Well-designed buildings should be assembled using dry fixing methods, which is a big issue in the recyclability of the built product. The modern car industry already guarantees that all its components be fully recyclable. You might have expected me to draw my inspiration from nature, but that is not the case. On the other hand, I wish to design our buildings such that it feels natural, not by copying external characteristics but by a deeper understanding of how atoms group together to form molecules, how molecules act together to form proteins, how proteins flock together as to form cells, and cells to muscles and bones, how bodies flock to form swarms, how cars behave on the highway, and how buildings and infrastructure weave together as to form cities. I am looking at quantum theory just as well as natural physics on the galactic scale. I want to understand how smaller components work together to form a larger consistent whole, and how they read external environmental information. I see buildings as swarming complex adaptive systems, as information processing vehicles, rather than as imitations of end-products of organic nature.

MAS: *It is noticeable in our modern world the major retreats of local identities. Do you think it is important to consider local identity in the work of architecture? More importantly, is it achievable through this new design paradigm?*

KO: My answer to your question is a clear yes. Relational design methods must be open to local chunks of information, and local datasets, in other words to the very nature of local identities. I have always worked like that, trying to understand the spirit of the place rather than copying the obvious external appearances. We were very pleased with the name our client Abdullah Nasser Al Mansouri gave to our office tower in the Capital Centre district in Abu Dhabi. He named the tower after the LIWA desert, obviously because he was born and raised in the LIWA oasis, but surely also because he recognized the gentle curves of the ridges of the dunes as caused by the prevailing winds, and the color scheme, which we chose to avoid the building to look dirty after a sandstorm. So, there was a reference to basic climatic conditions and available technology but not a superficial resemblance. We did not look at the impressive reddish dunes at the Empty Quarter and thought that we should make the building look like that. Not at all, that has been the lucky marriage of intuitive empathy and rational reasoning in combination with applied mass customization technology.

MAS: *Is the use of the term "scripting" instead of "drawing" only a matter of keeping up with the modernization of the terminologies of technology; or does it hold an ontological underlying meaning to it? Can you explain more to us?*

KO: My team at ONL only produces drawings derived from a 3D model to get the approval of the authorities, since they need to stamp the drawings for approval. For all other purposes, the design development itself, the design to production procedures, and for the interaction with other experts, we do not produce drawings at all. We are very careful to avoid producing 2D drawings and fancy renderings since they are in essence a dead-end street, a derivative of the real information. You cannot make comments on a drawing and then feed it back to the 3D model since that is imprecise, creates double work, and too much gets lost in translation. To make it clear to you how we work, we are not clicking together 3D models either; we program them on a variety of software platforms like Visual Basic, AutoLISP routines, RhinoScript, MAXScript, Processing, Grasshopper, and Pro-Engineer. At Hyperbody, we programmed our interactive installations in the Nemo/Virtools game development program and used Max/MSP to control the sensor data. Using those more recently developed software platforms is writing code via a graphic user interface. Some of our nerds take it as a challenge to write code directly, in Python or Unity. So, you see how deeply rooted working on a digital platform is for us, and bear in mind that it is a pure requisite for being to realize buildings like the Liwa Tower for a normal budget of a commercial building. Scripting and programming are also a prerequisite for an open verifiable multi-player design process.

MAS: *You have built major buildings in the UAE such as the Capital Centre in Abu Dhabi and designed Dubai Sports City in Dubai. Besides the aspects of passive design based on the informed data of climate and location, how are these buildings different or distinctive from other projects of yours around the world?*

KO: The thing worth noting here is that we would never have the chance to propose such distinctive designs as the Liwa Tower or the Sports City tower (NB: not realized as of yet) in the Netherlands, since back home, we suffer from a combination of a typical Dutch attitude of flattening things out to "normal," with an absolute dominance of OMA's ironic modernistic esthetics. As you know, OMA/Rem Koolhaas is quite influential in my country; having produced so many cloned studios, many of the younger leading architects in the Netherlands have worked for a while in OMA's office and have been deeply influenced by their journalistic, modernistic, retroactive, assumed witty, and ironic attitude. As you know, Rem criticizes the iconic repeatedly, and frankly, that is not good for us. As a form of collateral damage, the logic of mass customization is not widely appreciated by Dutch architectural critics. It is exemplary that our A2 Cockpit building – although appreciated by almost anyone who passes by or buys a car there – has not received any direct follow-up in the Netherlands in terms of new commissions. I am not complaining; we do what we do, and we do it without any resentment, but this observation would serve well as an answer to your question. We have found fertile ground in the UAE, and hopefully in the greater Gulf Region as well.

MAS: *For many centuries, architects had given their interpretations of nature and the universe including the human body and the human input into the world around them. Today's architectural orientation is more about the way than the form; do you agree on that, and if so, why do you think this focus on the way of form has become more important than the form itself?*

KO: I see the design process to get to the finished product as a collaborative effort on a level playing field by a swarm of experts, exchanging lean datasets in real time to achieve the higher goal of improving our built environment. Collaborative open design systems are process-oriented by definition. This does not mean that each individual may not have explicit opinions in their field of expertise; I require them to have such strong opinions. I request from a structural designer or a climatic designer that s/he has an opinion on the structure. I strive for a balance between a bottom-up process and top-down input into the process. I think this is the very nature of any natural body as well. Each organism receives top-down information from its environment, processes that information based on its own genetic rules, and sends out edited information back to its environment. For us, it works similarly for the design process. Yes, we *design* the design process, we design the rules of the game, and then we play the multiplayer design game with other experts, we jump into the process, and we navigate the process like a gardener as was stated by the musician Brian Eno, but at the same time, we develop strong conceptual ideas, which function as external information pumped top-down into the

337

design process, to test it against all odds. In conclusion, I do not favor the process over the formal outcome, I am not interested in a process that produces indifferent outcomes, which is by the way the very reason we hate the term "blobs" as coined by Greg Lynn when referring to complexity in architecture. Blobs are indifferent in their shape, while my designs are highly stylized (think of car body styling), relational yet explicit, flexible yet distinctive, precise, and intentional, yet the outcome of a collaborative open design process.

MAS: What is your advice for young architects in the region?

KO: Educate yourself, learn programming skills, train your intuition, join discussion forums, think global, and act local. I think the secret to success for younger architects is to position themselves as team player, have explicit skills, preferably developing programming skills yourself, and be open to collaboration with other experts of other fields of expertise on a digital platform. Remember that complex geometry is inclusive of platonic and traditional shapes; parametric design includes the traditional. Position yourself as an information architect, controlling the datasets of the geometry of the project to establish a direct connection between the design and the CNC production procedures. Develop design strategies that rely largely on CNC production techniques and dry assembly techniques, as to secure a lean and green sustainable design and production process, and to secure the sustainability of the life cycle of the building, thus avoiding double work and waste in any phase of the project. Optimize for complexity and multiplicity, not for mass production.

MAS: Your recent project LIWA Tower in Abu Dhabi has quickly become a landmark in the area; was this a main concern of yours in the design process and how could it achieve this from your point of view?

KO: For some reason, many of our designs have achieved the status of a landmark, without the explicit intention to become one;

they stand out based on their strong internal logic. I think it is in the nature of our rare approach to the fusion of art, architecture, and technology on a digital platform that inevitably produces landmark structures. Sometimes people experience our work as hermetic because it represents a new paradigm and hence constructs a parallel universe in itself, rather than using a common language such as modernism, deconstructivism, rationalism, historicism, high-tech, regionalism, postmodernism, or parametricism. Another often-heard comment is that our designs need a lot of space around them, which bias we have effectively challenged with the design and the realization of the Bálna mixed-use center in the historic city center of Budapest. The Bálna building shows that we can weave the new nonstandard paradigm seamlessly into an existing city fabric without destroying the charm of the old while adding a strong new element to it.

MAS: *We at Arabic Gate for Architectural News have launched recently an architectural initiative called "Rebuilding Destroyed Cities" in response to the unfortunate destruction that many of the world's cities are witnessing whether by nature or man-made; and as you may know, many of the cities in the middle east have met sad cases of mass destruction lately. In your opinion, how should architects look at those cities; as a tabula rasa, that should be built from ground zero or should they built on a certain foundation, and if the latter is true, what should the main strategy for rebuilding be about?*

KO: I am not authorized to give well-funded advice for such urgent issues caused by very unfortunate situations of mass destruction. My hunch would be, but I would not be able to verify after having visited those ruined areas and having spoken to the people, that I would opt for a balanced mix between literally rebuilding the once existing and adding fresh new structures. Completely rebuilding the old may lead to boring cities that feel more dead than alive, while the tabula rasa is not an option for many societal reasons. Rebuilding the community is a huge task; one will need to bridge the gap between the city one remembers and an unknown future. I believe the most important issue that one has to deal with is the structure of ownership. Who owns the plot of land; is there enough diversity of ownership to guarantee the diversity of the newly built structures? If there is one dominant owner, for example, the city itself, then I would suggest dividing the larger plots into smaller ones to evolve the multiplicity and variation any city needs to become a desirable place to live and work. A balanced mix of houses to live in and places to work is mandatory for a livable city. Strong architectural characters can be imagined from scratch, and others can be rebuilt from memory; both attitudes should get a chance.

New Nomads

We have become *de facto* nomadic international citizens. For the last five years, Ilona and I lived in three different countries. We follow the international trend that people are less confined to the country in which they are born; they travel and stay for longer periods. The nomadic international citizen follows the work, and the challenges, not their roots. The nomadic international citizen is the privileged version of the fugitive – without pain, poverty, and humiliation. But the principles are similar; one goes where the opportunities are to develop oneself. Not only will the new nomadic lifestyle affect housing needs, but equally on the way of working in the near future. Future homes will offer temporary shelter, for as long as it takes, while the location of future workplaces will vary from time to time. Multinationals send out their staff to their branch offices all over the world and see to it that they as expats do not stay for too long to avoid getting rooted and start taking things for granted. Multinationals know very well that regular exchange of personnel secures the enterprise to stay motivated. Especially since the start of the current pandemic, there has been a growing market for flexibility where one lives and where one works. The two years in Qatar taught us how expat life actually works; we have experienced what the benefits and downsides are.

Instead of accepting a boring unfurnished compound villa that was offered by Qatar University, we rented a furnished luxury apartment in the Falcon Tower at a prominent site along the corniche. The Falcon Tower is perfectly located between the Hilton and the Four Seasons, with a splendid view of the yacht harbor, the Nobu restaurant, and the iconic Sheraton Hotel, designed by the American architect William Perreira, completed in 1982. It did cost 20% of my Qatar University fee in housing costs, but very much worth it. The Sheraton was the first building in West Bay, built as a generator for things to come. The positive side of renting is that you enjoy a high service level, in-house cleaning service, your private parking space in the parking garage, immediate response from maintenance, omnipresent security, an excellent coffee shop on the ground floor, in-house fitness room overseen by a fitness trainer from Kenia, a spacious pool, and a barbecue lawn at hand; rent is all-inclusive. Just short of being a serviced apartment, where the service would include cleaning up and making the bed every day. On the weekends, expats go out to explore the country and dine out with friends, being free from doing household jobs. One of the downsides of expat life is that expats typically do not invest in taking part in the local culture except for frequenting sports venues, bars, and restaurants; expats typically connect to other expats.

Expats seldom buy modern art; at best, they collect traditional memorabilia, which partly explains the absence of a lively art market in Qatar. There are only a handful of art galleries in Doha, and they have difficulties surviving, whereas they depend on the 12% of Qataris as their clientele. In general, expats live a poor cultural life, busy with careers and kids, mainly going back and forth to work, eating and sleeping, and visit bars, beaches, and restaurants on the weekends. Being an expat for a limited period in an exotic city does not exactly give you the peace of mind to establish an independent art practice, form a dance group, or start an experimental band. Typically, expats are employees, not independent business owners. Therefore, expats are not the new nomads that I have in mind when talking about nomadic international citizens. The new nomads are their own bosses; they are freelancers, and gig workers. They enjoy similar benefits as expats but without the downsides; their work and family life is one indivisible whole. The new nomads are not voluntary prisoners of expat life; they combine work, home, family, and leisure in a balanced mix. The new nomads are not necessarily rich; they earn a living by working from home or by working in open collective well-equipped working spaces with fast broadband Wi-Fi, a perfect cappuccino, and a choice of café-restaurants nearby to have their low-carb lunch. In the bigger cities worldwide, whether in democracies, dictatorial states, or socialist states, many such open networked work hubs have been established in recent years. The new open workplaces are inclusive of the new nomadic lifestyle that allows one to work at any time at any place for any client. The new nomads work by and large in the service industry, some of them in the new maker industry, representing an ever-growing part of world-making.

Connecting People, Spaces, and Machines

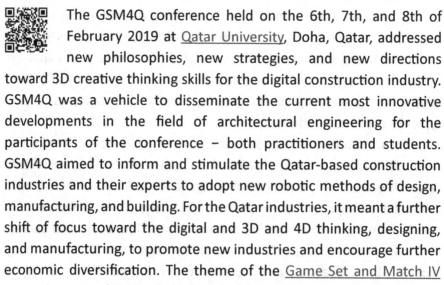

The GSM4Q conference held on the 6th, 7th, and 8th of February 2019 at Qatar University, Doha, Qatar, addressed new philosophies, new strategies, and new directions toward 3D creative thinking skills for the digital construction industry. GSM4Q was a vehicle to disseminate the current most innovative developments in the field of architectural engineering for the participants of the conference – both practitioners and students. GSM4Q aimed to inform and stimulate the Qatar-based construction industries and their experts to adopt new robotic methods of design, manufacturing, and building. For the Qatar industries, it meant a further shift of focus toward the digital and 3D and 4D thinking, designing, and manufacturing, to promote new industries and encourage further economic diversification. The theme of the Game Set and Match IV Qatar (GSM4Q) conference was Connecting People, Spaces, and Machines, with the subtheme The Informed Nomadic Monad. The GSM4Q conference covered a variety of specific yet strongly connected topics like Nomadic Swarms, Nomadic

Cities, Robotic Buildings, Informed Materials, Monads and Nomads, and Scalable Interactions. These and other topics were targeted by internationally renowned <u>keynote speakers</u> and challenged by hands-on workshops and <u>call for papers</u>. The GSM4Q conference topics were structured along the lines of the scalability of the connections between constituting monadic components. The leading themes for the conference sessions are

1) Informed Nomadic Swarms,
2) Nomadic Cities and Buildings,
3) Monads and Nomads, and
4) Scalable Interaction.

Day 1 was dedicated to the notion of the Nomad, and day 2 was dealing with the notion of the Monad. The two-day conference was framed into four discrete sessions: the day 1 morning session was structured around the theme of the Universe, the day 1 afternoon session dealt with the theme of the Community, the day 2 morning session addressed the theme of the Body, while the day 2 afternoon closing session covered the theme of the Cell. The <u>proceedings of GSM4Q</u> can be downloaded from QSpace, the institutional repository of Qatar University.

Where Are We Now

A compacted version of Where Are We Now is edited for the proceedings of the GSM4Q conference, published by Qatar University Press.

After 16 years of leading the Hyperbody research group as a professor of practice at the TU Delft, I looked at the Gulf region for further educational and professional activities. I was already familiar with the region, having lectured in Kuwait, Lebanon, Israel, and the UAE and eventually having realized an office building in Abu Dhabi. Our <u>projects in the Middle East</u> comprise design proposal requests for hotels (Flyotel) and office towers (Sports City tower, Al Wataniyah), and commissions for prestigious master planning projects, the redevelopment of Manhal Palace, the residence of previous ruler Sheik Zayed in Abu Dhabi, and a master plan for the relocation of the automotive showrooms to Automotive City, at the crossroads of major highways to Dubai and Al Ain. My continuous efforts to contribute to the building market in the UAE culminated in the realization of the LIWA Tower in Abu Dhabi (2014), commissioned by Abdullah Al Nasser, the CEO of Al Nasser Holding.

I was invited by the Department of Architecture and Urban Planning (DAUP) of Qatar University to educate the students, initiate state-of-the-art research, and innovate the curriculum, starting in September 2017, hired on a year-by-year basis. The challenging liaison to QU came to an unexpected end of service in the summer of 2019, after two years of enjoying living and working in Doha. Qatar was the most pleasant culture shock that Ilona and I experienced; it was more intense and more rewarding than the previous culture shocks when visiting the USA and Japan in the early eighties. No official reason was given; the dean of the DAUP assumed that the reason was my advanced age. Officially, the QU policy is retirement at the age of 60. Yet, I was hired when I was 67. The combination of being connected and unexpectedly being disconnected is why I chose to show the discomforting yet soothing video Where Are We Now by David Bowie as the musical intro to my lecture at GSM4Q. Where did we come from, where do we stand, and where are we going? My future has always coalesced with the here and the now. Many consider our projects futuristic, but for me, they represent daily reality, as a rule taking advantage of the actual technological and social achievements, as reflected in featured key built projects like the Saltwater Pavilion (Neeltje Jans, 1997), Web of North-Holland (Haarlemmermeer, 2002), A2 Cockpit (Utrecht, 2005), the Bálna Budapest (Budapest, 2013), and the Liwa Tower (Abu Dhabi, 2014). These internationally wide published projects are typically based on the paradigm shift

1) in design practice from 3D modeling to parametric design and

2) from mass production of a series of the same to mass customization of a series of unique building components.

ONL has been from the start in 1990 and has been since at the forefront of implementing the most actual design and execution methods into the built environment. Important sources of inspiration are complex adaptive systems, as I am fascinated by the swarms of starlings, trying to figure out what the rules are behind the ever-changing configurations of the swarm. I have endeavored to implement the principles of swarm behavior to the way building components are flocking together in dynamic complex adaptive design systems. The immediate file-to-factory link from parametric design scripts to robotic CNC-controlled production echoes the immediacy of how birds communicate in the swarm. With the experience of 16 years at the forefront of education and architectural practice, I was curious to see what impact parametric design to robotic production could bring about in the framework of Qatar University.

Abstract Thinking

The first project Ilona and I did together was the Design Methods design studio, open for third- and fourth-year students. Ilona's role as a visual artist was to lead hands-on workshops on abstract painting. Qatar University was, in those years, in the process of becoming NAAB accredited for a five-year Bachelor's curriculum, whose target has been achieved the following year in 2018. As a forward-looking full professor, I was instrumental to get the DAUP to pass the ongoing American NAAB accreditation. Eventually, the NAAB accreditation was rewarded at the end of 2018. The Department of Architecture and Urban Planning, under the auspices of the College of Engineering, is open to female students, Qatari students, and those from the region. Qatari students are without exception wearing a black abaya and a hijab; students from other Arab countries choose the colors of their abayas and hijabs freely, although many prefer to adopt the black style of the Qataris. It was a cinematic experience to see the students in their black abayas smoothly floating around the campus, proudly walking upright, wearing sneakers or high-heeled shoes of fashionable brands. A memorable event was the site visit with students that we organized in the spring of 2018 to the Singing Sand Dunes.

The blockade imposed by KSA, UAE, Bahrain, and Egypt, instigated by the interference of Trump, went into effect in June 2017, just a few months before we started in September 2017. The effect this had on the student's mindset cannot be underestimated. It meant a serious wake-up call for both Qatari and expats; they immediately knew that Qatar had to become more resilient and independent from the import of basic products. Seen in hindsight, the blockade that has been resolved by now was the best that could happen to Qatar. Qatar became morally, economically, and politically stronger, and that reflected positively on the students. They were quickly becoming militants in a peaceful way, being self-aware and increasingly interested in their historical roots. The current education framework at the DAUP is by and large traditional; there is little space for experiments, and much is framed in the brief and the assessment criteria. Plans and sections are the norm, and projects are presented on the mandatory A0 boards, which practically rules out the use of videos and animations. In the assessment criteria, one earns points for complying with the rules, not for being imaginative. Complying with fire safety measurements has the highest priority. We decided to turn that around 180º and asked the students to start from a purely abstract expression, without reference to nature or existing objects, without any functional program of demands – no portraits, no trees, no animals, no houses, no cars, and no coffee cups. We believe that abstract design thinking is essential to become an architect or artist since one needs to construct a reality that has not been there before. We asked them to put their paintings together to form a larger painting, stimulating teamwork and creating shared values. Exercising abstract thinking in painting brings along the capacity for conceptual design in architecture. Their ability in conceptual design thinking was further developed in a series of subsequent architectural design studios. My ultimate goal was to prepare the students for parametric design to robotic production methods, which is based on a similar abstract expressionist way of thinking. Students need to abandon the traditional method of coding architecture with two-dimensional plans and sections. Instead, they must be trained in processes, spatial relationships, and performativity of constituent components. Conceptual thinking is the first necessary step to understanding the principles of parametric design. The QU students needed to restart from scratch – from abstract thinking and expressing oneself in pure gesture, pure color, pure light, and pure relationships between the gestural components, in collaboration with their fellow students. By asking to put their abstract paintings together with a group of four to five group members to form a bigger picture, they are made aware of the performance of their immediate neighbors. As in the swarm, one must find a strategy for how to relate one's imagination and performance to that of their immediate neighbors. Abstract thinking and doing prepare them for the structuring of architectural design in terms of relationships between the components, the performance of the components, and the personal styling of the components. As per definition, a spatial concept is expressed in three dimensions, by parameterizing relationships. Navigating 3D models may show spatial relationships, while plans and sections remain poorly informed cut-outs from the 3D model. After full abstract and spatial thinking has been achieved, the traditional coding, which is eventually necessary to get a building permit, is embedded in the 3D model. Plans and sections are dead-end streets, meant for printed-out presentation boards and bureaucratic approvals.

Plans and sections are not essential to the design process; they are deviations from the process. There is no feedback possible from a 2D drawing to a 3D model. The 3D model is what it is, while 2D plans and sections at best suggest a spatial relationship but are otherwise inaccurate to describe the spatial experience. A design process needs to be bidirectional; one always needs to be able to go one or more steps back and choose another road to proceed. Parametric 3D models are dynamic design engines, allowing the design to gradually grow into full detail, fit for communicating with the production machines directly. Conceptual design thinking forms the basis for parametric design, which in turn forms the foundation for interactive and proactive structures that

347

live their lives beyond the design process. The brief I gave for the sustainability design studio for fourth-year students that I conducted was to design a self-supportive diamond-shaped city on an artificial island off the coast along North Beach in West Bay. I gave each of them a triangular plot, a small maximum footprint that would urge them to go vertical, and the requirement to physically connect to their neighbors higher up in the air, to make them negotiate with their neighbors in an abstract new context. The exercises that we had executed with them before were paying off.

Interactive Tower

In the meantime, in the fall of 2017, I wrote several research proposals, internally for Qatar University for small 20K USD research projects, and nationally for 600K USD 11th cycle of the National Priorities Research Program (NPRP) grants. Three proposals were rewarded: the GSM4Q conference, the QU MANIC research for Multimodal Accommodations for the Nomadic International Citizen, and the national research project Qatar Robotic Printing, which research eventually started in November 2019, after my end of service. I had to find a colleague at the College of Engineering (CENG) to take over my role as Lead Project Investigator and continue to be involved as an external consultant. Meanwhile, I developed, together with a team including the then Head of Department Fodil Fadli and his teaching assistant Sara Zaina, a concept for a 16-m-high interactive tower, named ICE-Café, to be located in front of the entrance of the College of Engineering, functioning as a coffee shop and an iconic public art structure to represent the dynamic nature of the college. The interactive tower swings and rotates in its entirety, having a life of its own. The 16-m-high tower bends itself in the direction of prevailing winds to catch a cooling breeze and bring the breeze down the coffee shop on ground level. The structure consists of nodes and edges, an interior and exterior skin. The nodes adjust passively to changing rotations, and the edges have a small electric piston embedded, and they adjust their length actively according to the programming. The interior and exterior skin is flexible and adjusts passively to the changing rotations and bends. The programming of the programmable components is open to external data taken from weather stations. The programming responds in real time to the wind direction and the speed of the gusts of wind. When there is no perceptible wind, the tower gently waggles back and forth, just for the pleasure of moving. The interactive tower is meant to be an urban orientation point, being twice as high as the CENG building, in front of which it stands. The design was presented to the dean of the CENG as a demonstration of the real-time behavior of parametrically designed and robotically produced assembly of programmable components. Although I secured the financial support of a steel manufacturing company that would deliver the materials for free, it was not considered a feasible project. Is the world ready for interactive structures yet? When I presented Trans_Ports back in the year 2000, I expected to proceed to realization within five years. Twenty years later,

nothing of the kind happened yet, despite many attempts from my side to inspire decision-makers, up to a meeting of the Dutch Minister of Infrastructure and Environment Melanie Schultz van Haegen. She was impressed by the concept of the interactive Dynamic Sound Barrier, but her staff did not follow up.

Maidan Monument

Together with a team of then-Qatar-based artist Diogo Esteves and architects Yassmin Alkhasawneh and Najeeba Kutty, Ilona and I joined an international competition for a public art project in Kyiv. The brief was to design a monument for the Hundred Heavenly Heroes – those who gave their lives in the 2014 revolution. The images of the

Maidan Revolution made a deep impression on the world; photos of the dramatic sceneries were propagated all over the world. Especially the image of the thousands of people and the Ukrainian flag rising from the crowd against a background of fire is deeply imprinted into the collective memory. The images of burning barricades formed the backdrop of the humanitarian drama that unfolded on Maidan square and its immediate environment. We wanted to capture these iconic images in a monument that represents a combination of abstraction and literal representation of the concepts of democracy, open society,

and enlightened citizenship. We chose to take the flag as the virtual carrier of the geometry of the Maidan monument. An imaginary three-dimensional volume of a waving flag is randomly populated with exactly a hundred points, each of them representing one of the hundred heroes. The lasting impression of the image of Ukrainian people holding the flag in a joint effort is translated into the dynamic wave shape of the constellation of the hundred nodes. The nodes form a spatial three-dimensional network, thereby representing the mutual connectivity and the robust collaboration between the heroic people, whether they belong to the heavenly hundred or the thousands of their fellow protesters. The constellation of nodes resembles a collective brain with brain cells and synapses, operating in sync to get things done. The Maidan monument represents a network, a network of nodes and edges, whereas the main

component is the node. Basically, following the Exoskeleton Grasshopper plugin that we used, the node component includes half of the spike; the node and the spike are the same thing, the nodes growing toward each other via its tentacles.

9th Architecture Day Exhibition

I was invited in May 2018 by the head of the department to design and organize the 9th Architecture Day Exhibition, the yearly exhibition of student work. Probably as a sort of ability test to work under high pressure, I was given a mere 2 weeks to conceptualize, design, organize, and execute the show. My colleagues told me that typically the head of the department distributes last-minute challenges to the professors and the assistant professors, to test how they respond, and use it to create a faculty dossier for the yearly appreciation cycles. With the help of the tall Qatari teaching assistant Abdullah Al Nuaimi, like all Qatari men wearing a white <u>thobe</u> and ghutra or keffiyeh, and a bunch of students, I managed to create a unique setting in the main hall of building C7 using wall elements from a previous exhibition within that short time. Being bored with the regular A0 posters students have to produce for critiques and reviews, I wanted something different. I decided to show as many models as possible and also included some of the abstract paintings that my students produced the semester before. The show had a different feel to it than earlier and later shows, more personal, more architectural, more artistic, and more liberal. Somewhat contradictory to the task of doing abstract painting that we asked our students the semester before, we invited them to hang their often excellent, realistic portrait drawings on the curved wall, such that they could show off their traditional drawing skills as well. After the show, the portraits were removed, presumably while one student filed a complaint against the somewhat liberal nature of some of the drawings. Who exactly is the higher management, you never get to know. As they say in Qatar University, this is how the Arab culture works; one does not question top-down decisions. The middle management of universities in general, and in Arabic universities in particular, has mastered the technique of avoiding responsibility, on the one hand by shifting responsibilities downward in the hierarchical chain, and on the other hand by living with instructions from higher up in the hierarchy. As a consequence, it is sometimes hard to get things done; forward-looking initiatives typically die in the bureaucracy that self-protects the middle management. One colleague conveyed to me the meme that faculty personnel in Qatar must have two pockets: one pocket for the generous tax-free income, and the other for bottling up bureaucracy; once one of the pockets is full, you will leave Qatar. Living and working in Qatar is an interesting mix of freedom and restriction. Freedom because of the open desert, the services offered, and the economic potential you feel every day, and restriction because of the institutional bureaucracy that is slowing down new initiatives. And some things that we are used to having in Europe are simply not there. There are hardly any hardware shops, which is fully understandable since neither Qataris nor expats spend their time doing odd maintenance jobs at home.

Two Legs

Working and living in Qatar is experiencing the future and the past simultaneously. In the past because many sheikhs are *de facto* feudal masters who build castles for their extended families, as in feudal Europe was done by kings, princes, and dukes. Knowing Qatar from the inside out, I do not criticize the palaces of the sheiks, while we as Europeans have no problems with the many castles in Europe. Qatar is a society undergoing an accelerated transition from a tribal community via the current feudal society toward a new form of democracy, to which we do not yet know where it will lead. Living in Qatar means living amid a fast and intense transition, which feels like living inside evolution in many ways. The article that appeared in the Guardian about the death toll of migrant workers at the FIFA 2022 stadiums is tendentious, and therefore misleading. In the first place, most of the workers who died did not work in the stadiums but outside in the desert on the infrastructure, and many deaths are not directly work-related. Second, there is a lot more construction work going on in Qatar as compared to Europe, meaning more deaths. Third, Qatar is improving fast; the position of migrant workers has substantially improved since then, and new compounds are being built. In comparison, in the Netherlands, there are on average some 25 work-related deaths per year in the building industry. But it is true that most low-skilled laborers still live in poor conditions, often with many people in one room. Poor working conditions for the lowest-paid workers were considered normal during the building of Europe's palaces in the past centuries. This aspect of the still-feudal societal structure in Qatar is what it means to live with one leg in the past. My other Qatar leg lived in what Europe's future might eventually look like. A future that adopts new technologies faster and more efficiently is feasible in a Europe that unfortunately more and more tends to adopt regressive nationalist policies.

Qatar is a society that develops at a greater speed than can be maintained in our history-rich environments. Huge sums of oil and gas money are spent on education, healthcare, science, sports, and tourism. Like Saudi Arabia and the UAE, Qatar is also determined to diversify its economy, transform into a green economy, become a destination for highly skilled workers, become a hub for new scientific developments, and become a destination for professionals and tourists. The Gulf region is quickly becoming what is the Mediterranée for Europe. Qataris and Emiratis do not take history as something to defend; they take it as pride instead, but even more so as a driver for success, progress, and happiness. It feels like being inside the future and simultaneously inside the past in Qatar. Living and working in Qatar stretches your mind to the extremes of comfort and confusion.

Robotic Workshop

One of the most rewarding experiences during our stay in Doha was the conceptualization and execution of a robotic workshop.  Previously, we found new ground for art and research with the robotic painting project Machining Emotion, instructing large industrial ABB robot arms to paint on canvas, merging emotive gestural marker strokes with algorithms that execute the robotic paintings. At the time, nowhere in Qatar could a large industrial robot arm be found, but, fortunately, I met John-John Cabibihan of the College of Engineering at Qatar University, and we teamed up to offer a robotic workshop at Doha Fire Station, open for students of Qatar University and artists in residence of Doha Fire Station. Cabibihan brought in his small humanoid NAO robot; the only industrial robot at QU was a medium-sized very slow pick-and-place robot. We bought two vacuum cleaning robots to play with, an LG HOM-BOT and a Roomba by iROBOT. These cute little ground scrapers soon became our favorites. For Machining Emotion Qatar, we set up the boundaries for two large paintings: the first one is an additive painting for the vacuum cleaning bots to move on the canvas inside the 2 × 10 m boundary, and the other one is a subtractive painting of the same dimensions. The additive painting was the domain of the two vacuum cleaning robots with wide-tipped acrylic markers attached to their back, adding acrylic paint to the large canvas, following predefined linear and circular paths, bouncing into the wooden boundary frames, and rebounding back into the arena, occasionally bouncing into each other and changing course again. It was a truly joyful experience to manipulate and watch the robots make their moves, producing a 2 × 10 m work of art. Watching the process of making the subtractive painting meant another mesmerizing experience. For this one, we scattered some kilograms of white flour onto the floor inside the 2 × 10 m working area. The vacuum cleaning robot did what it is designed for and sucked up the flour, thereby leaving traces of its itinerary. The best results were achieved by having the robot circle in spirals. Once one subtractive painting was done, the vacuum cleaner was full of flour; we reused the flour, wiped it out onto the flour, and made another one. The robotic workshop at Qatar University taught us that even with a very limited budget one can do meaningful projects with simple robotic devices. Using equipment for other purposes than designed for opens up endless possibilities for innovation in art and architecture.

Earlier at Hyperbody, we typically worked with a modest 3000 EUR budget for each hands-on design studio, which is a very small budget compared to the most impressive results of the interactive student projects. Students learn how to become socially intelligent to find material support from outside of academia, how to develop practical skills to 3D print custom-designed components, and how to connect, tinker, solder, and program the Arduinos and the sensors to bring the interactive installations to life. The same holds for most ONL projects, where we always had to find ways to give something more within standard budgets. Where others have become masters in raising the budget and "talking" money out of the client's pocket, I specialized in maximizing innovation within modest financial frameworks. Not a single project went over budget; I always found it my social responsibility to deliver within predefined mutually agreed conditions. On the other hand, modesty is not the best strategy to get things done; it may well be that my acceptance of modest conditions in combination with a lack of arrogance has stood in the way of further growth of our endeavors.

QRP

One of the hottest research topics of some years ago is robotic 3D printing. I worked with Cabibihan on a Qatar-University-based research project in an attempt to develop a hybrid additive–subtractive all-inclusive component, using 3D printing with a local concrete mix for the additive part, and 3D robotic milling for the subtractive parts. I acquired the NPRP 11th Cycle support for a research project called Qatar Robotic Printing during my years as a full-time professor at the Department of Architecture and Urban Planning at Qatar University. After my second retirement, this time from Qatar University, I handed the Lead Project Investigator role over to John John Cabibihan and changed my role to that of an external consultant, working from our new base in Nagymaros, close to Budapest, Hungary. We aim to build a series of one-to-one prototypes for bespoke building components, featuring a full integration of building physics, structural strength, and compelling esthetics. The concept for the hybrid component combines three basic materials:

1) limestone-based concrete for the compression forces,
2) polystyrene foam for the insulation, and
3) CNC-cut steel for the reinforcement.

The limestone-based concrete mix is a hybrid in itself, with limestone cement, polymers, glass fibers, and grained rubber in the mix. Qatar cleared a mountain of waste rubber tires stored at landfills from the numerous SUV vehicles. The 3D-printed components are connected using magnets that are embedded in the edges of the 3D-printed components. Collaborating robots ensure the exact positioning of the magnets and thereby the exactness of the position of the components in relation to each other. Using magnets guarantees easy assembly and future disassembly. Speculating on the future of 3D printing and robotics on the building site, we applied for a grant for an innovation program Expo Live in Dubai. Our proposal comprised the robotic strategy to 3D print a complete complex building employing an army of <u>on-site robot arms</u> to produce a series of unique components with respect to building physics and finishing surfaces with well-performing structural components using locally mined materials. This research project was not awarded. The ultimate logic of 3D printing is to bring the printer as close to the final destination of the components as possible, by establishing temporary on-site factories and minimizing the transport of raw materials and heavy prefabricated components. For 3D printing in the Arabian deserts, we are considering limestone with polymer binders and fibers for the reinforcement, together with grinded, used car tires for creating insulation cavities in the concrete. The concept for 3D-printed components is to put the material only where needed and as needed, as in animal bones, meaning a higher density of concrete along the edges and surfaces, transforming toward the center of the components into an open structure with ubiquitous insulating voids.

There is a world to win using 3D printing but only when going all the way toward all-inclusive components. Not just a roof, not another pavilion, not just a 3D-printed cement wall. Virtually, all 3D-printed examples I have seen in the latest years are at best partial solutions, not integrated solutions for a complete building, including floors, walls, roofs, insulation, climate control, and esthetics. A building is a complex amalgam of different and sometimes conflicting performative requirements that must be resolved in the constituting components. The actual state of 3D printing today is by far not sufficiently developed to form an alternative for bigger buildings. And it might as well remain an illusion that it will ever be. A more drastic change is needed in how buildings are designed, produced, and assembled. Dubai may be boasting that by 2030, 25% of all new buildings will be 3D printed, but that is not a realistic goal; I have not seen any real progress in the first five years after it was proclaimed. When concrete pouring pumps and cranes are robotically teleoperated, which is very well possible by then, that might be regarded as 3D printing. Then Dubai can boast that it surpassed its initial goal. But in fact, nothing will have changed really. Traditional technology will be automated, but no real innovation will have taken the stage. Real innovation means transforming the polluting cement and concrete industry into a clean distributed and ubiquitous operation, transforming mining into a form of landscaping, leaving behind an environment with more flora and fauna than was found, building smart by parametric design to robotic production, utilizing swarms of robots, building only when needed and as needed, and designing, producing, and using just that, just there, and just then.

MANIC

The Multimodal Accommodations for the Nomadic International Citizen (MANIC) research, for which I got granted a 20K USD budget at Qatar University, builds upon earlier studies by the Dutch architect John Habraken on an open building strategy providing flexibility in the use of the homes, and on the visionary art project New Babylon by the Dutch Artist Constant Nieuwenhuys, and a follow-up of our Pop-Up Apartment initiative. In New York in the eighties of the last century, the concept of living and working in spacious lofts in abandoned industrial buildings became popular. In our first culture shock visit to Manhattan and Brooklyn in the early eighties, we visited an architect friend who was living in such a spacious loft. We were immediately impressed; it resonated with how we have lived and worked most of our career in abandoned school

buildings, enjoying the large multifunctional spaces both inside and outside the former school building. We lived and worked in schools up to around the year 2012 when the crisis and our debts at the banks became unbearable and we had to sell our domain at the Essenburgsingel 94c in Rotterdam, a former gymnastic hall. Typically, an NY loft is a spacious abandoned industrial

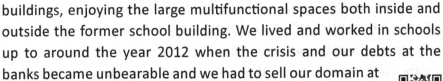

storage floor, hired or bought by double-income citizens to live their fashionable lives, free of the constraints of traditional ways of living. The new citizens are the harbingers of new internationally oriented metropolitan culture. Building lofts as new buildings have become increasingly popular in the last decades, allowing for the tenants or owners to organize the interior layout and define the use of the spaces themselves. The spaciousness though is by and large

357

compromised by traditional constraints in building technique and by lack of cultural agility. The MANIC research takes the loft concept and the New Babylon nomads to a new level, introducing the concept of fully programmable furniture and fit-out to transform the available space in real time into different types of usage. I chose a basic unit of 50 m^2, equivalent to the size of a standard schoolroom, column-free thanks to a span of 7.20 m, larger than is common practice in housing, where the standard span is 5.40 m, in more luxurious homes extended to 6 m. The span of 7.20 is common practice in office buildings. The choice for the larger span brings homes and offices on the same denominator; it unifies the idea of what an office is and what a home is, and thus facilitates quick conversions from home to office and back. The idea of MANIC is this: all furniture is programmable and pops up only when and as needed, popping up from a double floor with a usable depth of 1.20 m. The 50 m^2 unit can be instantly (re)programmed to become a 50-m^2 bathroom, a 50-m^2 bedroom, a 50-m^2 living room, a 50-m^2 kitchen, or just a void space of 50 m^2 to leave behind visual and spatial stimulation. Any combination of these functional pop-up components can be programmed. Multimodal accommodations facilitate the migratory life of the Nomadic International Citizen, unbound in time and space. International nomadic citizens feel at home whenever and wherever they are. They feel at home in any place that they can make their own by programming that place in real time. The concept of programmable multimodality is a logical step beyond mixed-use developments, whereas the mixed-use is democratized and customized, leveled up to the grain size of an individual unit of 50 m^2. Each spatial unit may be booked for any activity, leisure, and work alike, and for any period. The design allows for the possibility to book for the short term or the long term, to book for lodging, living, office, workshop, dining, or hanging out, to book for rent or to buy. In the context of the QU-based research, I developed the conceptual framework for a Ubiquitous Booking App, booking.com, and airbnb.com to program the configurations of one's future temporal and local home in advance. The ubiquitous booking system facilitates the migratory lifestyle of the nomadic 21st-century international citizen.

MANIC means Habraken's open building taken to the extreme, and Constant's New Babylon brought to life.

West Bay North Beach

I taught at the Architectural Association in the years 1988–1989. In my Intermediate Unit #13, I had one female student from Qatar; she was in London together with her husband. Later, I understood that female students are not allowed by their family members to travel alone, let alone study abroad. Although the Qatari law does not explicitly forbid independent travel by women, traditionally, the men are overly protective; they reportedly do not want to expose their spouses to possible insecurities, inside and outside Qatar. When walking in the park with their spouse and children, the men always walk a few steps behind, supposedly to be ready to act when danger is coming their way. Although much has changed in the last decades under the liberal regime of the former Sheikh Hamad Al Thani and his (third) wife Sheikha Mosa, the traditional relationships between men and women are deeply rooted in the still predominantly tribal Qatari families. Tribal is seen as the equivalent of what is known as feudal in Europe. After 30+ years, I met Badria again at the excellent Helmut Jahn Doha Exhibition and Conference Center (DECC) where she had a stand with her own Badria Kafood architectural office. When I reached their stand, she was not there; she was out praying. Typically, Qataris pray five times a day, which is another deeply rooted cultural custom that is bound to stay, whatever technological progress Qatar will be going through. When she returned, I saw a different Badria from that in London. In Qatar, she dressed traditionally in the black abaya and the black hijab; in London, she dressed more colorful, yet wearing the hijab. Badria is the only female architect that runs her own architectural practice in Qatar. Later, I heard from another leading female architect that they wanted to run their own offices but were discouraged by their husband and she gave up that idea. Badria conveyed to us that she is regularly asked to spend much of her time on community services for jurisdictional disputes, to downplay her work as a leading female architect. We visited her office, which she runs with a liberal partner from Lebanon, discussing how we might work together. When there was one of those rare design competitions announced in Doha, I decided to ask her to join forces to submit a design for the West Bay North Beach competition published by the semi-governmental institution Ashghal that oversees all infrastructure works including public parks in Qatar. To be able to enter the competition, registration in the Qatar engineering community is required; so we registered under her office's title. I asked a former Hyperbody student and parametric design programming specialist Arwin Hidding to join the design team along with the well-established consultancy firm Northcroft Middle East, by then led by Nigel Couper, to check the Bill of Quantities to estimate the total project costs. Our concept for the West Bay North Beach master plan aims to provide a publicly accessible beachfront along the whole stretch, while the commercial target of building 2,000,000 m² real estate is achieved by interweaving generic towers with iconic ones. The generic towers are meant to be able to be designed and built within a very short time to be ready by 2022. The iconic buildings on the triangular new artificial islands are meant to form tourist attractors to comply with the governmental document Qatar Vision 2030, to be developed between 2022 and 2030. The generic towers, the triangular treasure islands that contain the leisure parks, and the iconic landmark buildings are connected by an extensive network of smoothly shaped elevated traverses, providing shade and comfort for pedestrians and bicyclists. It is not so much the high outdoor temperature that is the reason for Qataris and expats to live an indoor life but the lack of shade and natural cooling. If only Doha planted a million trees and narrowed the city highways to give way for trees and cafés, the city would already be much more friendly for pedestrians and bicycles. In our West Bay North Beach master plan, numerous covered traverses are added, hosting cafés and terraces in a shaded but well-lit environment. Perforations in the roof and the sides of the traverse network are based on solar diagrams, to let in the sunrays wherever functional and not cause excessive heating up. Ample parking and service areas are foreseen under the park areas on the triangular islands. The spacious traverses form an extensive network; they link everything to everything and everyone to everyone, and they connect the generic towers and the existing area of West Bay North Beach to the new developments extending into the bay of Doha. Inserted between the existing towers in the West Bay North Beach area, our master plan foresees the realization of several 200-m-high urban farming food towers, providing for food to be produced and consumed locally in West Bay. To the west end of West Bay North Beach, a very dense urban forest will separate the new developments from the indicated sensitive area, where the family of the emir presumably owns some palaces. We have not heard any official response to our competition entry. Just recently, I found out that they changed the brief for the <u>Vision Competition for the West Bay North Beach Development</u>, without the over-ambitious target of building 2,000,000 m^2 of real estate.

Palaces

One of the palaces adjacent to the North Beachfront has a particularly intriguing shape, definitely not traditional as other palaces that are built in a blend of Arabic and Romanesque styles. Seen from above, it looks like a giant octopus, with five curved legs rotating around a dome-shaped center. I was told that it might be the <u>summer palace of Sheikha Mosa</u>. No information can be obtained about the design of palaces; it is kept strictly secret. It is forbidden to take photographs in the proximity of royal palaces. Designers of the palaces, Massimiliano Fuksas and Thomas Leeser among them, are not allowed to publish anything about their palace designs. As Leeser told me once, they are contractually not allowed to take pictures when visiting the building site either and are typically not invited to see the end result. Time will tell though when the feudal Middle Eastern leadership style eventually will have evolved into a more democratic system, as happened in Europe, future generations will enjoy visiting the palaces, and stand in awe of the rich gold-plated decorations, as we do in Europe. Some interior designers of the palaces have adopted a modern version of baroque, enhanced with oriental motifs, and sometimes with some explicitly modern additions. Mohanad Barakat, a befriended Doha architect of Syrian background, who runs the office Baroque Design & Build, specializes in the design of such new hybrid style interiors, while his brother Nashaat runs the company that produces the artifacts. Syrians are famous for their exquisite craftsmanship. There are only a few exceptions to the dominant classical style. Along the D-ring road in Doha, I spotted an abandoned palace in Paul Rudolph style, or perhaps actually designed by Paul Rudolph without the public knowing it. I asked many of our Qatari friends if they knew anything about the palace, and one said that it is probably abandoned because of a disagreement on one of the many less prominent branches of the royal Al Thani family. That palace is an excellent piece of modern heritage; it would be a huge loss when left to decay further.

HyperCUBE

My interest in the Food Energy Water Nexus was triggered by the

Belmont-supported international research project called M-NEX, which was co-initiated by Anna Grichting Solder, who had been a long-time Assistant Professor at Qatar University when I came in. Ilona and I had met her before; we went to see her two years before I was appointed professor at Qatar University, city hopping with the 45 minutes flight from Dubai. That was during the summer break; the university was almost empty, which did not give the full impression of the campus that we had later when crowded with female students elegantly floating in their black abayas from their cars to the department building, from the department building to the female food court. The Qatar-based part of the M-NEX project aims at turning the campus of Qatar University into a green field laboratory, growing food in permaculture, to be consumed in the food court. I took over the Principal Lead Investigator role from Grichting after her contract was unexpectedly terminated, after six years of well-appreciated educational work and excellent networking at the university, but terminated without a reason given.

Almost immediately after I took over the project, after a one-year-long bureaucratic process, not in the least from the side of the Belmont grant, unexpectedly, I was confronted with the unexpected termination of my contract, before I could even take a proper lead in the project. I could only guess what was going on here. Was there a relation between the terminations of two liberal professors who were both popular among the students? In the short period of leading the project, I aimed to turn the project into the direction of a more encompassing master plan for the campus, incorporating the rather soft ideas of the original M-NEX ambition into an ambitious master plan featuring two long green boulevards with thousand large shade-providing trees, banned for car traffic, fit for transportation by electric vehicles, and including a prototype HyperCUBE building that foresees in its own consumption of food, energy, and water. The design principle for the HyperCUBE is in concept straightforward and game-changing for future buildings. The HyperCUBE produces as much food in a reasonable diversity for the population of researchers inside the structure. The main volume of the HyperCUBE is a compact cubical form, wrapped in a lightweight open structure, shaped as an outward projection of the cube, and hence the name HyperCUBE. Solar panels and windmills, glasshouses for urban farming, and fine steel meshes to harvest water from the moist air are attached to the lightweight structure. The growth of food is designed to be based on vertical aquaponic and aeroponic farming, growing algae, mushrooms, and fast-growing plants like tomatoes and peppers, whereas every single plant is monitored and provided with exactly as much water and nutrition as needed for optimal growth. The production of electricity is provided by solar panels distributed over the external structure to catch the power of the sun wherever most effective. For harvesting the water, the external structure is equipped with a fine steel mesh that functions as a water condenser. Although largely self-sufficient, the HyperCUBE is not meant to be an autonomous structure but an active hub in a network of other food, energy, and water-producing buildings in the CU campus. This radical FEW Nexus design concept may be applied to existing buildings as well. It comes down to the addition of a spatial second skin, loosely wrapped around the five faces of a building. The first idea for the concept was born in a conversation with my colleague climatic expert Djamel Ouahrani, with whom I shared an office at the department. We discussed the idea of covering existing mosques topped up with an additional lightweight structure with solar panels.

The HyperCUBE building, located at the armpit of the east-west and the north-south boulevards, is meant to become the home of the Center for Sustainable Development, with whom I partnered for further research.

The key issues that are addressed apply to any existing or new building. In 2020, I was approached by befriended architect Mohammed Al Sabbagh of LC Consultants in Doha to join forces in a competition for the completion of the Corniche Tower, a development funded by Emiratis but was stopped due to the financial crisis of 2008, and never restarted. The four-year-lasting economic and cultural blockade by the UAE, Saudi Arabia, Bahrain, and Egypt against Qatar, euphemistically referred to as the diplomatic crisis, was not helpful either to get things going again. The raw concrete 22-floor tall structure of the Corniche Tower stands there on its prominent location for more than a decade unfinished, at the side of another even bigger tower, the Burj Al Mana, that was left unfinished back then but recently got refurbished in anticipation of the FIFA 2022. I proposed to top up the concrete structure with three lightweight structures: the first for harvesting solar energy, the second for harvesting food, and the third for harvesting water. The highest addition that I proposed is a more than 100-m-high solar tower, high enough to generate the energy needed for the refurbished existing building, on top of the central core. On top of one of the two wings that are designed to be repurposed into multimodal housing units, a structure is projected to grow food using the most advanced urban farming methods in robotically controlled environments. On top of the other wing, an open structure is foreseen with an extensive network of steel meshes to harvest water from the other, as to supply the building's water consumption. Detailed calculations will eventually show how high the additions in reality will have to be. The design is flexible to calculations and applied technology; there is a direct dynamic relationship between the need for space and structural components, and the need for food, energy, and water, when and as needed.

The HyperCUBE stands as a prototype for future buildings that are self-sufficient in terms of food, energy, and water, but not autonomous in the sense that they are not connected to the network. In Another Normal, buildings will become production facilities, active components in a local network of similar performing components. As such, the hyper-buildings behave like the birds in the swarm. They see each other, they stay in a dynamic relationship with each other, they listen, they think and talk to each other, and they alternately follow or lead each other to find the optimal balance for their local neighborhood. Similarly, neighborhoods will communicate with other neighborhoods to find the perfect FEW balance for the city as a whole. And cities communicate with other cities, within one country, but just as important are networked to cities in neighboring countries, balancing, and negotiating the production of food, energy, and water. In Another Normal, cities are the hubs that communicate with each other; cities manage the world, not countries.

Little Babylon

Rezone, a design and event studio based in Den Bosch in the Netherlands, led by Tessa Peters and Rolf van Boxmeer, invited us to design an inflatable and interactive sculpture, in a team with Erik van Dongen of the Air Design Studio, and the app developers Marijn Moerbeek and Thomas Rutgers for the interaction. The inflatable sculpture functions as a mobile installation at festivals to collect data about the city it has landed in and interact with the citizens of that city. Originally, we called it the Mothership, as it was designed to host a flock of drones that would fly out to collect specific samples, sounds, and images, from the immediate environment. The concept has changed since and was redesigned to retrieve data from Twitter feeds; the project was renamed Little Babylon, while

based on words and language from Twitter feeds instead of sounds and images. The mothership Little Babylon communicates the Twitter data with the audience in a dynamic audio–visual performance. The behavior of the inflatable sculpture adds a digital, emotional, and social layer to the urban condition. The soft sculpture personifies the digital city; it mines Twitter for specific topics and translates the data into movement, light, and sounds, thereby uncovering specific feelings and emotions of the city's inhabitants. On a smaller scale, the public interacts with Little Babylon by posting Twitter messages using the required hashtags. The sculpture repeats the Twitter post after some 10 seconds as if to say that it has understood the message and adjusts its emotional behavior, depending on the emotional content of the message. This unique blend of digital data and dynamic architecture portrays, in the form of a big three-dimensional data visualization, the moment-to-moment emotions of the citizens in real time. The real-time performance of cute Little Babylon stands as a prototype for interactive components in a network of similar real-time operation processes with which it is communicating. Little Babylon, as an art installation to be extrapolated to complete buildings, stands for an active component in a living city lab. A city is a laboratory, where digital data inform physical structures that are linked to each other in a dynamic network. Implementing interactive art installations is a first step toward a city that evolves not only by adding new structures to change its appearance but also by developing the behavior of the constituent city components in real time. Acting structures assign a new component to what it means to be a dynamic city; buildings are becoming the acting components in the cityscape. Little Babylon was first installed in the courtyard of Doha

 Fire Station in November 2018, and later in February 2019 in front of the Main Library at Qatar University during the Game Set and Match for Qatar (GSM4Q) international conference, which we initiated and were awarded by a national grant. Getting initiatives approved by the QU authorities, to have things transported and delivered in time, turned out to be a challenge in itself. All possible bureaucratic thresholds have to be taken to get things done; at first, it seems impossible to get through, but, typically, by last-minute improvisation, it miraculously ends up well in the end.

Components for Public Art

In the fall of 2018, I was asked to lead Design Studio I for first-year students of DAUP Qatar University. One of the numerous committees of the department had concluded that to modernize the curriculum, for which I was hired, I should teach first-year students and lead them in subsequent years toward parametric and interactive thinking. They believed that fourth- and fifth-year students were already too framed by the rigidity of the system. Tutoring staff at Qatar University typically complains that the students are lazy, cheating, and copying each other, but they do not realize that it is them who fail to inspire the students. After one year at QU, it became clear that what they expected from me was a better version of their teaching. However, what I delivered was beyond their framework of validation, using extensive spreadsheets with detailed predefined deliverables that are dealing with mainly technical drawing aspects, with no rows or columns about design thinking or spatial development, let alone imagination or vision. The narrow interpretation of the standard assessment criteria of the American NAAB, the National Architectural Accrediting Board, a private organization that facilitates the accreditation of Middle Eastern universities, is imposed by the College of Engineering, of which the Department of Architecture is a subsidiary. Having studied the NAAB student performance criteria shows that it would have been very well possible to adopt a much more creative interpretation of the student performance criteria and still pass the test of accreditation. One only has to check the student outcomes of the SCI-ARC in Los Angeles, which also is subject to the NAAB accreditation, to see how much room for interpretation the NAAB gives. To break the rigid assessment frame, I designed a challenging brief for the first-year students, stimulating the students to think abstractly and parametric from the beginning. I challenged them to think as designers, not draftsmen. The first brief I gave them was to choose a household object, coffee machine, vacuum cleaner, or mobile phone holder, and dissect it into its constituting parts. Then I asked them to draw these parts as if they were the designer. Only when thinking as the designer of the parts, one will fully understand the why and how of the dissected parts. I asked them to show how the parts fit together and how they work together to perform their function. And then I requested them to describe how the parts were made, what materials were used, and how these were produced. The next task I gave them was to analyze parts of a car and the context they were in, and how they are related to the immediate neighboring parts. I took the students on an excursion to the exorbitant car collection of Sheik Faisal Al Qassim, in the middle of the Qatar desert, yet easy to reach via half an hour drive on one those ultrawide eight-lane highways that are under construction all over Qatar. The collection consists mainly of American cars that were imported in the beginning years of the oil and gas boom in the sixties and seventies, but some older models, including a variety of trucks, from the fifties, are present too. I asked my students to focus on the headlights, to study them from all angles, from the top, from the side, and from the front, and find out how they are related technically and stylistically to the neighboring parts of the car. Then I asked them to make a physical model of the headlight in its context and make a drawing and/or 3D model of it. Some students proved to be surprisingly good at drawing the car in its totality as well. By this method of taking things apart and understanding how the parts act in their context, they self-taught how a complex object like a car is an assembly of different components that are designed to fit exactly together. From assembling parts to cars to components being assembled to form a building is then a comprehensible step. By this hands-on analog teaching method, they readily understand the importance of building relationships between the constituent parts; in other words, the principles of parametric design.

The final part of the semester was dedicated to the design of a large object of interactive public art. The only requirement I gave them was that they had to design the artwork as an assembly of different parts, whereas not a single component is allowed to be the same and design the interaction with the public. For the presentation,

I prohibited them to present on A0 boards, as in the other design studios, it is usual practice and mostly mandatory, but just a 1:20 scale physical model on a given 60 × 60 cm ground plate. I gave them hints to use base materials and banned cardboard or other typical model-making materials. The choice of materials is half of the work. In this design studio, I tested how to maximize student performance, by some simple rules of play, since only a set of simple well-chosen rules will lead to diversity. The student's outcome was above and beyond; the results were imaginative and diverse, and no one even thought of copying examples from browsing the internet as so often seen in other third- and fourth-year projects. There were innovative approaches, and each of the 25 students found their unique abstraction and analog parametric components. These first-year students were clearly on to it, enjoying their acquired abstract thinking, and exceeding their own expectations. These first-year students were acting as mature designers and were proud of their achievements. In general, when student outcomes are somewhat traditional, it is not because of the mindset of the students or of the capacities of faculty, but by and large because of an all too strict assessment of the students' and staff's performance. Students and staff are getting demoralized rather than inspired, because of the demotivating bureaucratic aspects of grading and assessment that are imposed on students and staff.

Without involving me, a selection of the sculptures was moved to the College of Engineering to show off the department's creative power in front of the dean's office. My efforts to have the designs showcased in the corridor at the faculty building, in a nicely organized row as it was prepared for, ended up in bureaucracy. No one is authorized to approve; everyone including the head of the department typically points at someone else whom I supposedly should have turned to for approval. But that someone almost per definition would say that it is not their competence to give such approval. According to Arabic customs and traditions, it is not polite to say no directly; they prefer to keep it undecided, without having to bear personal responsibility. Yet, how different is that from the bureaucracy in European universities?

White Majlis

In the two tropical years of our stay in Doha, Ilona and I initiated many 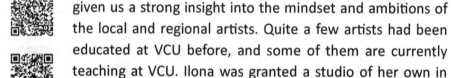 activities in the fields of architecture and art. We established connections to the Doha Fire Station leadership and Ilona was asked by curator Bahaa Abudaya to become a mentor for the 18 artists in residence for the 2018–2019 batch. Since its establishment in 2014, the Doha Fire Station has become a major hub for artists and culturally interested Doha citizens in the few years after its foundation. The Doha Fire Station is a repurposing of the fire station in Doha that was built in 1982. That timeline gives an impression of how fast developments are going in Qatar and how recent governmental investments are building a cultural community. Fifty percent of the selected artists are Qatari; the other 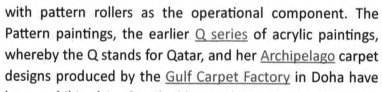 25% are residents from mainly neighboring countries and teaching staff from the VCU, the Virginia Commonwealth University branch in Education City, which is a renowned art school in Qatar. Discussions with the artists in residence have given us a strong insight into the mindset and ambitions of the local and regional artists. Quite a few artists had been educated at VCU before, and some of them are currently teaching at VCU. Ilona was granted a studio of her own in the guest wing of Doha Fire Station, where she produced a new series of acrylic paintings called the Pattern series. The Pattern series are based on intuitive gestures executed with pattern rollers as the operational component. The Pattern paintings, the earlier Q series of acrylic paintings, whereby the Q stands for Qatar, and her Archipelago carpet designs produced by the Gulf Carpet Factory in Doha have been exhibited in the Sheikh Faisal Bin Qassim Museum, abbreviated the FBQ Museum, located on a large piece of private property populated with some huge limestone horizontally stretched castle-type buildings, palaces, horse stables, a lush park, and a zoo. All buildings are clad in the same local limestone, both inside and outside, *in situ* cut into manageable building blocks, right out of the ground of the FBQ property. The main extra-large building complex, the FBQ Museum, extends every year to house his ever-growing collection of art pieces and memorabilia. His vintage car collection is famous, as is his extensive international carpet collection. Sheik Faisal Bin Qassim, assumedly the richest man in Qatar, even collects characteristic items from various religions, whereas each religion has been assigned a separate room. The former director of the FBQ Museum Kees Wieringa invited Ilona to show her work in the White Majlis, the abandoned summer palace of the Sheikh. Ilona asked our architect-artist friend Yassmin Alkhasawneh to contribute her salt crystallization works for the show titled Intuition & Procedure. Later in 2019, Ilona exhibited solo at the Doha Fire Station, showing her Pattern series and Q series. In her own words as printed on the FBQ poster: "The speed of the operation is essential to my working procedure, since I want to avoid deliberate thinking, I want to surprise myself. And by that intuitive procedure, I take the viewers by surprise, orchestrating an immersive form of serendipity." This statement is exactly what we wanted to achieve with the students of the Department of Architecture and Urban Planning. We wanted them to liberate themselves from conventional learned preoccupations, from regimented institutional framing. We wanted them to see themselves as grown-up women choosing their direction in becoming an architect.

Intuition & Procedure

Accompanying text that I wrote for Ilona's solo exhibition at Doha Fire Station Workshop 4 in March 2019

Looking at Ilona's paintings and sculptures, one feels a tension between dimensions. Working on a flat surface as a painter, she creates a tension between the surface of the painting and a three-dimensional world. Working in three dimensions as a sculptor, she typically sets up a reverse tension with flat surfaces. Ilona builds her universes in the voltage fields between the two-dimensional and the three-dimensional. The titles that are given to the paintings reveal their intention: the Polynuclear series, the Tangle series, the Omniverse series, and more recently the Q series. However, Q stands for quantum, as well as for Qatar, where the recent paintings were made. The paintings are the opposite of metaphorical; they do not have the intention to even remotely look like a universe; they are the direct expression of a series of gestures, intuitively smashed on the canvases, using wide-tipped acrylic markers. The speed of the operation is essential to the working procedure of Ilona, since she wants to avoid deliberate thinking – she wants to surprise herself. And by that intuitive procedure, she succeeds in taking us as the viewers by surprise, therewith orchestrating an immersive form of serendipity. Ilona works with large canvases on the ground, circling the canvas and even stepping into the canvas while marking her fast and furious gestures around her, thus building her personal universe. There is no up and down in her universe, and no front or back; her paintings are universal in the true sense of the word. Only after the paintings are made, the decision is made on how to mount the painting on the wall. Meaning and metaphors are not intended by the artist and are not imposed on the viewer. Now it is up to the eyes of the beholder to jump freely into these rich multi-layered universes, surprising in their levels of detail. Standing in front of the large paintings, walking toward them, and zooming in on them is a rewarding experience, as one starts to float freely in the evoked three-dimensional space, feeling the power of the strokes. Not accidentally, Ilona refers to her intuitive gestures as powerlines, which are put on the canvas with force and speed. The power she has forced upon the continuous gestures is gratefully received back by the viewer when navigating her universal paintings. Ilona's paintings give us the feeling of being infinitely small and unlimitedly powerful at the same time. One is happily caught in the voltage field between the dimensions.

East-West/West-East

During our stay in Qatar, there was one specific art project that we repeatedly had to see over and over again, the East-West/West-East project by Richard Serra, located in the Brouq Nature Reserve in the middle of the Zekreet desert, in the western part of Qatar near Dukhan. The first time we went there was in 2014 when it was installed; the only clues for its location were the coordinates: 25.5171°N, 50.8710°E. To get there, one has to take dirt roads across the rocky desert, by four-wheel drive. It took us 10 km through the desert without any clear signs of where to go; the only indication was our mobile phone's GPS. The young Sheikha Mayassa Al Thani, the younger sister of Tamim Al Thani, the current Emir of Qatar, suggested Richard Serra to do a project in the desert, and her father, the Father Emir Hamad Al Thani, suggested the location. After turning away from the highway to Dukhan at the location of the Cuban Hospital, which is famous for the high quality of their doctors while granting them only low wages, one drives 10 km deep into the desert, losing contact with telephone networks and GPS signals. The project is half hidden between two 14-m-high plateaus, the leftovers from millions of years of erosion of what once was the bottom of the ocean. Visiting East-West/West-East is a mesmerizing experience. According to Richard Serra, it is the most fulfilling thing he has ever done. Four huge chilling steel plates that are 12 cm thick, 2 m wide, and 14 m high at varying distances are placed between the winds eroded limestone formations, matching the top of the plateaus. The four steel plates rigorously stand as marks from an alien civilization. Four-wheel drives are silently crossing the desert around the corrugated steel plates. East-West/West-East is my favorite destination in Qatar. At least four times, we took our expat friends and visiting architects there to share the experience, including picnics in the shade of the plates. People step out of their four-wheel drives, walk around, endure the heat, come closer, take distance, and climb the plateaus for the overview. In the first encounter, you walk alone because you want it to be a personal experience, only later to share your experience with the company you are with. People are wandering around at proper social

distances from each other, as in the Continuous Monument beach scene collage by Superstudio. The experience feels as if the four steel plates are a sign from a remote civilization,

perhaps hiding underground, communicating in a precise, abstract visual language. The Black Monolith from 2001, A Space Odyssey, comes to mind, an alien object that you are tempted to touch, yet prudently; scanning the unknown, eager to learn from it. As for many of our friends in Qatar, the East-West/West-East means nothing less than a pilgrimage site. It is a perfect example of the power of simplicity, a form of simplicity that triggers a stack of emotions, and a form of simplicity that we like to match with our strategy of designing simple rules that trigger diversity, which makes it a perfect example of what I have called simplexity. In the case of East-West/West-East, it is not diversity in many differently shaped components as in nature, but diversity in the emotional responses it triggers, and in the social patterns it creates around the art installation. On a larger scale, I see it as an inspiration for a new form of urban planning that becomes a form of landscaping. Similar to the way that I have peeled off from architecture the traditional elements like doors, windows, columns, and beams, and have replaced them with a performative fusion of art and architecture on a digital platform, also urban planning must be freed from the traditional elements as there are streets, building blocks, squares, center versus periphery, and replaced by a fusion of landscape and architecture, in a blend of the known natural and the new synthetic nature. The East-West/West-East installation does exactly that.

8

Proactive Components

Proactive Architecture

The design of components for static structures is one thing, and the design of components for interactive and proactive structures is another challenge. Designing for interaction means another level of thinking, informed by sensor and actuator technology. What exactly is the relation between nonstandard architecture and interactive nonstandard architecture? Nonstandard architecture is the design and construction of a series of unique components for complex geometries. Interactive nonstandard architecture is the design and construction of components that change shape and content in real time. When the interactive installation is built up from mobile components like drones or crawling robotic creatures, the interaction resembles the component's performance within a swarm of birds. Drones typically are a series of the same type, in terms of shape and dimensions not different from one another; they just need a unique digital identity. To form nonstandard configurations where the playing components are bound to their immediate neighbors requires flying, floating, or crawling components that can adjust their shape, dimensions, and content to fit in that particular place at that particular time, in real time. Nonstandard interactive components communicate with their immediate neighbors, not only to adjust their position but also to synchronize their shape with the neighbor for a perfect alignment. In the communication between two adjacent shape-changing nonstandard interacting components, they will need to negotiate how to change shape, and how and when to connect when both shapes have been adjusted to fit together. In theory, this can be one; yet, in harsh reality, there are many hurdles to be taken. The first hurdle is to choose the physical properties of the acting component. Real-time shape-adjusting components tend to lead to complicated mechanisms. Typically, one does not want to depend on complicated physical mechanisms as in <u>Jean Nouvel</u>'s interactive facade of the Institut Du Monde Arabe in Paris. The many <u>complicated diaphragms</u> in that facade are prone to failure. We have learned along the way that it is hardly ever the software that fails, but the hardware executes the instructions given by the software. Hardware is vulnerable, and mechanisms that are delicate like clockworks certainly are fragile and expensive. Once the software for interactive installations works as intended after iterative tweaking and testing, it will just keep doing what it is meant to do, until a mechanical part fails.

To overcome the hurdles of unreliable hardware, there are several options to be considered. One option is to use flexible materials in combination with programmable electric pistons, which are proven technology. For the behavior of the <u>Trans_Ports</u> pavilion, I chose large electric pistons that can adjust their length in real time, connected to flexible tubular sliders that are moved up and down by the pistons, and we applied flexible materials that stretch between the sliders. The sliders allow for changes along the length of the pavilion, while the 3D harmonica-shaped rubber skin allows for changes in the lateral direction. In <u>Trans_Ports' interior,</u> I planned to have a point cloud of thousands of LED points, attached to the flexible interior skin, visualizing almost anything anywhere, immersing the public. The visuals would come from all around the public, a fully immersive environment. The interior is the flexible version of the programmable visitor experience we achieved in the Sensorium, the upper level of the lemniscate track inside the Saltwater Pavilion that we completed a few years earlier. With the Trans_Ports project, we wanted to explore the logical next step from the programmability of light and sound, toward a fully immersive environment that physically interacts with the public.

For the exterior, we envisaged a rubber skin that is folded to allow it to be pushed upwards by the electronic telescopic cylinders, while elastically folding back when the pistons would retract. The maximum stroke of the pistons is 1.4 times its initial condition, meaning that the rubber skin must be able to extend with that same factor. Pistons can take both tension and compression forces and are thus fit to replace static structural steel members. The programmable pistons with position feedback effectively act as the members of the swarm of pistons. In 2000, I imagined Trans_Ports to be built anytime soon; I expected to find a sponsor within five years. Together with Ole Bouman who was involved in the development of the concept, we talked to the Rotterdam Harbor Authorities, who liked the concept and also suggested a location in Hoek van Holland. Structural engineer John Kraus substantiated our vision. The extra costs of replacing static steel columns with programmable electronic pistons are known, as are the

costs of the other constituting components of Trans_Ports. According to our calculations, the realization of the programmable pavilion would require twice the budget when compared with a standard budget for smaller pavilions. The return on investment comes from the multiple uses of the pavilion. Since the structure can adjust its shape and content, the structure can be scheduled to be used for many different functions during one day. Trans_Ports has many possible emotive modes of operation: Trans_Ports may perform as a research facility during the morning, as an educational experience in the afternoon, and as an entertainment facility at night. The down-to-earth mentality of the Dutch culture turned out to be a bit more resistant to my provocative ideas than expected, and more than we were hoping for. Twenty odd years later, I still have not been asked to do an emotive structure like Trans_Ports 2000 yet.

At Hyperbody, we found ourselves limited to smaller-scale interactive installations. I developed with our Hyperbody students at TU Delft a dozen interactive installations between 2001 and 2016. Building upon my experience with Trans_Ports, we discovered in 2002 the 2001-launched invention of pneumatic Festo muscles. We used the Fluidic Muscle for the first time as an actuating component in a network of 72 swarming artificial muscles wrapped around the flexible light blue balloon-skin of the NSA Muscle project, exhibited in Centre Pompidou in early 2003. Subsequently, Hyperbody students re-used the same muscles for the programmable Trans_ Ports section, the Muscle Tower, and the Muscle Body. Structures with Festo muscles follow quite different logic as compared with structures based on programmable electric pistons. Festo muscles do not take compression forces; they take tension forces when they contract by inflating the flexible tubes, becoming shorter, by the innovative diagonally woven reinforcement structure. Festo muscles may be used in combination with rigid members of the structure that resolve the compression forces. Every Festo muscle structure is a form of a tensegrity structure, whereas the (sub)components to take the tension and the compression are working together by division of labor. The interactive projects typically feature a sort of wobbling massage, which is best experienced when inside the structure. Using proximity sensors, the kneading movements are triggered by the

public. The Muscle Body, designed by Hyperbody students, performed a perfect translation of Marshall McLuhan's original title. The <u>Medium Is the Massage</u> is the original title of <u>Marshall McLuhan</u>'s famous book. The original title is not the Medium Is the *Message*, which was an accidental mistake of the printing company, which has since become one of the most frequently cited quotes. McLuhan readily accepted the faulty interpretation and did not try to reverse the message anymore.

Another way into facilitating dynamic structures, besides structures based on the incorporation of electronic pistons and Festo muscles, is to consider the structure as a whole as an inflatable that can be inflated or deflated to form the different spatial configurations on command. The same Air Studio that made Little Babylon manufactured 20 years

earlier the ParaSITE interactive sculpture, designed for R96 Festivals in Rotterdam. <u>ParaSITE</u> is not a programmable sculpture as is Little Babylon; the inflatable gently gives in a bit when people enter through a gill on the side of it. Opening the gill lets air out of the sculpture that is kept under constant pressure by a ventilator; the sculpture inflates again when the gill self-closes. The beauty of inflatable sculptures is that they are packed on a euro pallet and travel the world. We sent ParaSITE on a European Tour, visiting Graz, Dunaújvaros in Hungary, Helsinki, and <u>back to</u> <u>Richard Meyers' Town Hall in The Hague</u> in the Netherlands for its final performance. A similar fabric is used for the Little Babylon project, with multiple air chambers, programmed to alternately inflate and deflate to change its overall shape. Little Babylon was first inflated in Den Bosch in the Netherlands, later once in Sao Paulo in Brazil, and finally twice in Doha, Qatar, once at the Doha Fire Station cultural center in November 2018 and installed again at Qatar University during our GSM4Q international conference in February 2019. Little Babylon is driven by data that is harvested from the internet. A special software application picks up words from tweets with the Little Babylon hashtag and maps those words into the program to match the four basic emotive categories: happy, afraid, angry, and sad. The emotional states, imagined and executed by Rezone, are accompanied by sound and light. Having sent a tweet, any tweet with the hashtag #littlebabylon would read that tweet out loud for the local public. The Little Babylon sculpture moves smoothly up and down along with the different emotions, alternately inflating and deflating the smaller air compartments at the top of the inflatable. Each of the three legs of Little Babylon behaves individually, allowing for its smooth and seductive body movements. Especially at night, Little Babylon shines convincingly, mumbling, occasionally shouting, and always colorful. Other solutions are imaginable for programmable interactive sculptures and buildings. One could theoretically think of incorporating phase-change materials and memory materials. But, as the spatial effect of phase-changing materials on the grand scale of architecture is minimal, I have not touched upon this field of research. My leading design principle is to create a maximum effect with minimal means. The use of phase-changing materials does not fall into that category.

Game Theory

The very concept of interactive architecture naturally transforms the traditional task of an architect, which is to design buildings as static interfaces for human activities, into that of a game designer, who designs dynamic environments that unfold by interaction with the players. Besides a theory of architecture, the new generation of information architects must develop an understanding of game theory. Game theory is the study of mathematical models of strategic interaction among rational decision-makers who have agency as in adaptive systems. Game theory builds upon numbers, modeling, building blocks, agents, patterns, configurations, formulas, correspondences, and mapping. I use similar terms to describe design interventions. However, I do not design theories of play as in Huizinga's Homo Ludens, but rather games that are designed to design. Serious design games follow a set of strict rules, as do social games. And just as with social games, design games thrive on a few simple rules that generate complexity. Chess is the ultimate example of how a few simple rules create an endless number of possible configurations. Game theory is at the heart of our understanding of architecture as a complex adaptive system, based on rules and mechanisms, generating perpetual novelty. Simple rules and gaming strategies create the unknown, the alien, emergent from the critical mass of numerous possible configurations. I introduced the idea of working in game-like environments early in my Hyperbody era, which was for me the logical step beyond complex yet static geometry to interactive architecture that operates in real time. In my second year at TU Delft, I crafted a design project called The Emotive House. The goal of the design studio The Emotive House is that each student designs a part of the house in a real-time dynamic relationship with other parts of the house. A change in one room, for example, the kitchen, informs a neighboring room, for example, the living room, and urges to respond immediately. I encouraged the students to find a procedure to exchange information in real time, such that the changes occur instantaneously. Students enthusiastically plunged into this challenging new concept of architecture. Notably, students Chris Kievid and Michael Bitterman picked up the idea and dived deep into the architectural concept of programming emotions. The title of my TU Delft inaugural speech that I delivered one year before in 2001 is E-motive

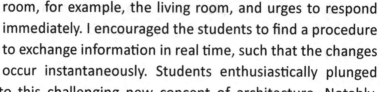

Architecture. Not emotional, but emotive. The word emotive means arousing emotions, while the word emotional refers to showing emotions. Emotive programming is the acting component that triggers emotions. While static architecture frames and guides the spatial behavior of people inhabiting the structures, emotive architecture is the active form of framing, influencing the behavior of people through real-time changes in the environment. I spelled the title of my inaugural speech as e-motive, like an e-bike, to indicate a digital, electric, and programmable motivation. The Emotive House design studio is aimed at a house that proactively proposes emotions. The Emotive House is a digitally augmented environment for the interacting rooms in the house, not in the sense of what is commonly known as a smart home. A smart home simply involves the automation of the tuning of the house equipment and of automating actions that one would otherwise do physically and by analog proceedings. An emotive house is a home where the rooms are actively proposing emotions of their own, rooms that propose spatial and emotional configurations for the inhabitants to respond to and to start a dialogue with. Already back then, I realized that the future of design is in the programming of bi-directionally communicating environments, for people whose lives are enhanced by the Internet of Things and people. Both the people and the things are the players in the game of life. The bi-directionality of the process is important to understand the design direction I have chosen to go. Bi-directional communication between immediate neighbors is the opposite of the one-directionality of a Big Brother that is watching you. Swarming bottom-up interactivity is the opposite of top-down crowd control.

At Hyperbody, we, as early adopters, discovered the French game development program, Nemo, later renamed Virtools, which is the perfect instrument to perform architecture as a design game. Nemo/Virtools features a graphical interface, which is comfortable for visually oriented designers to understand how programming works. I intentionally use the word instrument instead of a tool, since the program is used to compose new configurations, process data in real time to develop design environments, and unfold the serious game. I consider the building to be a vehicle that processes information in real time with the purpose to change its shape and its behavior. The Virtools game development platform program is not used for its intended purpose, which typically is to design web-based interactive applications for businesses to involve the public in making otherwise predefined choices, such as choosing the color of the car, choosing accessories, and viewing their choices in three dimensions by rotating around the object. Game development platforms are used to construct games for entertainment, like the popular shoot-them-up games, or the mystery games where you have to find the key to the secrets. But Hyperbody uses the game development interface as an instrument to design programmable environments that have a proactive life of their

own. Environments as living diagrams that continuously feed themselves with fresh data process the information and reconfigure themselves in always new yet unknown configurations of the self. Applying game theory by playing the game development platform as an instrument quickly became the default state of Hyperbody. The theory of complex adaptive systems became deeply embedded in all of Hyperbody's education and research projects.

ONL from the nineties, and Hyperbody from the zeroes, have been early adopters of a truly programmable architecture. Only in recent years, the paradigm of interactive architecture is taking off internationally through the establishment of interactive architecture courses at universities worldwide, a prime example of which is the Interactive Architecture unit at the Bartlett and the subsequent establishment of the Interactive Architecture Lab, established in 2014 by Ruairi Glynn, who visited Hyperbody back in 2006. I embraced the inevitable logic of Interactive Architecture while designing our early interactive projects, catching butterflies in the Sculpture City CDROM (1994) and synthesizing sound samples in the R96 ParaSITE inflatable sculpture (1996) and the VR environment of the Saltwater Pavilion (1997), what we back then thought was becoming the new normal. Interactive architecture is like being *in* the game. One of my favorite films of the time is eXistenZ (1999) by David Cronenberg.

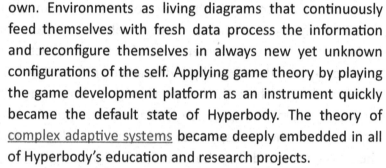

The gamification of life and hence of architecture is bound to become the default mode of the <u>Metaverse,</u> the digital twin of Another Normal.

Hyperbuildings

Foreseeing what the future will bring can only be a personal view based on one's personal experience with building. My concept of the future is per definition rooted in the here and the now. I am emphatic with available social devices and technologies and feed that knowledge into my design concepts. What I design is never speculation but firmly rooted in today's social and technological state of affairs. As I prefer to design real-world constructs, that is, not as some hypothetical construct for a distant future, much of it depends on whether I will have the pleasure to find an enlightened client, who will allow me to do what is needed to further develop my vision on the practice of architecture. My work is by many seen as futuristic, but nothing is less true; my work is always firmly based in the here and now. My pragmatic attitude is for the first time realized at a larger scale in the bespoke Saltwater Pavilion. The world was taken by surprise by a radically designed sculpture building with an interactive programmable interior. The logical next step will be to realize a fully programmable building body, allowing the body itself to interactively be tweaked in the interaction with its users.

A programmable building is a hyperbuilding, a structure that is an active player in a real-time operating global network. Hyperbuildings are "hyper" in the sense of being linked to other structures in real time, as a hypertext is linked to other text, wherever and whenever, as the images on hypersurfaces are linked to other images or videos, as hyperbodies are linked to other building bodies, both in the physical real world and in virtual aka hyper-real universes. Hyperbuildings are an assembly of acting components that interact with each other, with their immediate climatic environment and with their users to continuously shape and reshape their dynamic spatial environment. Hyperbuildings are living hyperbodies. Working in cross-disciplinary teams with interaction experts, all the necessary expertise is available to realize a hyperbuilding at any scale. Although I have proposed upon my initiative many different versions of hyperbuilding, not a single enlightened client has come forward. There is nothing futuristic about a programmable hyperbuilding because the concept is based on proven computational technology in design and execution. Many of the necessary interaction techniques are tested in a great number of Hyperbody's master design studios over the years, in the design and execution of one-to-one scale prototypes for interactive environments. Hyperbody's first batch of Ph.D. candidates, notably Nimish Biloria, Henriette Bier, and Hans Hubers, has been involved in the design studios. We educated a game-changing total of about 5% of the student population of the Faculty of Architecture at the TU Delft in the years between 2000 and 2016.

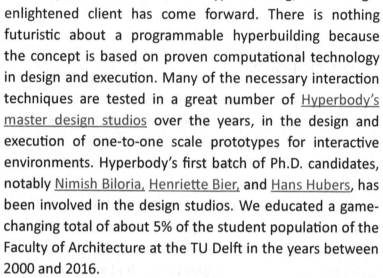

Venice Biennale 2000

The earliest proposal for fully programmable built constructs is the programmable interior for one of the capsules of the International Space Station (ISS), especially modeled for the Dutch daily newspaper Trouw on invitation by the Dutch journalist Robbert Roos, the current director of the Kunsthal Kade in Amersfoort. I imagined the interior skin of the ISS to be composed of a dense network of hundreds of small electronic pistons. Changes in the interior configurations are made by programming the length of the pistons. Since electronic pistons take both tension and compression forces, the network will be stable at all times. Networked pistons can only act when they work in sync with their immediate neighbors. If just one single member would want to change its length, the others would withhold it from moving when not informed to work in sync. The individual members inform each other of every change they are going through, like the birds in a swarm. The neighbors give in or move along with it. Programming a network means programming the flock, the swarm of individual members to work together. The Space Station was the first design of acting components in a networked environment – components that operate like the connected birds in the swarm, like the connected brains of the human population. A single component in a networked environment cannot do anything unless it cooperates with its immediate neighbors. This has become the leading theme for all of my subsequent designs, whether resulting in static buildings or interactive installations. The electronic pistons of the ISS capsule configure four radically different modes of operation:

1)   the meeting mode,
2)   the game mode,
3)   the home mode, and
4)   the sleeping mode.

In the configuration of the meeting mode, the central element that pops up is the meeting table, which is formed by pushing up the interior skin from what is temporally considered the floor. In a weightless space, there is no such thing as a floor, a wall, or a ceiling, as the sides form a continuous interior skin. Electric pistons can change their length by the telescopic stroke of a maximum of 1.5 times its most compact size. Suppose the default situation is somewhere in between maximum and minimum sizes, the pistons can either shorten or elongate from there. To shape the rounded-off square block of the local and temporary meeting table, the programmable components near the rounded corners will shrink, while the components in the relatively flat fields will expand. Programming the networked skin means instructing certain groups of pistons to work together to change their length to reshape the interior skin. The concept for the programmable interior skin of the International Space Station turned out to be *the* game changer for using available digital technologies to build truly interactive environments.

Soon after our ISS proposal, I got an excellent opportunity to develop the idea further. The curator of the Architecture Biennale 2000 in Venice Massimiliano Fuksas, together with his spouse Doriana Fuksas, invited us to participate in an installation in the Italian Pavilion. Fuksas was familiar with our design for the interactive environments of the Saltwater Pavilion and found that it would fit well in his concept of Less Aesthetics More Ethics for the Venice Biennale 2000. Fuksas gave us a prominent central space in the Italian Pavilion, the 15 × 15 m² room upstairs from the other central space on the ground floor. Fuksas installed a mock-up of the International Space Station, hanging from the ceiling. From there, the direct view via a large opening between the two spaces built a direct link between the ISS and our interactive installation Trans_ Ports 2000. We conceived the installation Trans_Ports as a circular interactive arena, in which the public interacts with projected virtual building bodies around them. Stepping into the sensor-equipped arena surrounded by the 360º of virtual reality panorama, the public deeply immerses itself in the three different modes of operation. By moving around in the area, the public tweaks the content of the unfolding virtual worlds that are projected onto the three curved screens. The public is inside the game. The glass-fiber-reinforced PVC projection screens leave three openings for the public to enter the arena. The display on the backside of the 3D-printed screens has an illustrated explanatory description of the multimodal installation. The interactivity of the Trans_Ports installation is unique in many ways. Trans_Ports is the first interactive installation of its kind within the realm of architecture, even anticipating similar interactive installations within the arts. We were already acquainted with smaller responsive art pieces by media artists like Jeffrey Shaw. In his interactive piece called Legible City (1988), the public sits on a bike that navigates the streets of a virtual letter-city. Such early interactive art installations formed the basic available knowledge before the design and execution of the real-time interactive light, sound, and virtual reality behavior of the Sensorium of the Saltwater Pavilion. Trans_Ports introduced a new level of interaction. Inside the Trans_Ports interactive arena, the public adds information to the ongoing process of manipulating the geometry of the projected virtual worlds, by moving around inside the arena, inside the design game. At the Venice Biennale 2000, there was another large interactive installation, a highly technical wall piece titled HypoSurface, designed by Marc Goulthorpe. The hundreds of small components of this wall are driven by small pistons pushing the constituting triangular parts outward and pulling them backward again. An array of sensors that are built into the wall piece register the presence of people and respond to their immediate presence by pushing the pistons toward the spectator. Walking along that interactive wall gives the impression of it being alive, having a magnetic relationship with the surface, morphing the surface without touching it, a great responsive piece by Goulthorpe. However, HypoSurface is responsive to the public, immediately responding when someone comes closer, but otherwise stand-by. Trans_Ports is alive when there is no one in the room. While interaction requires two active parties, I consider Trans_Ports an interactive installation, not a responsive one.

The public at Trans_Ports gathers in the central arena that measures 10 m in diameter. Moving around the public triggers infrared sensors that hang from the ceiling. The sensors are organized in three concentric circles: the inner circle triggers background colors, the middle circle triggers images that are projected on the surface of the virtual 3D model, and the outer circle triggers the changes in the geometry of the 3D model. The public becomes a co-designer of the projected 3D model, by manipulating the default geometry while moving around in the arena. By trial and error, the public finds out which movements trigger what response. The default 3D model is a triangulated abstraction of the Saltwater Pavilion, whereby the structural members are virtual pistons instead of steel profiles, virtual pistons that can change their length up to a stroke of 1.5 times its most compact size, as physical pistons. By orchestrating the individual pistons, this virtual version of the Saltwater Pavilion rotates, twists, and zigzags like a snake, pumps up its volume, and shrinks back again. Each sensor in the outer circle informs one of those pistons to change shape, while that piston informs its neighboring pistons to move along with it. As in the concept design for the ISS module with the programmable interior skin, in a dynamic 3D structural network, none of the members can change its length alone. If one component changes, at a minimum, its immediate neighbors must change with it. While there are always more people in the arena, many of these movements happen simultaneously, delivering complex changes in shape, content, and perpetual transformations in basic atmospheric conditions. The changes in shape and content feel like a smooth massage of the virtual body. The massage is the message. Trans_Ports never has a dull moment. Even when there is no one there, the installation does not sit idle; it executes a program to entertain itself.

Trans_Ports is designed to operate in three distinct modes:

1) <u>Self-explaining Trans_Ports mode</u>,
2) <u>North-Holland pavilion v1</u>, an interactive virtually augmented path through the first (rejected) version of the Web of North-Holland, and
3) <u>Handdrawspace mode</u>, an interactive abstract painting by Ilona Lénárd.

The programming is done by the ONL team member <u>André Houdart</u>, assisted by the interns <u>Rich Porcher</u>, <u>Nathan Lavertue</u>, and Thomas Müller. None of what we did had been done by anyone before, nor by ourselves; we took an adventurous calculated risk that could just as well have failed. Along the road, we realized when executing the project that history was in the making that we were living inside evolution, and we were overcome with emotion when the plan eventually came together. Sensor technology expert <u>Bert Bongers</u> tinkered and soldered the connection of the sensors to the virtual reality software. Bongers has been working with us for many years, and he managed the programming of the behavior of the Saltwater Pavilion. Bongers designed the sensor boards inside the Saltwater Pavilion, with which the public manipulates the interior light and sound conditions of the <u>Sensorium</u> space. Together with Hyperbody's researcher Chris Kievid and performance artist <u>Dieter Vandoren</u>, Bongers later played a crucial role in setting up the first version of the hands-on interaction protoSPACE lab at Hyperbody.

Like Trans_Ports and Handdrawspace, the interactivity is programmed using Nemo, later renamed as Virtools, a French software platform with a graphic interface to program the behavior. Nemo/Virtools has been a key instrument in the development of our take on parametric design, interactivity, and real-time behavior. We were among the first, if not *the* first, early adopters worldwide to apply this game technology to the design of architecture-scale installations, buildings, and cities. Until then, game development programs were typically used for web-based interactivity to boost the sales of products by user interaction, and for educational purposes. Programs like Grasshopper are inspired by the Nemo/Virtools graphic interface, only to be introduced 10 years later. Yet, Grasshopper is not a game development program; it is limited to parametric design for otherwise static environments. Robert Aish, who developed the ingenious Generative Components (GC) software, claimed that he could set up similar interactions between components as in Virtools, but the big difference remains the user-friendliness of Virtool's graphic interface compared with the technical engineering approach of GC. The huge progress though is that parametric thinking has now invaded the more advanced architectural design practices as the preferred modeling tool for even the most generic designs. The parametric design method is quickly becoming mainstream, but not necessarily as a style, and not as parametricism, which Patrik Schumacher has declared the dominant style of our epoch in his Parametricist Manifesto (2008). Parametric design builds continuously different relations between the components. Parametric design is not a style; it effectively orchestrates the variable relationships between the building components, whether simple or complex in shape. While parametric design is a key method for the construction of nonstandard constructs that are otherwise static, Virtools is the key instrument to orchestrate relationships between building components that act in real time. After having taken over Nemo, Virtools in its turn is taken over by Dassault Systems in 2005 because it had some useful programming features to enhance their proprietary software, but Virtools is no longer supported as a platform for interaction design; what was left is a 3DVIA/Virtools web player until it was permanently closed down in 2009. Big international giants seem to be more interested in bringing advanced developments to a halt instead of investing in state-of-the-art products that are pushing the limits. Nemo/Virtools is badly missed; today, Unity is a pretty good alternative for game design.

In the previously realized art and architecture projects, we defined the passive constituent components. Now, what are the active constituent components in the Trans_Ports installation? Which are the components that act in real time? First of all, one group of acting components is represented by the people, the public that moves about in the interaction space and triggers the sensors that hang from the ceiling. Second, the system that includes the sensor field, the computers running the Virtools design game, and the projectors beaming the 360º panorama represent the other component. Sure, the mass-produced individual sensors are standard components on a smaller scale, but in the bigger picture of the interactive installation as a whole, they are subcomponents that make up the system. The array of sensors communicates in real time with the interactive design executable file that runs on the server. The sensor field builds the connection between the flock of people and the software that projects the virtual world on the surrounding screens. The three projection screens are the physical components of the installation. The two acting components, the group of people and the interaction arena, work together to build the interactive experience.

Inside the three modes of operation of the Trans_Ports 2000 installation, the acting components are the avatars and the virtual environments. The main components of the above-mentioned three modes of operation of the virtual worlds are

1) the virtual pistons of the virtual Saltwater Pavilion, of which the changes in length are triggered by the movements of the public,
2) the immersive events in the series sectors of the North-Holland pavilion; while the avatars are hopping up and down the winding trajectory, the public moving about in the Trans_Ports 2000 arena triggers the immersive events to the left and right of the avatar, and

3) the swarm of dots that are in real time emitted from the 3D virtual
    sketch in the Handdrawspace universe.

In the three worlds of Trans_Ports 2000, navigating the arena
manipulates the virtual worlds to create an enchanting experience.

The visual appearance of Handdrawspace, the Trans_Ports
2000 art mode, is based on an intuitive spatial 3D digitizer sketch by
Ilona, whereas dots are continuously emitted from the sketch that
is in itself made invisible. The presence of the dots is set on a timer;
the dots are gradually dying out, making place for new emissions,
triggered by the movements of the public. The sensors in the middle
ring of the interactive arena trigger the size of the dots, occasionally
blowing them up to gigantic proportions, causing the larger dots in the
Handdrawspace world to cluster together to form large black stains.
The dots in Handdrawspace are continuously changing their shape,
appearing in the scene, and fading out again. Some dots live longer
than other dots. The movements of the public in the outer ring of the
arena change the background color, while movements in the center
trigger the number of dots emitted. Sound samples, short
cuts out of a piece by the Italian composer Luigi Nono, are
linked to the triggering of the sensors, celebrating the fact
that Nono actually was a native inhabitant of Venice and
lived his whole productive life there.

Our first proposal for the North-Holland pavilion, which forms the
basis for one of the three Trans_Ports virtual worlds, turned out to
be too ambitious for the Province of North-Holland; the design was
rejected and was redesigned into the nonstandard summer pavilion the
Web of North-Holland as it was eventually built. I decided to keep the
North-Holland version 1.0 alive as an interactive experience. Moving
about in the arena of the Biennale installation triggers responses from
the virtual electronic LED surfaces of the pavilion to the avatars that hop
up and down the path. Sound samples from the adrenaline-
boosting Three Little Birdies track by the Chemical Brothers
enhance the immersiveness of the environment. We had
a lot of fun designing and executing the Trans_Ports 2000
installation; I consider the interactive installation a major achievement,

and we, ourselves, were as a collective emotionally overwhelmed by the mesmerizing experience. It was supported by a modest 30K budget from the Dutch Stimulation Fund, which was in Dutch florins at the time, one year before the public introduction of the EURO in 2002. The impact of the interactive experience is by and large the outcome of an innovative combination of game software and a spatial experience, eventually leading to a completely new vision of spatial constructs with real-time behavior, anticipating what is now popularized as the Metaverse. When we would reinstall Trans_Ports now, 20 odd years later, it would still stand out as revolutionary. The times are changing, but not as fast as I thought they would; I still expect history to go full cycle and conjure up an enlightened client for a full-sized structure like Trans_Ports.

Ground Zero

 The most radical concept for a large programmable structure is the Ground Zero (2001) proposal. Among a few dozen American and international architects, I was invited in early 2002 by the Max Protetch Gallery to respond to the demise of the World Trade Center towers. The designs were exhibited in his New York Gallery first, and a few months later at the Venice Architecture Biennale in the American Pavilion. The radical erasure of the Twin Towers required an equally radical response. My designer's response was equally critical of the apocalyptic act itself as of the counterproductive response by American president George Bush Jr. The tragic event recalibrated the global political climate into irreconcilable opposing positions, without in-between choices. The slogan "You are either with us or against us" is counterproductive for resolving political tensions. After having condemned the vicious attacks, I criticized the USA for striking back disproportionately. I noticed that, as in virtually every Hollywood movie, the self-proclaimed good guy typically kills more people in revenge than the stereotyped bad guys. Max Protetch called me and kindly requested to soften my tone in the project description for the exhibition catalog A New World Trade Center: Design Proposals from

Leading Architects Worldwide. He said that his governmental sponsor would not allow for such a critical voice. Eventually, I gave in. I did not want to add more oil to the fire but took the opportunity to develop my ideas of a fully programmable architecture one step further. With the inconvenient truth of the inconceivable destructive act and the disproportionate revenge in the back of my head, I proposed a fully programmable building, to be programmed by international users of all faiths and beliefs, to be modified after their own local and temporal preferences. We visualized our concept in 12 radically different scenarios, one different mode of operation for each month of the year, published in full color in the opening section of my book Architecture Goes Wild (2002). The built structure takes on radical different configurations using actuators for the internal structure and the flexible exterior skin, embedded with sensors and programmable LED lights. Ground Zero's default state is a programmable cube, a HyperCUBE, of $500 \times 500 \times 500$ m^3. The HyperCUBE responds to events taking place at that particular time of the year. Imagine a structure where the horizontal parts to walk on are becoming gently sloping parts to connect to another level. Imagine vertical parts become twisted and inclined to become a part to climb up to, as one can climb up stairs. While the constituting components self-transform, a myriad of possible configurations can be spatialized. As a result of the horizontal and vertical slabs extending and shrinking, the internal spatial structure and hence the outline of the HyperCUBE change dramatically. The base programmable components of this imagined structure are the slabs, both horizontal and vertical in their initial condition. My team, which included Michael Bittermann, Chris Kievid, and Michaela Tomaselli, modeled 12 extremely different configurations of the programmable HyperCUBE and photoshopped for each of the 12 configurations a seasonal collage. Using images from the internet came with a price; Getty Images came after us and billed us for a substantial amount for illegal usage of one of their images. However, we only used a small, hardly recognizable part of the image and had put it in a completely different context. For January, we proposed the United Nations Mode, February is for Valentine's Day, March for the transformation economy, April for International Art, May for the Love Parade, June for Doomsday, July for Independence featuring the structure disguised as the American Flag, August is for the invasion of the Body Snatchers, September for the 911 Memorial reconfiguring the structure into two separate twin towers, October for Theater, November for the New York Marathon, and inevitably December for X-mas.

The Ground Zero HyperCUBE can be any other mode at any other time, to be decided by the movements and informative input by the indoor and outdoor public. We designed a multitude of buildings rather than offering one Ground Zero with one fixed spatial design concept, and as such stood out amongst our peers. I believe that the integral concept of programmable multimodality will become a key factor in the future of architecture, not just in the Metaverse but in the physical reality as well, and intimately connected to its digital twin in virtual reality. No longer do we want static office buildings that are only half occupied due to COVID-19 or other reasons, and no longer do we want to build single-use buildings. No more monocultures. Even when the structure itself is not programmable, buildings should be designed to accommodate different functions at the same time. Any office-space would be able to be converted into an apartment, a workshop, a private practice, a classroom, a meeting place, a dining room, or a lounge in no time. The perfect spatial unit for such multimodality has the size of a classroom with a high ceiling, suited to build a mezzanine into that space.

ProtoSPACE 1.0

Soon after my installment as a professor from practice at the Faculty of Architecture at the TU Delft in February 2000, I established Hyperbody's first protoSPACE lab. Building upon our experiences with the virtual worlds of the interactive installation Trans_Ports, the Hyperbody team set up an interactive design environment, by and large following the technical design that was developed by sensor expert Bert Bongers. Bongers managed the various modes of digital communication, soldered the electronic elements together to form the interacting components, and made things work, above and beyond expectation. The details of the subsequent versions of Hyperbody's protoSPACE labs are elaborately described in the 624-page 4-cm-thick book titled "Hyperbody, first Decade of Interactive Architecture." The thick-as-a-brick book is conceived as a visual relational database by graphic designer Niels Schrader. ProtoSPACE is designed as an instrument for multidisciplinary and collaborative design in real time. In a traditional design process, much gets lost in translation, while the people who are involved in the process do not actually work together but usually take some discussion points back home and respond in the next meeting, sometimes first weeks later. What if the response of the collaborating experts can be processed instantly? That is the challenge we set for ourselves. We set up a digital working space that is the digital equivalent of a dialogue between people, not as a replacement for human interaction but as a digitally augmented version of the interaction. People talk and exchange ideas, values, and opinions with others, and so does the digital protoSPACE working space. Digital models are modified on the spot by the collaborating experts, by their actions and movements in the interaction space. We learned from Trans_Ports how to set up an immersive interaction space. For the development of protoSPACE 1.0, we took further steps to enhance the interaction, to become an interactive *design* environment, a decision-making machine. The physical environment of protoSPACE 1.0 that we installed in a part of Hyperbody's office space has a 6-m-wide wall covered by two beamers, and an interaction space of 4 × 6 m in front of it. In protoSPACE, the additional acting (sub)components of the interaction system are

1) the pressure-sensitive floor that triggers specific user commands,
2) the RFID tags that identify the designer who is in the design game,
3) video tracking to define the position of the players in the interaction field,
4) infrared sensors to switch on and off different functionalities,
5) voice recognition to swap from one scene to another,

6) rotation sensors to enhance hand movements inside the interaction space, and of course

7) the rules of the design game, written in the Nemo software.

The subcomponents together form the protoSPACE instrument. The main components are the human players and the protoSPACE instrument. ProtoSPACE 1.0 is a customizable digital instrument, a sleeping beauty that comes alive when being played. The triggered sensors and trackers send signals to the server that hosts the game design program. The definition of an acting component is that it reads, processes, and sends information in a dialogue with other components in real time, not as a response but as an act of will. An acting component is permanently alive and does not wait for orders; it *self-acts*. The instrument protoSPACE is played by a group of acting components, the designers when they enter the scene and intentionally move about in the design space. The other acting component is the running virtual world using protoSPACE as its interface, a world that is navigated and tweaked on the spot. ProtoSPACE is the interface, the environment, and, above all, the instrument that facilitates the two main components to interact. The executable software and the human players, by playing the instrument, are proactively working together to form a local dynamic adaptive system.

I consider every design concept to have a unique entity that develops from a single idea aka an acting cell into a complex model aka a live organism. Every design concept and every conceptual idea should be entitled to get a birthright certificate; it acquires a unique identity from the very beginning of the design process and upholds that identity until the project is disembodied. Step by step, the identified, and hence the identifiable, idea evolves into a coherent complex of constituent components of the project. The acting components are assigned a unique identity, such that they can interact individually and be addressed as a group of collaborating individuals as well. This radical, self-evident concept forms the basis for Hyperbody's protoSPACE design software,

 subsequently named protoSPACE Demo 1.0-1.4, culminating in a fully operational collaborative design instrument that we used for the early design phases of several ONL and Hyperbody projects. The basic idea of the <u>Swarm Toolkit</u> is that every design concept starts with one single volume, the mother cell. This single pre-identified mother cell is positioned in the meant-to-be building site. The mother cell divides into as many children, grandchildren, etc., as needed. The mother cell and the offspring are subject to global parameters and local parameters. Global parameters, imposed on the cells, are the total gross floor area (GFA) and the boundaries of the site, aka the playground. Local parameters are the shape of the cells (rectangular or round), the area of the footprint, the number of floors, and the level, color, and transparency, assigned to one or more selected cells. The toolbox of the protoSPACE Swarm Toolkit demo lets the mother cell divide into the components, regroup a selection of them, and set point attractors or linear attractors, attract neighboring components with gradually diminishing attraction forces, and set the area of influence. The components always are subjected to global parameters. When the GFA is set to a lower value, all cells shrink in size proportionally, unless their values are fixed values; in that case, the shrinking is only imposed on the non-fixed components. After a cell division, each new cell is assigned a new identity, while the dependency on the global GFA remains intact. Two divided cells together have the same volume as that of the mother cell. A divided cell can be divided and regrouped again until an endless number of smaller cells populate the digital canvas. Eventually, the single mother cell is divided and regrouped to create the desired spatial configuration of all the constituent components. Each cell is an acting component in the design, parametrically related to neighboring components and subject to external influences like the GFA. Exactly how a swarm of birds has a consistent internal organization and is at the same time subject to dynamic external circumstances that inform the swarm to move. Similar to cell division in organic bodies, the cells are specified in a specific direction to fulfill specific functions of the spatial design. Groups of cells congregate to become an internal organ, and other cells the skin. In the analogy of the natural cell division, protoSPACE's silicon-based cell division specifies specific groups of components to fulfill specific functions as constituting parts of the building as a whole. The divided specialized cell, the single acting component, is considered a subcomponent of the protoSPACE software, while a family of similar – but not the same – components forms the basic building parts, aka the "organs," while the different groups of components are considered to be the super-components, aka the whole "body." Individual subcomponents interact with other individual subcomponents and as a group interact with the base component. A group of subcomponents, a base component, interacts with other base components, and as a group interacts with the super-component, which is the body as a whole. More levels of components are possible, but in ONL/Hyperbody projects, it is typically limited to a small number of categories of components. The Nemo/Virtools software supports the existence of components and groups of components that are actors in a dynamic system, internally coherent while responding to external stimuli, in the form of a dataset, informed by a logic that resides outside of the group of interacting components. ProtoSPACE, as Trans_Ports, is a local intranet of things and people. The interaction space functions as a giant keyboard that the players walk upon, a spatialized keyboard that is operated by a group of people rather than by a single user. The different design experts use this room-sized spatial keyboard to manipulate the design model in real time, from the point of view of their authorized unique expertise.

With the protoSPACE interaction space setup, we anticipated what later became known as collaborative working groups on other software platforms, usually directed at architects working in real time together with engineers to detect clashes between spaces, structures, and installations. At ONL, we already practiced collaborative design: we worked on the design development of the Bálna Budapest in 2007 with our two independent limited companies, ONL Hungary kft based in Budapest and ONL (Oosterhuis_Lénárd) BV based in Rotterdam. We used the parametric Revit software to work in collaborative working groups, almost instantly refreshing the models for concrete, steel, and renovated existing structures, checking their mutual consistency. For that innovative collaborative achievement, we got in 2008 the Revit BIM Experience Award.

Virtual Operation Room

 In 2002, Hyperbody was commissioned to contribute an interactive piece to an exhibition at the Delft Science Center.

Knowing that the exhibition was dedicated to innovative research on medical equipment, we decided to build a virtual game called the Virtual Operation Room, a time-based architecture for the augmented body. Without using the term, we designed an immersive Metaverse. At the end of the nineties, notified of its existence by Marcos Novak, we built our own ONL plot in the virtual city of Active Worlds. It is interesting to see how the plots were claimed. While the AlphaWorld citizens are free to choose their plots, the most popular plots of the city are located along the X- and Y-axis, and along the diagonals, as can be seen from a satellite image of Active Worlds, which looks like a radial version of Los Angeles. For the Science Center exhibition, we decided not to think like an architect looking for building opportunities but to look inward into our physical bodies. Together with bright Hyperbody students Christian Friedrich, Michael Bittermann, Chris Kievid, and Sven Blokker, we designed a virtual world that is accessed by the patient for a process we called self-healing. The design concept of the Virtual Operation Room is that the patient virtually accesses their own body, as in a computer game, navigating one's organs and adjusting the organs to cure the problem. The Virtual Operation Room (VOR) is a spatially self-organizing structure that self-develops according to running algorithms. The patient inhabits an avatar that navigates the body. Game consoles control the navigation of one's avatar and trigger events in the meta-world of the body organs. We designed four distinct worlds, the Body Portal, the Kidney Purification Plant, the Growing Brains, and the Peristaltic Lymphatic System. The Body Portal, designed and programmed by Christian Friedrich, is the initiation module, from which the self is catapulted to one of the three locations in the body to heal. The Body Portal is a soft ball-shaped, dynamic self-massaging triangulated mesh with a transparent membrane, into which the avatar must find a way to penetrate, and from there start the self-healing journey through its own body. We have introduced game elements in the navigation to trigger the user's imagination and sharpen their attention. In the Kidney Purification Plant, designed by Chris Kievid, the patient tries to shoot kidney stones, like in a laser game. When unsuccessful, it is game over, and one has to give it another try from the beginning. The score of the healing process is in the Peristaltic <u>Lymphatic System</u> instantly recorded, and visually given back to the patient. The Peristaltic Lymphatic System, designed by Michael Bittermann, offers a journey through the tubular system, where numerous small bacteria are floating, some of them malicious. To heal the self, the patient must identify the malicious ones and collect as many of them as possible to improve their health. Soothing sounds accompany the patient's journey through the circulation system of the body. The patient gets instant feedback on the score. The third and probably most advanced virtual world is that of the <u>Growing Brains</u>, designed by Sven Blokker. In the Growing Brains, there is an algorithm running that grows malicious brain cells in real time, and the patient must shoot them off. Growing Brains is a deeply immersive shoot-'em-up game, highly dynamic by the alarming rate of growing cancerous brain cells, enhanced by aggressive sounds. One strongly feels the need to shoot them to protect oneself from further destruction. Having never done something like the Virtual Operation Room before, I knew that we could do it, simply based on trust in each other's dedication and emerging skills. We worked intensively together to make things work; we accepted the challenge given to us by the Science Center and finished ahead of time, within one month.

Mapping the Virtual Operation Room concept into the world-making of architecture, one practical application is to build algorithms into those active components of the design that form a potential threat, especially for structural capacities and climatic performance. For example, not matching the design criteria for CO_2 neutrality or circularity can be built in as a red alert into the design platform, or not fitting within the bandwidth of structural requirements. In our design for the Climbing Wall, we have built in a red alert for exceeding the maximum angle between the components. Whatever design one proposes, one may opt to secure the targeted performative and structural requirements. The performance is monitored by virtual sensors and informed by global data coming from external systems that feed into the local design system. If not performing according to the criteria for soundness, the red alerts will go off and one will need to reconsider some design options to further develop the design. Hypothetical design situations as in the Virtual Operation Room game advance the profession of architecture, as they sharpen the collective mind. It was only a matter of time before we applied some of the findings applied in the physical world.

NSΛ Muscle

After the Trans_Ports project, a second important step toward fully programmable bodies is our bespoke NSA Muscle installation at the Centre Pompidou in 2003, a physical embodiment of the Trans_Ports paradigm of a programmable body. NSA stands for nonstandard architecture. On the day of the opening, Frédéric Migayrou, the curator of the Nonstandard Architectures exhibition proudly showed me that the Trans_ Ports project appeared on the front page of the Libération daily newspaper in France. The Libération newspaper probably chose that image of our project to represent the supple moldability of projects shown at the NSA exhibition. The NSA Muscle is ONL's first 1:1 scale physical translation of the Trans_Ports 2000 project. Together with co-curator Zeynep Mennan, Migayrou organized the game-changing exhibition on the mezzanine floor, including the works of many of our colleagues, notably Greg Lynn, Hani Rashid/Lise Ann Couture, Bernard Cache, Sue Kolatan/Bill MacDonald, Servo, Lars Spuybroek, Mark Goulthorpe, Francois Roche, Dagmar Richter, Tom Kovac, and Ben van Berkel. The NSA exhibition is undoubtedly one of the most important milestones in the Nonstandard Architecture movement and, as stated before, was ignored by Rem Koolhaas in his Fundamentals show at the Venice Biennale in 2014. Before NSA, we participated in another groundbreaking exhibition titled Digital Real, Blobmeister First Built Projects, curated by Peter Cachola Schmal, at the Deutsche Architekturmuseum in Frankfurt am Main in 2001, and a series of smaller exhibitions titled Transarchitectures, curated by Odile Fillion and Michel Vienne started back in 1997/1998, just after the completion of the Waterpavilion, which was one of the exhibited projects. And more acknowledging my work than anything else, I participated in the groundbreaking Venice Biennale curated by Massimiliano and Doriana Fuksas in 2000, featuring Trans_Ports 2000 and some other NSA architects in the Italian Pavilion and the national pavilions as well.

 At the NSA show, we were assigned a <u>central space in Gallery 3 on the mezzanine of the Centre Pompidou</u>. We showed an 8-m-long 1:200 3D-printed model of the A2 Cockpit in the Acoustic Barrier, a 1:20 model of the nonstandard structure of the Web of North-Holland, and the four 1:1 scale models of the constituent components of the TT Monument. We installed the Cockpit/Acoustic Barrier model at eye level to have a realistic experience of the project when passing by. This was much to the dislike of our neighboring exhibitor <u>Lars Spuybroek,</u> who complained until deep into the night to Migayrou that his work would not be sufficiently visible, and requested to have our model lowered, fortunately in vain. We always present our models at eye level because all too many projects are designed to look great from a bird's eye view, thereby neglecting the view from street level. Besides the central position at the mezzanine, we were asked to install an interactive piece in the Forum at the entrance lobby of the Centre Pompidou. With the financial support of the Dutch Stimuleringsfonds, we developed an inflatable sculpture that is wrapped in 72 muscles, donated by the Festo branch in Delft. The blue balloon, made of a fabric that stretches in two directions, is engineered and executed by Rienk de Vries of Buitink Technology. The 72 muscles are connected to the ellipsoidal balloon in a triangulated mesh, leaving the two ends, the transparent front and back, open for the public to look into the muscular object. The 72 muscles are the actors with a collective hive mind that collaboratively massages the inflatable in real time. We made the NSA Muscle hop, twist, and shout, responding to touch-interactions by the public. The 72 muscles act together as one coherent swarm. The swarm of 72 muscles and their sensitive nodal points represents the acting sub-component of the installation. The balloon is the passive sub-component that gives in to the actions of the muscles. Besides the muscular body, the server computer is the third sub-component, orchestrating the movements. The public, triggering new movements by gently touching the sensor disks, and the NSA Muscle are the two main interacting components. The triangulated mesh, the public, and the server running the software consist of a variety of subcomponents. All components and subcomponents form the adaptive ecology of the interactive installation. The number of changes in the length of each muscle shows how much the installation is alive. Each nodal point of the triangulated mesh is connected to the balloon, and some of them, notably those at eye level, are provided with a disc with touch sensors and proximity sensors. As for the Saltwater Pavilion and the Trans_Ports installation, interactivation designer Bert Bongers designed the sensor discs and put them together. By subtle movements, the inflatable sculpture invites the public to touch its muscular body and thus interact with its behavior.

Looking out from the inside through the round transparent caps to the public in the lobby, one feels alienated and cocooned. Being inside the NSA Muscle feels like being on an expedition to unknown territories, like being in an otherworldly spacecraft, with stargates at both ends to look back at the old normal. The NSA Muscle is not designed as a habitat though but as a sensorial experience for the public around it and peeping into its operational heart. Inside the inflatable NSA Muscle sculpture, the central brain runs the Nemo/Virtools executable file, instructing the 72 valves to vary the pressure in each of the 72 flexible muscles and thereby make the body move. The techno-Gothic interior of the NSA Muscle is filled with 72 tentacles of the respiratory system, 72 thin light blue flexible hoses from the array of valves to each of the muscles, aligned with 72 data cables from the computer to the sensor disks. The programmed respiratory tubes are the lifelines that keep NSA Muscle alive. Pumping the pressure into the Festo muscles means shortening the muscles, and tightening the affected part of the mesh, thereby shrinking the affected part of the balloon. Letting the pressure drop means relaxing the muscle, letting it go. Each muscle is an actor in the hive mind, operating in permanent relation with its immediate neighboring muscles. When only one muscle would be programmed to move, it would cause no perceptible movement of the NSA Muscle. Only when a muscle works in a coordinated effort with its immediate neighbors, the NSA Muscle sculpture changes shape.

 The NSA Muscle informative screen shows in real time exactly which of the 72 muscles is in which state of inflation/deflation, and which one of the sensors is triggered by which proximity sensor and the touch sensor. The management, design, production, execution, and operation of the NSA Muscle is the result of an intense cross-disciplinary design & build team effort, working from scratch with programmers, engineers, artists, and designers, motivated by one strong shared idea and determined to make it work. The accomplishment of such a complex behavior of the interactive installation showed us the way to the inevitably dynamic future of what traditionally is static architecture. The future, which in our vision is the embodiment of the here and now, is to imagine and realize interactive, performative, parametrically designed and robotically built projects.

 Dynamic interactive architecture can only be achieved through a deep interdisciplinary exchange of knowledge and ideas, working as one coherent level playing field team, from the initial idea to mothering the working process.

 Invited by CCA director Phyllis Lambert, who was instrumental to get Mies van der Rohe onboard to design the Seagram building, and project curator Greg Lynn, the author of Animate Form, the NSA Muscle is archived in its digital entirety at the Canadian Center for Architecture (CCA) at the Archeology of the Digital, Media, and Machines exhibition.

Hyperbody

Working with Hyperbody master students has been a rewarding experience. Not only did we realize technological innovations, but we also implemented educational innovations as byproducts of the challenging design brief, for the students to put their teeth into. While my assistants would take care of the more regular teaching tasks, I would write the brief, define the challenge in the opening lecture, and join the weekly discussions on the work in progress. Each new Hyperbody master design studio brought about new challenges, building upon the achievements of previous projects. The educational innovations range from interacting working groups and real-time interacting project data to interactive lectures. I implemented the concept of real-time interaction between the nearest neighbors, whether people or things in all fibers of Hyperbody education. Usually working in five to six groups of three to four people on one project teaches the students to find their role in a dynamic hive of communicating minds. I typically request them to take on the role of a specific design expert, like the structural designer, the climate designer, the shape designer, the interaction designer, the stylist, and the project manager. Group work based on specific task assignments leads to bi-directional communication and a better understanding of the cross-disciplinary nature of design. Group work enhances interactive information exchange between the simulated experts, which principle is then mapped on the projects. I ask the students to link their projects in real time to the projects of their fellow students. I typically demand their projects to be linked in real time to neighboring projects, by exchanging information on their changes in geometry in real time. Thus, any chance in one project inevitably leads to a change in the neighboring project, which teaches them that no project is a closed system but as per definition acts as part of a local ecosystem, which in its turn is part of larger ecosystems.

As an example, I applied the dynamics of real-time interaction to some of my introductory lectures at Hyperbody. I invented the concept of the interactive lecture, where the audience participates actively by choosing images and twittering messages during the lecture, using a double screen, one for the images and one for projecting the Twitter feed. The projection screen with the images is a split screen showing one image chosen by the audience via a special purpose-designed mobile phone application, programmed by long-time Hyperbody computing assistant and Hyperbody wiki pages designer Vera László, and one image chosen by myself from a repository of images. I would make an improvised *ad-hoc* analysis that bridges the two images, by finding similarities and pointing at substantial differences. We involved the students actively via digital interventions, asking them to mine their contributions from the pool of images that we prepared for them, and from Twitter. The interaction between the speaker and the selected images resonates with how we, in the practice of designing buildings at ONL, typically handle the interaction between the constituent building components. We create connections where there were none before. Both ONL practice and the Hyperbody research are based on the paradigm of a real-time communicating swarm, creating a hive mind for the constituent components, people, and things alike.

My first teaching assistant at Hyperbody was Misja van Veen, one of my master students at that time. One year later, I invited the CAD experienced senior Hans Hubers to become my first assistant professor. In parallel, Hubers started his Ph.D. titled COLAB, Collaborative Design in Virtual Reality, in sync with the development of Hyperbody's protoSPACE lab. I continued accepting promising Ph.D. researchers, Nimish Biloria in 2002; his research is titled Adaptive Corporate Environments: Creating Real-time Interactive Spatial Systems for Corporate Offices Incorporating Computation Techniques, and Henriette Bier in 2003, whose research is based on System-embedded Intelligence in Architecture. Eventually, I promoted a dozen Ph.D. candidates at Hyperbody and stimulated them to embark on groundbreaking subjects for their Ph.D. research. Notably, Alireza Hakkak with Enhancing [Spatial] Creativity, Gary Chang with HyperCell: A Bio-inspired Design Framework for Real-time Interactive Architectures, Christian Friedrich with Immediate Systems in Architecture, Tomasz Jaskiewicz with Towards a Methodology for Complex Adaptive Interactive Architecture, Achilleas Psyllidis with Revisiting Urban Dynamics through Social Urban Data, Han Feng with Quantum Architecture (work in progress), Sina Mostafavi with Hybrid Intelligence in Architectural Robotic

Materialization (HI-ARM), and Jelle Feringa with Architectural Robotics: Bridging the Divide between Academia Research and Industry. All of them became experts in their field and found high-level jobs, most of them in academic environments, some starting a business. From the beginning of their research journey, I asked most of them to assist in teaching Hyperbody's master students and hired them as assistant professors during and after they finished their Ph.D. research. Hyperbody became a popular choice for architecture students and the Hyperbody Research Group was quickly growing into an expert group of 12 people at its peak in 2008. Unfortunately, after the global recession, we experienced serious budget cutbacks and were cut back to six people. In 2016, I left TU Delft due to having reached my retirement age, and that ended the Hyperbody project.

Muscle Tower

From as early as 2004, we started building large 1:1 interactive prototypes, with consecutive batches of bachelor students at Hyperbody TU Delft. We kept doing that until my retirement in 2016, and each semester, we chose a new subject. The early Hyperbody interactive installations, between 2003 and 2008, bear titles like Muscle Reconfigured, Muscle Tower, Muscle Body, Muscle Facade, Bamboostic, Muscle Passage, Interactive Portals, Bubble Pods, and Jealous Portal. Later, we also involved master's students to design more complex interactive installations as part of their master designs. In the last few years of my professorship, we cooperated with the Faculty of Industrial Design and Electrical Engineering, in a cross-disciplinary student Minor project.

The first prototypical installations were done by groups of bachelor students in a six-week elective course. After the NSA Muscle show, I donated the 72 Festo muscles and the exo-brain server of the NSA Muscle to Hyperbody to inhabit the interactive prototypes as actuators. In 2004, we built the Muscle Tower, a towering 6-m-high structure in the former machine hall of the faculty building of the Department of Architecture. The tensegrity-ish structure features fixed elements in combination with the programmable muscle components, allowing the tower to rotate, bend over, swing, and sway. Similar to tensegrity structures, the fixed elements ensure that the tower stands up, while the tensile Festo muscles facilitate dynamic movements. As in the NSA Muscle, I consider the swarm of muscles the main active interacting subcomponents, not the design of the node, as is the case in the static structures that we designed, while the static subcomponents around the node are hinged in a pretty generic way. The other subcomponents consist of the sensors that are embedded in the four feet of the tower. The behavior of the tower is programmed by the students to respond to the proximity of passengers. The tower and the passers-by are the two acting parties. Approaching the feet of the tower results in the tower bending over toward the passenger, as to bow in respect.

Applied on a larger scale, one could imagine a 60-m-high commercial advertisement tower alongside the highway responding to the traffic that is passing by. Nagging toward the automobilist or following the cars such as a Muscle Advertisement Tower would create added attention value and hence represents a higher commercial value for the advertisement tower operator. With our interactive prototypes, I was typically aiming at large-scale commercial applications. An advertisement tower that responds to passengers, catching their attention not only by the video content but enhanced by the interactive physical movement of the advertisement tower itself, would make a great candidate for a real-world project.

Muscle Body

The Hyperbody 1:1 prototype following the Muscle Tower using the same Festo muscles is the student-designed red Muscle Body, conceived as one long linear flexible black PVC tube keeping the orange stretchable fabric in place to create a closed envelope, at a dozen randomly chosen points connected to itself by the programmable muscles. The orange-colored stretched fabric is laced at the inside of the flexible black PVC rainwater pipes. The black tubes are 5 cm in diameter, loosely wrapped around the orange-red volume like flexible veins around the pumping heart. Proximity sensors are placed inside the body trigger sensuous massage-like contractions and expansions. The movements of Muscle Body evoke the idea of it being alive, like a slowly pumping heart. Muscle Body is an intimate space to be accessed through an interactive opening in the form of a gill. One of the sensors is designed to open and close the gill, which is a part of the continuous envelope at times to function as the door – a door that is not a door but a possibility to enter, to invade the soft body. The openable part of the skin invites people to step inside and quietly closes after them.

The design concept, the structural solution, the material research, and the assembly programming, that is, the whole project is designed, purchased, produced, and assembled by a group of six bachelor's students within the time of six weeks. All that was needed to inspire the students was the stack of Festo muscles, a working exo-brain server, and videos of previous projects. By and large, at Hyperbody, we created educational situations where the students self-learn and self-educate themselves. The educational principle behind Hyperbody's prototyping projects is similar to the simple rule, complex outcome strategy that we adopt in real-world projects. As Stephen Wolfram postulated in his book <u>A New Kind of Science,</u> out of many possible simple rules, only a special category generates complexity. Only when the initial condition has a rare combination of strangeness and normalcy, of the alien and the generic, the simple rule will reign.

What do we consider the acting components in this Muscle Body piece? Structure and skin go perfectly well together, and mobile programmable parts are included as part of the structural system. We cannot single out the nodes to be considered as the components or the Festo muscles; they are supporting subcomponents of the whole body. The flock of randomly placed muscles with the stretched fabric in between acts as one coherent component, actuating the subtle massage of Muscle Body. The whole of the muscular body could be considered one single complex component, similar to ONL's parametric one building one detail paradigm, synchronizing structure, and skin. There is no structural or styling difference between up and down, left, or right; there is a robust continuation of the same principle all over. There is not a floor, a wall, or a roof. Once inside, the visitor steps on the same fabric as is used for the upper parts. This structural and material unity makes one feel fully immersed in the body; it provokes a sense of wholeness and completeness. The subcomponents assembled to form the body are patches of stretchable fabric, muscles, sensors, and a single very long black tube that wraps multiple times around. Exterior to the body, the same exo-brain server that is used for the NSA Muscle manages the varying pressures of the muscles. The Muscle Body fully depends on this external respiratory system. The exo-brain component is one of ONL's old unused desktop computers, having enough number-crunching power to run the Virtools software that informs the valves to switch on and off and to read the signals from the sensors into the system, designed to evoke the varying degrees of massage.

Bamboostic

 After the Muscle Tower subsequently followed the Muscle Reconfigured, the Muscle Body, the Muscle Facade, Bamboostic, the Muscle Passage, and since the big fire that burned down the faculty building into a total loss and after we had to close down protoSPACE 2.0 in the iWEB as well, many more interactive installations in the context of Hyperbody's master design courses were materialized. Each of those interactive installations was innovative of its kind, and many of them are still accessible at the Hyperbody archive site and on my Youtube channel. The sheer multitude and inventiveness of the interactive installations, and the students' prospering geniuses that come with it, would deserve a separate big fat book titled "Hyperbody Part II, Interactive Installations, Master Courses and Ph.D. research of Hyperbody from 2010 to 2016." Quite a few video recordings of the installations are available on the Internet, ready to be found by googling interactive installations Hyperbody TU Delft. Special attention must be given to the interactive and proactive installation Bamboostic, led by the Hyperbody student Sander Korebrits, specially designed on the occasion of the bespoke three-day international Game Set and Match II Conference (GSM II), which we organized in 2006 at the old brutalist Faculty of Architecture at the TU Delft. Bamboostic not only proposed an inventive use of the Festo muscles but also introduced interacting components of a different kind in the interaction with the public and the muscles. Bamboostic consists of four main components:

1) the forest of 18 vertically bundled bamboo sticks, balanced by three Festo muscles each,
2) the people wading through the forest,
3) a living goldfish in a bowl of water, and
4) a video tracking system, suspended from the ceiling, that monitors the movements of the people in the forest and the movements of the fish and sends instructions to the bamboo forest on how to behave.

Bamboostic represents the three main themes of GSM II:

1) architecture as a real-time unfolding *game*,
2) *set* the parameters of the complex adaptive systems, and
3) *match* the exchange of information between data processing components in real time.

The forest, the public, the fish, and the tracking system are processing data in real time and influencing the behavior of one another. Together, the four components form a complex adaptive system, a synthetic ecology that includes the alien logic of the fish. The video tracking system records patterns in the movements of the public and the fish in the bowl and, based on data analysis from the patterns, steers the movements of the bamboo forest by varying the tension in the muscles. On the one hand, the forest self-moves without the interference of the public, and on the other hand, the muscled bamboo sticks respond to the presence of people by leaning toward those who are standing near the bamboo column for a moment, similar in behavior to the earlier Muscle Tower. Navigating through the

Bamboostic forest is a mesmerizing experience; one feels immersed in another ecosystem with a logic that is different from that of our daily lives. The Bamboostic forest is driven by an internal force and moves in response to external factors, which are two basic conditions for things to be considered alive. Not alive in the sense that organic plants and animals are alive, but alive as a silicon-based synthetic organism that digests data instead of food. Synthetic beings are a new species, emerging from a critical mass of installations that are augmented with ProtoSPACE 2.0.

ProtoSPACE 2.0

In addition to the budget for building the iWEB, the second life of the Web of North-Holland, Hyperbody was granted a 120K EUR budget to install the interactive design environment inside the iWEB, protoSPACE 2.0, the sequel. Building upon the Trans_Ports Biennale setup and the setup of protoSPACE 1.0 inside the faculty building that burned down, I designed a <u>pentagonal interaction arena</u>, embedded into the logic of the geometry of the iWEB. The iWEB is a freely tweaked six-frequent dodecahedron, molded around a regular pentagonal interior ground floor and roof plan. Five huge projection screens are suspended from a web of cables connected to the edges of the inflated roof, leaving 1-m space between the screens to enter the interaction arena. The setup is similar to that of Trans_Ports, but here with translucent projection screens, allowing viewers to see projections from both sides. The central area is the interaction space, and four of the five sectors outside of the projection screens are the group working spaces. The fifth sector is taken up by the entrance. The beamers can switch to mirror image and back to be used alternately in the interaction space or the working spaces. The iWEB creates thanks to its small footprint and cantilevering sides all around <u>a natural auditorium</u>, gently sloping upward from the central pentagonal space. The screens can be individually lowered and raised as needed, thus allowing the space to transform into a traditional confrontational lecture setting with one projection screen. At the exterior side of the screens, we furnished the working space with refurbished red polyurea-coated pieces of ONL's polystyrene hot wire cut 5 × 5 furniture series.

ProtoSPACE 2.0 is a digitally augmented group design environment for research, education, and commercial applications, for the professional fields of art, dance, architecture, urban planning, and leisure activities. Experts, clients, and the public work together on the participatory design platform of the Swarm Toolkit software. The main interacting components are

1) the protoSPACE software, supported by a powerful server PC,
2) the participants in the design game,
3) the array of sensors connected to the top of the screens, identifying the players, and tacking the patterns of the group of players, and
4) the virtual design worlds, the newborn virtual objects that evolve by the combined effort of the interacting components.

The revolutionary XiGraph meta-application protocol supports concurrent modeling via a semantic network, communicated in the open .xml format, thereby linking design software, hardware components, sensors, and actuators. The XiGraph-based application is developed by Hyperbody's Ph.D. candidates Christian Friedrich and Tomasz Jaskiewicz. XiGraph supports the handling of real-time interactions between a variety of professional software tools of a variety of users for a variety of purposes like interface mapping, design versioning, process modeling, conceptual diagramming, and semantic modeling of the design. In the Swarm Toolkit application, design objects function as smart active agents or as passive top-down controlled objects. Hyperbody's ProtoSPACE 2.0 software links dynamic families of components of different natures together in real time. The people move freely as in a swarm, the sensors monitor their movement, and the software communicates with the actuators that are embedded in the virtual designs. The swarm acts and is acted upon; there is nothing more natural than that.

413

The iWEB was confronted with its tragic death as the big fire that destroyed the old faculty building left the iWEB without power, short of one year after the opening. Until that moment, Hyperbody had been gaining weight, fulfilling its promise of being one of the most, if not *the* most, advanced digital design studio in the academic world. After the fire, together with the other departments, Hyperbody slowly went into a downward spiral, by and large, caused by severe budget cuts imposed by the general TU Delft management. The tragic downfall of the faculty coincided with the 2008 financial crisis, which harmed the level of academic research at the faculty. For example, from then on, Ph.D. candidates had to bring their own funding, material costs were reduced, short-time contracts of staff were not prolonged, and no new investments could be made, except for using the insurance money for rebuilding what was destroyed. Despite these serious setbacks and after two years working from temporary improvised locations at neighboring faculties – Hyperbody occupied a nondescript space in a corridor at the faculty of Industrial Design – we eventually set up a lab in the renovated building called BK City to embark on new developments in the still emerging area of interactive architecture, nonstandard geometry, synthetic, parametric, and robotic architecture.

ProtoSPACE 3.0

After tough negotiations, Hyperbody got assigned a 150-m^2 space in BK City, after convincing the relocation committee that we would use the same space in a multimodal way for education, courses, studios, workshops, and as the lab space for further 1: 1 scale interactive applications. We lost the unique environment of the iWEB; fortunately, we could save the computers and projection screens from destruction. At BK City, we got a regular square space back, with four columns in the middle that are standing in the way of the free movement of people participating in the interaction. The four columns are unfortunately too close to each other for mounting the projection screens in between; so we had them mounted on the surrounding walls instead, clearly a compromise since we could not use both sides of the screen in the same collaborative setting as in the iWEB. We decided to reintroduce the interaction space and enhance it with a new component, the interactive floor. The concept for the interactive floor is to have the floor act as a pressure-sensitive keyboard, a walkable keyboard. Under a one-time faculty recruiting program for hiring the so-called high potentials – Hyperbody was lucky to have <u>Marco Verde</u> and <u>Mark David Hosale</u> – I asked them to design and execute the interactive floor. The sensitive floor tiles of protoSPACE 3.0 behave like an intelligent swarm. The individual wooden pieces are parametrically designed to be unique in shape and in-house CNC fabricated. The wooden components are inhabited by pressure sensors and Arduino devices to communicate with their peers as the members of a swarm.

The components "talk" directly to their nearest neighbors, disseminating information on their actual status in real time. Stepping on one of the floor tiles means creating information about one's position in space and one's weight, which information is then shared with immediate neighbors, who in turn share it with their neighbors, thus spreading the news. By moving around on the interactive in specific patterns, the participants steer the interaction in a certain direction. In one particular setup, the interactive floor is directly connected to a point cloud of reference points for a spatial design. By taking steps, the players manipulate the reference points and hence tweak the 3D model. The interactive floor as a whole is a design instrument, played by walking on the giant keyboard. Besides the bottom-up creation of information by physically navigating the protoSPACE 3.0 lab, the array of individual tiles is directly connected to a central server to receive top-down instructions. The interactive floor is used for the interaction with 3D models, for public interventions during my interactive lectures, for dance experiments with dancers from the dance group of <u>Kristina de Châtel</u>, for Ph.D.-related research, and is used as a design instrument for student design courses.

What did we learn, on top of what we already achieved, from the components of the interactive protoSPACE 3.0 floor? Somehow, we reached the limits of our knowledge and capacity to make things work. We imagined how to move forward, but with every more advanced concept for the interaction, we realized that we could not progress without the cross-disciplinary participation of hardcore mathematicians, aka programmers, who would know how to further develop the self-organization of the swarm. One step forward – and sideways at the same time – was to join the European Commission-funded MetaBody project that started in 2013. From our side, MetaBody was by and large led by Nimish Biloria until 2015, to further explore the spatial interaction of the human body with adaptive "posthuman environments." The name MetaBody was invented by the Spanish project leader Jaime del Val; in his own words, the "non-human non-binary philosopher-artist-activist: meta humanist polymath, transdisciplinary media artist, performer and metaformer, ontohacker, meta-technologist, uncurator." To me, there is nothing more natural, and therefore not "post-human," to immerse oneself in imagined, designed, and executed environments, to surround oneself with things, gadgets, and technology in general; in other words, to merge the physical and the metaverse-type reality in real time. We involved our Hyperbody students in the MetaBody project; one project that stands out is the Reflectego interactive installation from 2014, which was reinstalled in Madrid during the MetaBody conference in the Prado Medialab. Reflectego is an interactive triangulated piece that is hanging from the ceiling. The triangular, acting components are equipped with sensors that detect the presence and proximity of the public underneath, looking up at fragmented images of themselves. People are seduced by the movements of the mirrored triangular pieces to move their hands toward the ceiling, which in turn triggers the sensors to change the configuration of the ceiling. The effect is that one sees different fragments of oneself, reflecting one's ego, and blending with the egos of the other players. The suspended triangular components hinge on their immediate neighbors, together forming a reflective cloud that never has a dull moment.

The ultimate goal of Hyperbody's research is to have interactive components all over a space, as we have in protoSPACE 3.0 the interactive floor, as the Reflectego for the interactive ceiling, and as in ONL's 1999 Space Station project to have an interactive interior skin that senses, reconfigures, and proactively proposes new configurations. That is, interactive components that interact with movements and actions of people inside and climatic conditions outside the project, as a living interface between people and their environments. What is more natural than architecture as a real-time acting interface between living beings and unfolding climatic conditions? An architecture based on interacting components is the default state of Another Normal. To complete the overall picture of interactive component-based environments, further Hyperbody research and education focused on self-organizing building structures, spatialized game environments, and Pop-Up indoor and outdoor installations, while ONL developed further practical applications as the Interactive Wall project in collaboration with Festo, the Dynamic Sound Barrier, and a participatory urban design instrument called Participator. One of the most outstanding student projects was protoSPACE 4.0, a collaborative design for a new iWEB to be established at the entrance of BK City, the new faculty building. A real option then back in 2010, but in the years ahead of 2010, turned out to be wishful thinking.

Interactive Wall

The interactive installations by Hyperbody evolved in terms of scale and professionalism into the Interactive Wall project, commissioned by and developed in collaboration with Festo AG in Stuttgart. Festo is a worldwide operating company producing components for the process industry; we applied their innovative pneumatic muscle actuators. The Interactive Wall (IW) consists of six 6-m-high programmable wall components, triangular in cross-section, equipped with sensors in the base and just one single Festo muscle down at the base. The shape is based on an in-house Festo invention called the Fin-Ray, a stretched triangular shape whereby the two long sides are connected by parallel plates, hinged to the long sides, which allows the wall component to move elegantly forward and backward. Festo developed the Fin Ray technique for the flying Air-Ray wings, and for the flapping wings of their swimming

<u>Medusa</u>. The way the Fin Ray element is actuated by one single muscle in the Interactive Wall project meant a new innovative combination of Fin Ray and muscle. Walking along the Interactive Wall triggers a wave-like behavior of the installation, building a dynamic relationship between the built environment and the users. Sensors in the base of the components sense the presence and proximity of passers-by and inform the IW brains to act. How the components of the Interactive Wall act is up to the imagination of the programmers of the behavioral choreography, fed into the Hyperbody exo-brains that instruct the components, while limited to the maximum stroke of the Fin Ray technique. The elegant movements of the components in response to passers-by are the default state. The basic setup forms the basis for further rules of play for the interaction between people and things, whereby both the people and the components are the actors who propose configurations. As postulated before, a meaningful interaction needs at least two acting parties, both of them with a will of their own, and both of them proactive.

After having realized the above-described groundbreaking interactive indoor installations, the focus of Hyperbody gradually shifted to smart environments and robotic design to production, while our practice ONL gradually shifted toward the realization of large-scale buildings like the A2 Cockpit in Utrecht, the Bálna Budapest, and the Liwa Tower in Abu Dhabi, none of them offering an opportunity for further developing concepts for programmable buildings that change shape and content in real time. My concept of real-time adaptive architecture, peaking around 2010, somehow lost its momentum, and only recently I encountered renewed interest, mostly from the side of academia and not (yet) from commercial firms. At the <u>Alive! international symposium</u> at the ETH Zurich in 2013, under the direction of <u>Ludger Hovestadt</u>, not surprisingly co-organized by Hyperbody's Ph.D. candidate Tomasz Jaskiewicz whom I promoted in 2009. We met many like-minded researchers who are in one way or another involved in building interactive installations. Although generally smart and delicate, the projects that were presented were disappointingly small-scale, and none of the other presenters catapulted to the larger scale of a building. One of my criteria for substantial change in the building industry is that the techniques that are developed, albeit parametric design, robotic production, or programmable components, must apply to the grand scale of architecture. The ultimate test is whether the technology applies to multi-story buildings, if not skyscrapers. If yes, that technology is bound to become a game changer. If not upwardly scalable, it is doomed to remain a fringe activity. As of now, parametric design and robotic production are widely acknowledged game changers, but programmable architecture, above and beyond kinetic architecture, is not there yet. Time will tell whether my efforts in this field will eventually bear fruit. The closest to a meaningfully applied interactivity that we have seen so far in architecture are adaptive facades. Some research works at universities today focus on facades that respond to changing weather conditions, which is just a simple first step in the direction of a fully programmable architecture. ONL proposed a programmable facade structure called the Adaptive facade back in 2003, based on an assembly of inflatable cushions and programmable Festo muscles, arranged horizontally as a second adaptive skin around the periphery of the floorplan, applicable to any existing multi-story building. The Adaptive facade responds both to changing external weather conditions and to changing internal usage of the space. The Adaptive facade is an interactive membrane, an active interface between the outside and inside.

Over the years, I have been spreading my design focus over a wide variety of ideas, instead of focusing on one single topic and exploiting that single topic to the max, as other designer-inventors respectively architects-entrepreneurs typically do. To integrate the different performative aspects of building, one simply has to combine efforts in a wide variety of knowledge fields. Mature programmable architecture involves much more than some superficially responsive elements; it requires a complete synchronization of structure and skin, built up from programmable components. The strategy of programmable architecture can only be successful when it is an essential genetic strand in the very DNA of the design. Full-blown programmable architecture changes the spatial conditions of the building, not only the surfaces. Full programmability is not just responsive, but proactively proposing

419

new spatial configurations, starting an interaction with the users and the weather. Full programmable architecture is a dialogue between people and their environments. Programmable architecture lives on the Internet of Things and people.

Digital Pavilion

Along with Hyperbody's progress in the field of interactive and proactive architecture, Hyperbody's bright collaborator Christian Friedrich took on the idea of Immediate Architecture for his Ph.D. research. The term immediacy is meant in the sense that all factors that in some form or another influence the design are in real time connected to the design environment, whereas design and fabrication are fused into one evolving system. To implement the paradigm of fully programmable buildings into the structural fabric of a substantially large building, one will need to rethink the way it is used and how it modulates. As in natural physics, every single component behaves in space and time; nothing is an isolated object, and constituent components live in intimate relation to each other, to their immediate neighbors in the first place. The condition to consider for a truly alive building is that many key components are programmed to change the spatial conditions of the environment. That must include the entire structure, the skin, and the overall interior spatial layout. What used to be static structural components must become actuators, possibly actuated by electronic pistons or phase-changing materials. There is ample existing technology available to make it work. When we design the structure to consist exclusively or partially of actuating pistons, the spatial conditions can be programmed to dramatically change over time, as we have demonstrated in the Ground Zero design for the Max Protetch Gallery.

Five years after Ground Zero, I have been commissioned by the South-Korean exhibition developer TNT to design a Digital Pavilion, as part of a larger ICT complex in Seoul. The Digital Pavilion is a programmable environment for a real-time educational gaming environment unfolding in an existing structure. The

Digital Pavilion is a large complex adaptive robotic system of interacting installations. Navigating the interior feels like walking in the interior of a living installation, like living inside evolution. The sectors of the Digital Pavilion installations are not only interacting with the public but also with other sectors. The output of one programmable sector provides relevant data that are used as input for another sector. The visitor gets a unique number using RFID tags and builds up its unique profile while navigating the four floors with interactive and interacting installations. Each visit is a unique experience; the installation will never repeat its exact shape and content, while in real-time adjusted by the movements of the public and by streaming content, facilitated by the 2012 new wireless broadband techniques. The public interacts with the installations using the then-just-introduced 4G smartphones. Thus, the building as a living installation becomes a showcase for the technological priorities set by the Korean government.

The structural components are mutually connected electronic pistons, including the surfaces that are spanned between the triangulated structure, made of a stretchable fabric to move along with the pistons. The surfaces are designed to be infused with numerous LED lights to broadcast any possible image or movie. The pistons change the spatial conditions of the interior space by interaction with the public, either via the app on their cell phones or by their physical presence. For example, approaching one particular sensor causes one or more pistons to extend and thus create an opening through which the public enters the adjacent space. Along the way of the design evolution of interactive architecture, I have unlearned myself to label such an opening as a door, while the word "door" has too many traditional connotations. It is more proper to talk about an opening since that relates directly to the act of becoming open. I stopped referring to the object and referred to the action instead. The opening appears for the visitor and disappears again after passing through. An alternative term for Interactive Architecture would be *disappearing* architecture. This is because the traditional static architectural elements like doors, canopies, ramps, stairs, floors, walls, ceilings, roofs, etc., tend to disappear, making way for programmable components that transform from one configuration into another. Much to my disappointment, the

Digital Pavilion design was not selected for realization. The road ahead for interactive, immediate, and disappearing architecture is clear though. The technology is here, and the design concepts are here; yet, the societal urgency to build programmable buildings at a larger scale, financed by motivated clients, has not arrived yet.

Self-moving Components

When mobile programmable parts form the constituting components of an interactive construct, one of the design questions is who or what causes the changes in the configuration? By what external force are they informed, and what drives the internal dynamic of the interacting components? The ultimate situation is when the components move proactively by themselves, as self-moving identities, executing a will to move, to change the spatial arrangement, informed by a dynamic interplay between the users and environmental data. Not unlike self-driving cars but applied to components that built spatial three-dimensional constructs. Only some of that external data would come from human users of the construct. When the behavior of the components is no longer directly linked to the directives of people, then the construct is no longer responsive, but proactively making its choices. We consider the construct to self-act when it organizes its spatial behavior, based on an internal drive to move forward while processing external data. Just as humans respond to environmental changes, an acting construct also responds to changes, in its unique way. It must be noted that most human responses in daily life are utterly predictable and not that unique. Many human responses are by and large pre-programmed and often quite repetitive and robot-ish, albeit with a personal touch to it, like a parametric variation on an otherwise pre-programmed behavior. How many free choices do we humans make in one day? Even creative professionals like artists, architects, composers, and writers accept to live a life by and large organized according to a pre-programmed pattern. For example, my daily morning routine is to empty the dishwasher, squash a citron, and make a cappuccino every morning before breakfast. Each day, I repeat the same action, over and over again. Yet, not necessarily in the same order. Sometimes I put the milk container before refreshing the water in the coffee machine. Sometimes I empty the dishwasher before making the hot citron and the coffee, and other times thereafter. Different roads lead to the same destination. That is one aspect of freedom of choice, which can be randomly programmed into interactive and proactive environments. Every time the same actions, in the same space, using the same ingredients and the same kitchen equipment, but in a different order. When one writes the script for a robotic operation, some actions must be performed in a linear order, and some do not. Such partly flexible routines can be robotized by programming the relationships between the individual actions, by introducing (a seemingly) random order of actions, and by if–then statements. Scripts to operate robots typically are hundreds of pages, since every single change, move, gesture, or act from either the actor and/or the things that are acted upon has to be described in minute detail. A typical limitation is that most acts must be foreseen in detail. Scripting means breaking down acts and activities to the smallest detail of the design and describing their mutual relationships in time and space. Modern computers do not worry about the vast number of operations. The script of making hot citron and coffee may be in itself a well-defined case, but every process always unfolds in an environment; it is never a stand-alone process. When we consider a robotic system to self-make the hot citron and the coffee, then we need to position it as an actor in a dynamic environment and have the robotic setup make choices responding to information from outside the citron- and coffee-making process, even if those external circumstances hardly ever have a decisive influence on the internal logic of the process. In real life, only a few quirky actions or thoughts occasionally cause a drift from normalcy. It is exactly these not pre-programmed situations that define the capacity of a being or a vehicle to self-act. Self-aware vehicles look around and process information from the world around them. Advanced programmed Boston Dynamics robots are trained to respond to unpredictable situations; the robots are AI trained to interpret patterns of behavior instead of exactly pre-described situations. Intuitive and creative acts seem counterintuitive and sometimes absurd and contradictory; they feel like the proof of a non-programmable free will. Yet, being unpredictable and perhaps even the absurd can be a programmable form of behavior. Self-driving vehicles and self-moving components of a building construct must be able to cope with unforeseen circumstances, and they should act themselves a bit weird, as to educate their peers to respond in a way not to harm anyone. This applies to humanoid robots as well as to interacting components of dynamic real-time behaving structures.

I have a keen interest in developing interactive installations that act by themselves, in a mix of functional responses and unsolicited spatial behavior. An excellent example of such a possible larger-scale installation is the Hyperbody student project POD⁶⁴ (2016). POD⁶⁴ is just one out of many other equally relevant student projects. On the Hyperbody archive www. hyperbody.nl site that I am keeping in the air after my mandatory retirement in 2016, many more proactive environments can be found. POD⁶⁴ consists of 64 cubes that can self-arrange in an endless number of different configurations, like a four-sided Rubik's cube. Each cube can move by itself; it can move horizontally on the ground or the top of another cube and is equipped with vertical cog-wheel sliders to crawl up along the sides of one of the other cubes. Each of the 64 cubes can do that. Moreover, each cube can rearrange its interior to become the stair to units above or under. Elevators are not needed, because the cube elevates itself. The cube self-moves and self-elevates. In its most compact form, the 64 cubes are neatly stacked in a spatial 4 × 4 × 4 = 64 arrangement. Each of the cubes can be in any position. Together, they form a building with 64 connected rooms that can be used in many different ways. The cubes can open or close in relation to each other, thus forming bigger spaces or smaller cells. The configuration of the cubes will never repeat over time, neither in its most compact form nor in its most distributed form. The cubes inform each other in what state they are, and the others know how to respond to that. There is no fixed arrangement of the cubes, only endless ways to (dis)connect.

Typically during nighttime, the cubes would go back to their most compact form, and during daytime, they would go out and distribute themselves in public space on the quay of the Rijnhaven in Rotterdam, to interact with the public. Each one of them operates alone, harvesting energy for its operation by plugging into base stations. They can move freely along the quay and other public areas and adjust to a function that is desired at that movement or to a self-created function. The pods look at each other so as not to bump into each other, and to form larger configurations, all based on the information they receive and the information they send out to their neighboring pods. Users can rent the space the way they want. A pod has the proper size for any kind of kiosk, street food unit, private working space, group meeting room, or minibar – basically anything that would be functional at a given time in a given space. The units are multimodal in themselves and can group to form clusters of units and cater to a larger clientele.

The main component in the POD⁶⁴ project is the basic unit itself. The individual units are the players in the real-time reconfiguration game. Proactive architecture orchestrates the components to act as the members of a swarm, to act individually, yet follow a set of simple rules that keeps the flock together. The units of proactive architecture have a hive mind. The units operate by communicating with their nearest neighbors in the first place. They communicate to avoid bumping into each other and stay close enough not to lose contact with the other members of the swarm. Beyond that, some external attractors steer the swarm as well. There are base stations where the units go for a refill, just like robotic lawnmowers and household robotic vacuum cleaners do. And in the POD⁶⁴ project, there is one central attracting point where the pods gather in their most compactly stacked format.

This project works well because, besides having a unified way to connect, all the constituent components have the same standardized dimensions; they always fit, in whatever configuration. The cubical pods represent a traditional form of modularity, whereby all units have the same dimensions and shape. Now here is the thing. When we aim at a combination of nonstandard architecture, whereby all constituting components are unique, and only fit in one position to form the greater whole, with the behavior of proactively interacting components the idea of modularity, needs to be rethought. Nonstandard units of a fixed geometry would only fit into one position, like the pieces of a 3D puzzle, meaning that there is only one possible form of the clusters of components. Hence, in proactive nonstandard installations, the spatial components must be able to reshape themselves to fit into multiple spatial layouts. Self-shaping of the constituting components is relatively easy when the components are one-dimensional, like pistons or memory materials. The challenge to reshape itself increases with proactive components that are two-dimensional slabs, yet still a surmountable challenge. You might want to think of telescopic slabs that can change their length and width and can rotate one-dimensionally to change the spatial layout of the building, as I imagined for the Ground Zero project. Or one might think of an elastic skin stretched on the actuators that give in to the changes in geometry, like in the Digital Pavilion project. The capacity to reshape itself in three spatial dimensions in real time as a consistent volume was proposed in the Trans_Ports project back in 2000. Yet, Trans_Ports 2000 is bound to the ground, and not designed to connect physically to other Trans_Ports pavilions, not on the same level, and not in a three-dimensional arrangement either, although, theoretically, this might be possible. In a thought experiment to have multiple programmable volumes acting together to form a stack of unique units, we would need Trans_Ports-like volumes that can move by themselves, climb up onto each other, and adjust their shape to their immediate neighbors.

 Hyperbody's Ph.D. candidate Christian Friedrich, who led the 2009 protoCOLOGY workshops with Hyperbody MSC2 students, came close to this ideal imagined situation by exploring the idea of immediacy. Multiple units in the form of unique irregular space-filling polyhedrons, based on the Weaire-Phelan structure, made instantly after being designed, flock together in the protoCOLOGY construct. Noteworthy is that the Weaire-Phelan

 structure formed the basis of the space-filling geometry for the Water Cube Olympic Pool in Beijing in 2008. The protoCOLOGY workshop meant a brave instance of the paradigm of immediate architecture, whereby the individual units are designed and produced instantly, thereby facilitating the flocking together of uniquely shaped units on the fly. Yet, after the production of the components, the process is frozen. Is there a way to keep the process of 3D formation alive?

Until today, neither Hyperbody (2000–2016) nor ONL has developed further concepts, nor have we seen many proposals from interactivity or proactivity colleagues for a spatial programmable flock of free-form units, where the units themselves can change shape and content in real time. A project that comes very close and which I appreciate is the MORPHs roaming robots project by Ruari Glynn with William Bondin at The Interactive Architecture Lab at The Bartlett. This octahedral robotic MORPHs structure can change its shape, and by doing so, it can walk and could adapt its shape to neighboring components.

We invited Ruairi Glynn to visit Hyperbody by the end of the first decade of the third millennium, and we shared our ideas on programmable building blocks with him at that time. He took the visionary step to establish the Interactive Architecture Lab (IAL) at the Bartlett in 2013, basically taking over where we had to stop. They have done wonderful projects since, although predominantly modest-sized indoor projects, and not yet applied to the larger scale of architecture. In many ways, the IAL can be seen as the sequel of Hyperbody. While we were losing support and budget at that time at TU Delft and were forced to gradually close down Hyperbody, pending my retirement, The Bartlett funded Glynn's interactivity Lab. Without further speculation on the idea of self-assembling programmable spatial units, self-shaping and self-moving proacting components certainly are the logical choice for the advancement of proactive architecture.

Pop-Up Apartment

What about programmable interior components? What about programmable content? The aim of Trans_Ports 2000 is to be a spatial construct that changes shape *and content* in real time. In Trans_Ports, the content is projected on the flexible LED screens that cover the whole interior surface. The content

427

also includes the way the programmable hard space is lived by its users. Participating by interacting with the projected content in an exhibition setting is one thing, and participation by living your life inside the programmable environment is another level of interaction. I imagined back in 1998 how a programmable interior might change using actuators in the interior skin of the Space Station project. In 2014, I proposed another unsolicited interactive design, the Pop-Up Apartment. Thinking architecture is not per se dependent on a client, I tend to think ahead anticipating possible future commissions and extrapolating the potential of the here and now. In those cases, operate like an autonomous artist; I do it because I feel an internal drive to do so. That also explains why I am familiar with proposing constructs that are not born from a functional program but from extrapolating developments in society, art, and science. That awareness forms the basis of the symposiums I have (co)organized in my early professional years, usually supported by academia and by generous national grants, events like

1) Artificial Intuition (Galerie Aedes and TU Delft 1990),
2) The Synthetic Dimension (Museum De Zonnehof Amersfoort, 1991),
3) Sculpture City (RAM Gallery Rotterdam, 1994), and
4) Genes of Architecture (Academy of Architecture Rotterdam, 1996), all of them together with Ilona Lénárd, and subsequently, academic events like
5) GameSetandMatch I (TU Delft, 2002),
6) GameSetandMatch II (TU Delft, 2003),
7) GSM III (TU Delft, 2016), and
8) GSM4Q (Qatar University, 2019).

All of them were initiated to further develop the fusion between science, technology, art, and architecture.

The Pop-Up Apartment study explores the idea of full programmability of interior space, while the envelope of the space itself remains static. Imagine a void space, like an empty classroom, as the initial condition. A generic classroom is a generously sized space that can become anything. Its emptiness intends to evoke that same feeling as one has when entering a vacant artist studio before moving in and embarking on a highly personal occupation of that space. I have chosen a base unit of 7.20 × 7.20 m, indeed the standard size of a classroom, and by and large the standard module for generic office buildings. With this generous size, the spatial unit, the voxel, accommodates all sorts of living and working conditions. For the Pop-Up concept, I imagine such a vacant space to become a fully programmable space. The programmability includes furniture objects popping up from the floor, coming down from the ceiling, folding out from the walls, and using sliders, pistons, and hinges to operate the built-in pieces of furniture. The temporary inhabitants, the nomadic international citizens, would use an app to program the desired configuration of the pieces of furniture in any given period. The users inform the Pop-Up unit from a distance while web-based. The nomadic international citizens (NICs) feel any of such programmable units as their home; their temporary home because it is laid out according to their individual preferences. Home is where you are. Or the NIC comes with friends and wants a kitchen and a dining table, and 2 hours later, the couch for casual chatting. The basic design concept is that there are only those pieces of programmable furniture operational in the room *when and as needed*. Only that, there, and then. If not in use, it sinks back into the 1.20-m-deep floor, which contains the other programmable objects. The Pop-Up Apartment's deep floor includes a variety of programmable components, a seating area, a dining table plus chairs, side tables, a double bed, a kitchen, a bathtub, and a toilet, while the deep wall hosts the closets and a home cinema screen, and the deep ceiling contains the light fixtures. Every component is available on call through the app at any time for any time, in any combination. The obvious added value of a fully programmable spatial unit is that one feels as if in a large space, whatever configuration one has opted for. One has a 50-m² bedroom, a bathroom of 50 m², or even a generous 50-m² toilet. Thus, the relatively compact unit of 50 m² feels like a much larger house. From the project developer's point of view, this concept offers an interesting option for future developments. On the one hand, the developer can build more units in a given volume; on the other hand, they can sell a compact unit as a luxury home. Earlier, we have seen proposals from architect Gary Chang from Hong Kong for partial multimodality, not as integrated into the building fabric though as the Pop-Up Apartment. In the Suitcase House at the Commune by the Great Wall in China, Chang has projected a kitchen that is hidden under a hinged piece of the floor, and some additional recessed private areas, also hidden under the wooden hoods. Another example from Chang, the Domestic Transformer in Hong Kong, born from pure necessity, is a tiny 32-m² apartment with sliding closets and kitchen, and a bed that folds toward the wall. Chang's multimodality is manually driven, and the sliding closets feel like living in an archive system. For the Pop-Up Apartment, we have chosen for full programmability, popping up from a completely vacant space, beautifully empty like a Zen space, more Mies than Mies; "Mies Is Too Much," as I tell the audience when explaining the One Building One Detail paradigm. Together with the then Hyperbody tutor Chris Kievid, we approached the Rotterdam-based project developer Blauwhoed, to support a hands-on Hyperbody MSc2 Pop-Up Apartment design course. Interestingly enough, Blauwhoed understood the market potential for our concept of programmable multimodality. Blauwhoed generously donated 10.000 EUR for material costs for the students to build their prototypes and arranged a place to exhibit the results. Inspired by the irrefutable logic of programmable environments, a bright batch of Hyperbody students built their own 1:1 prototype for a Pop-Up Apartment, each proposal imaginative and valuable in its own right.

At ONL and later at Qatar University, I proceeded with the concept of programmability and imagined a multimodal tower with hundreds of pop-up programmable units. The steel structure to host the pop-up furniture is based on an earlier original idea by engineer John Kraus of D3BN. Kraus invented a column-free office building with a floor span of a generous 14.40 m in one go for the Nissan Headquarters in Amsterdam, using honeycomb trusses to support concrete floor plates and concrete ceiling plates, floor plates, and ceiling plates together having enough mass for proper sound insulation while absorbing the transmission of contact-sound. The space between the two layers is 1.20 m high, one-twelfth of the full span. In between the hexagonal openings in the trusses, there is plenty of space for air ducts, electricity, and data cabling, in all directions. A similar spacious construction would be the perfect solution for the facade as well and for loosely distributed climate control instead of fixed positions for heating and cooling systems. Kraus' system intelligently merges floor, structure, and ceiling into one integral component. I adopted the strategy of ample space reservation inside deep floors and walls, for structure, ducts, and cables for our TORS sculpture building proposal at the beginning of the nineties. In many ways, this TORS sculpture, exhibited at the L,v exhibition in 1993 stands as a predecessor for all of our future proposals and executed buildings, while it offered the perfect componential framework for the integration of art and architecture into buildings that are habitable sculptures and for programmable multimodality as well.

My Qatar University-based research titled MANIC (2018–2019), Multimodal Accommodations for the Nomadic International Citizen, brings the concepts of the Pop-Up Apartment, the double column-free honeycomb truss floor in combination with a diagrid facade structure, and further social aspects of programmable multimodality, together in a proposal for a 20-story rectangular-shaped building. The MANIC building is a very lively community of short-term guests and long-term guests, and every possible period in between. I developed the idea of a ubiquitous booking app to make ultra-short-term, short-term, weekly, monthly, or even very long-term reservations to program the space in one particular personalized way. The same 50-m^2 unit transforms *on demand* into a hotel room, a working space, a workshop space, a classroom, a yoga workout, or a dining room. The extra costs for the actuators and for the technology required for popping up and disappearing of the furniture are easily compensated by the structural compactness of the MANIC spatial scheme and by the almost continuous 24-hour usage of each m^2. The spaces are programmed to be used all day and all night, vastly surpassing the currently optimal 80% or 90% hotel occupancy rate. MANIC calculates an occupancy rate of far over 100% daily, while spaces are booked two, three, or more times per day. The MANIC scheme is a somewhat restrained version of the possible merge of nonstandard architecture and programmable environments. MANIC is explicitly meant to appeal to project developers in Qatar; we left out the otherwise exciting combination with nonstandard design to production procedures. In the MANIC proposal, the proposed components are standardized from unit to unit. We restricted ourselves to one single commercially compelling aspect of actuation. A logical step for the foreseeable future is to apply the principles of parametric

 design to robotic production of the unique constituting components to the programmable components. But, the die is cast, in the words of Antonino Saggio in his preface titled

 "Alea Iacta Est" to my book *Towards a New Kind of Building*; he stated that now that I have crossed the Rubicon, there is no way back to normal. The conceptual logic is there, the social urgency is there, the technical feasibility is beyond doubt, and the commercial profitability is obvious; it is about time for programmable architecture to land between the ears of visionary investors.

Participator

 A recent effort to move forward in the field of participatory interactive design, based on the principles of parametric design, is the urban design instrument called Participator, developed for the project developer Steven Manhave in Rotterdam, in a consortium with New Citizen Design to represent the new lifestyle, Cepezed Projects as the circularity expert, and the engineering firm Royal Haskoning DHV. Based on the initial situation of a three-dimensional point cloud of thousands of reference points projected on the site, the Participator design game connects the dots between the different stakeholders. The serious game is programmed in the visual interface Grasshopper for Rhino, by ONL's gig worker at the time designer-programmer <u>Arwin Hidding</u>. In its essence, Participator is a connection machine, connecting experts to laymen, designers to users, and politicians to citizens. As of now, the program is designed to be managed by someone familiar with Grasshopper and Rhino. It requires experience and expertise to be able to work in a meaningful way. Adding new features requires even more expert knowledge in programming in Grasshopper. To allow an unskilled expert player to work with the design game, eventually, a user-friendly interface to the program must be developed. The user-friendly interface of the Participator app might look like a simple questionnaire, with quantitative and qualitative questions on sliders, as in ONL's attractor Game. Where is the location/target area? What are the boundaries of the working space? How many square meters will be planned? What is the maximum building height? What is the resolution of the point cloud of the initial condition? What is the imagined target group, and what is their lifestyle? Each question shall have an info button explaining the how and why. The main interface that follows the questions, as to set up the initial situation of the design game, will look like a screen with sliders, a different slider for each parametric component. The constituent components of Participator equal the components in the Grasshopper interface, processing input values while connecting their output to related components.

As in MANIC, in the initial condition of the Participator urban design instrument, the size of the modular unit is 7.2 × 7.2 × 7.2 m^3, roughly the size of a double-height classroom. The generic unit bears any function from home, office, workshop, and outdoor terrace to open hanging garden. The spatial layout of the units is based on the chosen porosity value of the whole urban structure. The higher the porosity value, the more open the structure. At the start of the design exercise, the total desired volume of the development is chosen. The budget can be set to a fixed value. The players are asked to draw one or more lines to define the approximate imagined location of the masses. The lines function as the backbone of the structure, assembling the voxels around them. The lines may be straight or curved, horizontal, diagonal, or upward moving. The drawn curves represent the centerlines of the foreseen development and perform as a linear attractor for the number of scheduled multimodal units, as in the protoSPACE toolkit software. Choices are made, one by one, quantified using sliders. The effect of each choice is immediately visualized, and open for discussion and feedback at the workshop table. The Participator urban design instrument is designed to guarantee that, whichever choice is made, the result will match the set budget. Thanks to the parametric relationships between all acting components in the graph, the chosen values can be changed in any phase of the design process, and the whole structure is immediately recalculated and updated. To work in a higher resolution, the distance between the points of the point cloud – aka the size of the basic units – is adjustable. Infrastructure like elevators, stairs, and ducts is included in the total gross volume taken up by the modular units. As in virtually all ONL projects, the elevators, stairs, and ducts are included in the main volume, which is characteristic of the inclusive nature of my design strategy. In my take on architecture, all infrastructure must be integrated with the main components of the design.

On top of the deliberate choices, the design game is enriched with a randomized porosity factor, functioning as the serendipity factor, defining the level of openness of the generated three-dimensional voxels in beforehand unpredictable positions. The Participator design game allows players to play with constraints and values, with quantities and qualities, to find an optimum between the choices. Shared green spaces will enhance the overall Q-factor, whereas Q stands for the quality of the design proposal. Open spaces will come at a cost but will, as per definition, fit within the set budget. The stricter the budget, the less porosity will be allowed or must be balanced by other choices. An attractor line that is positioned closer to the ground implies a lower price per unit, and thus some more units can be placed within the allowed budget. A higher porosity factor will stir up building costs. The porosity value in combination with the intuitively drawn attraction lines generates often surprising outcomes.

The algorithms behind Participator generate configurations that the project developer and the spatial designer could not have imagined beforehand. The results of playing Participator are used as a quantifiable and qualifiable framework for a future master plan.

434

Experience with real-life design sessions at Manhave has proven that a three-dimensional Participator conceptual structure is looked at more open-mindedly. The outcome of a Participatory process triggers the spatial imagination of the players and serves as a reference model for the future master plan. The commercially valuable innovation of Participator is – whatever configuration is produced – that it always remains faithful to the chosen financial constraints.

Playing Participator draws the future inhabitants of the built environment into the design process of parametric and interactive architecture, giving them an active role in the design process that unfolds in real time, according to the principles of inclusive design. The principle of inclusiveness applies to interactive installations, where the main goal is to create strong experiences, like in the Trans_Ports installation. At Hyperbody, the principle of inclusiveness is applied to educational settings, where the students are playing the design game on a spatialized keyboard, as in the protoSPACE installations. Inclusiveness applies to 1:1 scale, to the built prototypes like the Interactive Wall, and potentially to buildings as a whole adaptive system. In this phase of development, participatory inclusiveness is still in the realm of design proposals, as in the Digital Pavilion project, as imagined at the largest possible scale in the Ground Zero project. Thus, the principle of inclusiveness is applicable in virtually all phases and at all possible scales of a design project, from sketch to completed project. In the earliest design phase bringing a diverse group of stakeholders together to explore a variety of possible configurations, in the group design of 1:1 prototype, in visionary proposals for programmable environments, in the shared experience of interactive installations, in multimodal hospitality buildings, and in interactive lectures as the inclusive alternative for one-directional teaching. An architecture that is an assembly of interactive components enhances diversity, inclusivity, multimodality, and participation, and thereby connects to current forward-looking movements in society and naturally lives in parallel in the Metaverse. Nonstandard and interactive architecture are fully in sync with actual social developments concerning diversity, equality, justice, climate, and inclusiveness.

9

Where Do We Go?

Pragmatic

So far, I have discussed the nature and performance of passive and active interacting building components, parts that are acting to shape their bodies. I have touched upon how the swarm of constituent components feeds on external data from their immediate environments. I will anticipate how the chosen strategies of component-based simplexity, nonstandard and interactive architecture, and proactive architecture possibly relate to current societal developments, especially in the context of a post-COVID-19 society. The next normal, aka my Another Normal, has to come to terms with the effects of the global pandemic, with the global food crisis, the energy crisis, the water crisis, ever-increasing inequality, and ultimately with the climate crisis. The climate crisis is the crisis of unrestrained economic growth; in other words, the crisis of the current out-of-control capitalism. I already mentioned the vision of the Greek economist Varoufakis in his book *Another Now*, in which he describes a participatory world, where people are the co-owners of the enterprise they are working for/with, where people have a guaranteed basic income and participate directly in every local and global decision-making process. The dominance of the current shareholder system that has financial profit as its highest priority is seen as the main cause of the crisis. There will remain differences between people with respect to how much money they accumulate, but, other than what is the case in the here and now, based on personal positive contribution to the success of the company, rather than on speculative investments of shareholders. Varoufakis remains critical to trend-following believers in naive utopian visions, while at the end of the book, he prefers the messiness of the current world over an all too sterile ideal world. He prefers to fight for a better world from the dire reality of the here and now over a flight into an ideal yet sterile society and Metaverse. I work from a similar point of view; I do not indulge myself in utopian visions but take advantage of available social and technical achievements of the here and now to realize my forward-looking goals. I tell my students and my team: you can do anything as long as it works. Whatever one pursues, it has to work, for you, and the good of society. In a scientific sense, my point of view argues for transparency, for verifiable validation of data, for taking responsibility for one's own produced data, and thus for being credible. In the professional field of nonstandard architecture, the scripts have no other choice than to work; otherwise, they are useless, and nothing gets built. Interactive prototypes simply have to work in real time; otherwise, there is no interaction. Building prototypes is almost the opposite of fact-free storytelling. I do not tell stories, I do not create illusions, nor do I use metaphors. On the contrary, I make sure that things work, and these things are what they are, without a hidden agenda, without the need for justification by storytelling. Making things work on the scale of realized projects, installations, and prototypes, I am keen to find out how such a visionary and, at the same time, pragmatic attitude might apply to global societal issues. In art, artists are responsible for the concept, the execution, and for the budget. In nonstandard architecture, where parametric design relates directly to robotic production, architects are responsible for the exactness of their data. For a society to advance into a fair and just Another Normal, a direct link must be established to the performance to verify that things work, with a well-defined responsibility for participating individuals, communities, and cities. I will refer to these three levels in society when it comes to the decentralized and distributed production and consumption of food, energy, and water.

Simplexity

Simplexity is the art of having simple rules that facilitate the emergence of a benevolent complexity. Will simplexity work on the larger societal scale as it does on the architectural scale? Could a benevolent form of simplexity apply to cities, neighborhoods, and individual citizens? Rules of law result in specific spatial or societal patterns. Navigating satellite images and discovering surprising patterns in the landscape, both natural patterns and patterns as a result of human habitation, virtually all patterns we see on the map are the result of laws, natural laws, and human-invented laws. The patterns I consider collective works of art are instigated by a simple rule. Being a secular world-maker myself, I appreciate the Islamic rule that requires any mosque to point its main axis in the direction of Mecca as a great example of such a global work of art that is based on a simplexity rule. This simple rule results in a magnetic pattern that is distributed over the entire globe. To appreciate the beauty of it, imagine drawing a world map with only Mecca and all the mosques in the whole world, and one sees the magnetism. Mecca, more precisely the gray granite almost cubic Kaaba, a super magnet, creates a virtual magnetic field that attracts almost one million mosques worldwide, the biggest artwork of all. The Kaaba dates back to the 7th century and is still operational, perhaps a stronger magnet than ever. In every Islamic city, there is approximately a 1-km-grid of mosques as the building guideline is that one should not walk more than 500 m from work or home to the mosque. Not unsurprisingly, a similar distance is currently advised for local services in 15-minute walkable cities, as coined by Carlos Moreno in 2016. On a super-local level of detail, in each hotel room in predominantly Islamic countries, there is an indication on the ceiling of what the proper direction towards Mecca is. Apart from the visual and spatial effects of the rule, there is also the omnipresent soothing sound of the muezzin, calling for prayer from the minarets. Five times per day, the muezzin sings his way through a prayer, which I have experienced as a pleasant direct personal touch. I felt invited to sing along with the elementary melody, and perhaps since I do not understand the meaning of the verse, only the sound and the pitch remains, just pure music. Besides the fixed places of worship and indications of direction, there are mobile versions too. When there is no mosque near, people take their prayer carpet with them. Car drivers spread their praying rugs along the highways at the roadside approximately in the direction of Mecca. In their winter camps deep down in the desert, traditional Qatari spread their rug in the proper direction and say their prayer aloud into the open. The global rule affects both traditional nomadic life and settled city life. Although less obvious, a similar rule governs ancient Christian cathedrals. In medieval church architecture, the main axis orients itself towards the east. The word orientation comes from the Latin word for east, the Orient. The east is where the sun rises from, or rather, toward where the Earth is turning itself to, eager to let the morning light pour in through the colored stained-glass windows. The east end is where the altar is placed, often within an apse. Similar to the position of the mihrab in the mosque, the altar in the church represents the shortest distance to self-proclaimed religious truths, mediated and guarded by the priests and the imams. The facade and main entrance are as a consequence at the other end. The orientation toward Mecca of mosques and the east-west orientation of ancient European cathedrals explain why they are often placed at odds with the fabric of the city, which is based on a different set of rules. Similar but not the same as the magnetic positioning of mosques, cathedrals are more often than not placed at an angle to their immediate environment, thereby setting themselves apart from the generic. This is the art of the rule, one simple rule, which adds a layer of complexity to the city fabric. It is a world-making process of what I call simplexity: one simple rule that generates a complex outcome. Similarly, in response to the Coronavirus, societies respond by adopting new behavioral rules that function as meme components, dynamically redefining the existing social structure.

Social Patterns

In the era of the spreading COVID-19 virus, it is fascinating to see how new rules are meme components that are forming unprecedented spatial realities. The 6 ft distancing rule changes the way people behave toward their nearest neighbor. Instead of almost bumping into each other, as citizens are used to in crowded environments, the people are requested to keep a modest distance of a minimum of 6 ft to avoid people breathing directly into your neck, making you inhale their droplets. Recommended by the

World Health Organization is a 2-m distance; yet, most European countries have reduced the rule to 1.5 m, while in the USA, the 6-ft rule was adopted. In practice though, it usually comes down to probably 1-m distance though or less. Two meters is more than one thinks it is, when in a conversation with one another, the distance is 1 m maximum. When the 6-ft distance is put into practice, the effect of this Corona parameter is outlandish; it heavily affects how people populate parks, streets, museums, and restaurants. People that are close family may sit together but must keep social distance from other potentially infected clusters. Even today, in the aftermath of the pandemic, I have adjusted my spatial behavior, even when walking in the wide-open space of a park. When I pass oncoming hikers on the path, I step a bit aside to maintain a proper distance; I might hold my breath for a while. Nothing like that has happened before, I have become much more aware of the other's near presence and have become sometimes overly cautious. Imagine an opening for an art exhibition; how will we organize ourselves when obeying the 6-ft rule? Usually, gallery openings are very crowded; one stands so close to the other that one almost directly inhales portions of the air that other people are breathing out. During openings, the air gets so dense with droplets that one can almost cut the air, as the saying goes. We survived that in earlier days, but now I follow the new guidelines for my own good and for that of others, I take more care, and I keep a proper distance, meaning that the capacity of simultaneous presence in galleries had been drastically reduced during the peak of the pandemic. I have seen galleries with dots painted on the ground to indicate where one can stand next to one another. In shopping centers, the advised distance before the cashiers is painted on the floor. At the entrance, guards are counting the number of people that are allowed to enter, proportionally to the size of the interior. In small shops, only one at a time.

Distancing from other people has become subject to a rule, a formula with a given set of parameters. We still behave as the members of a flock; we are the herd, and we are the swarm, but temporarily the parameters governing the herd had been changed. Obeying the rules, we still have that hive mind, like the birds that are landing on a power line, we are instantly and instinctively changing our distancing parameters. While sitting on the wire, the birds are still flocking, albeit in a relatively static configuration. It is more accurate to state that the rules are not changing, but only the parameters operating on that rule. Thus, changing essential parameters has a huge effect on spatial planning. In our temporary new normal, people need much more space per person, which reduces the population density in venues like art exhibition openings, shopping centers, and theaters, which in its turn requires a new take on the operational economy of exhibitions, shopping, and leisure activities.

Imagine again the pattern of dots 1.5-m apart on the gallery floor. To be able to move at all, not all dots may be occupied. There should always be one dot free, so that one person can move to give way to another person, like in the simple sliding puzzle game. If there are more visitors than capacity in the art gallery, then people will wait outside, still obeying the 1.5-m rule. Patterns in the streets will change dramatically as a result of the new rule. Following and maintaining the new rules is facilitated by Corona apps on one's cell phone. Apps may tell you when you have been too close to someone that is infected. The apps work a bit like the minesweeper game. A number tells you if, and if yes, how many mines are close to you. COVID-19 apps may create an invisible protective shield around you, in much the same way as automated cars are looking at each other to avoid bumping into each other. Apps may assist the public to arrive at the gallery just in time so as to avoid queuing. And here is the thing, real-time information is similar to the "just there, just then, and just that" paradigm of nonstandard parametric design to robotic production.

The digital invades not just the building industry but also our social lives; the digital orchestrates our social patterns. Even when the COVID-19 crisis will be under control worldwide, the digital techniques that have been developed are here to stay and are bound to have a permanent influence on how we as a human species behave. Not only will app-driven social distancing be useful to control the sneaky Coronavirus, but it will also apply to more regular influenza viruses as well, and to crowd control in general. The West might take the example of the Japanese and the Chinese to wear facemasks and keep their distance when infected, not egotistically to protect oneself but to altruistically protect others from getting infected. Social distancing inspires the creation of new patterns for daily goings in densely populated cities. New social patterns emerge out of the simple 6-ft rule. As in the noughts and crosses game, aka the Tic-Tac-Toe game, an individual in a crowd may have up to eight immediate neighbors. Moving around in a crowd means that people behave inclusively toward others, while one needs to take the positions and movements of others into account. Indeed, people are aware of one another like one of the 15 components in the number sliding puzzle, aware of each other like the birds in a swarm.

Scalable Distancing

New forms of relationships between the individual and the collective arise from the social distancing rule. Inhabitants behave more nomadic and more collectively at the same time. Keeping proper distance is a scalable behavior, but keeping a distance is quite different in respective cultures. Specific cultures develop specific behavioral rules simply based on the specific social distancing parameters they have adopted over longer periods. As an example of how local cultural parameters shape public life, life in Qatar gives some clues. In the culture of Qatar, men and women do not touch each other in public, and men and women do not shake hands. But women and women do, and they may hug each other quite strongly, as Ilona has experienced often when meeting with friends and students. Qatari men are considered the guardians of women. In parks, the men walk a few steps behind the women and the children to protect them from dangers from behind. According to tribal rules, women cannot travel alone without the guardianship of close family members, although traveling alone is officially allowed by the state of Qatar. Sometimes tribal customs are stronger than the law.

Without exception, Qatari women wear black abayas, sometimes also "open" abayas, topped with hijabs, with only a few women wearing niqabs, while many women do wear niqabs in Saudi Arabia. It all depends on the

443

strictness of the interpretation of the ruling culture. Wearing a black hijab is an expression of a culture, more than it is an expression of faith. In other Islamic countries, the hijab and sometimes also the abaya dresses are in color. Along with the general acceptance of black abayas, a vast new fashion industry has emerged in Qatar. The simple cultural rule for women to wear a black abaya when outdoors creates a wealth of abaya differentiations. Not a single (young) Qatari woman wears the same abaya; all their abayas are different, from high fashion with fanciful floral stitches to punk-style abayas decorated with needle pins. Summer abayas are very lightweight and thin and do not absorb heat. Under the abayas, they wear anything fashionable or comfortable, from stone-washed jeans to sluggish leggings. Students allow themselves to wear their abayas half open, but officially this is not recommended. In every governmental building, roll-ups are visualizing the mandatory dress codes, for men and women. Men are not supposed to wear short sleeves either or wear Bermuda shorts in public. In private, the women wear whatever they like; every family member, male and female, at home is casually dressed.

The spatial effect of a simple cultural rule is one of the takeaways from our two-year stay in Qatar. The simple social rule to wear a black abaya has an enormous visual impact; it is an art project in itself. It is impressive to watch the choreography of Qatari students wearing their black abayas at the colleges, moving about swiftly, as if lightly floating just above the pavement, with always ample space around them because of the distancing rules. Cultural rules are scalable; they vary from country to country. The culture-specific rules guarantee to keep a culture-specific distance. The cultural and temporary variations of the rules inform the members of the swarm how much distance to keep at that particular moment and in that particular place, when needed and as needed. Corona has added new parameters to the already existing distancing rules, only to make clear that the complexity of people's behavior is subject to simple rules.

Science Rules Art, Art Rules Architecture

For the Food Energy Water Nexus design studio I did with third-year students at Qatar University, I told them to consider art to become a dominant economic force. I gave them examples of large-scale art projects like the Running Fence of Christo, and the East-West/West-East installation in the Brouq nature reserve in the desert of Zekreet in Qatar by Richard Serra. These large-scale art projects bring about movements of people and trigger economic activities that turn them into economic forces themselves. Bespoke large-scale art projects clearly show that economy equals social activity and that new social activities, aka economic values, can be created from the production of art. I asked my students to imagine how art-driven concepts could shape the framework for a new form of sustainable use of the land. The students came up with wonderful, enlightened design concepts. One group, who proudly called themselves the Rebels, was inspired by the Singing Sand Dunes that make a rumbling sound when the wind blows and when sliding down from them. They proposed an inhabitable series of sand dunes covering an area of roughly 2 × 2 km, similar in size to a typical Doha neighborhood in Doha. There would be no visible infrastructure, only man-made and nature-enhanced sand dunes that are crystallized into habitable structures at the lay-side, away from the prevailing winds. Self-driving electric 4WD vehicles would cross the desert in between the sand dunes without the need for tarmac roads or traffic lights, just tracks in the sand and on the limestone rocks, as in the desert. The building structures embedded in the dunes would include urban farming, while the skin would be clad with arrays of solar cells.

I showed them a research project by NASA for a space colony of 10,000 inhabitants. Already back in the early seventies, NASA showed how this could be done in minute detail. If an autonomous yet connected settlement can be established in

445

space, it sure can be done on Earth. Another group of Qatar University students – we always work in groups and dedicated specific tasks to each group member – imagined a city that is cut deep into the desert to form a cool canyon, reaching for fresh water, and taking advantage of the natural cooling from the deeper layers of the earth. Therewith anticipating NEOM's Trojena – built in a canyon between the mountains – that is recently published by NEOM as part of Saudi Arabia's vision 2030. Seen from above and from within such city concepts that are driven by simple conceptual rules would be experienced as giant land art projects, they will not be experienced as a new form of landscape destruction, on the contrary as landscape making. These food, energy, and water nexus-driven landscaped cities will not produce pollution caused by fossil fuel burning, as they integrate solar power and urban farming. In every aspect of the performance of conceptually strong art-driven design, the original design concept must be maintained at all levels and scales, in their formal appearance, and their food, energy, and water production performance. THE LINE by NEOM, which was not yet published when I was teaching at Qatar University, is equally based on conceptual art-driven design concepts. Conceptual art-driven landscaped cities will be appreciated as an enrichment of the natural landscape, as a new incarnation of the natural. The future of building lies in a serendipitous fusion of science, art, and architecture. The future of design will be based on facts and be driven by intuition. Science rules art, art rules architecture, and architecture frames life.

Informed

Learning to live with the virus, social distancing has, by necessity, become scalable in time and space, leading to new forms of social patterns, and eventually to new forms of spatial layout, first new forms of interior layout, most likely to be followed by a novel arrangement of spaces in new buildings. We will certainly experience a more loose, less compact distribution of people on the streets and in the buildings. People will regain their space in the streetscape, repulsing the dominance of cars. Automated cars display a looser distribution on the streets as well and, at the same time, take up less space. Automated cars move more smoothly, maintaining regular speeds and mutual distances. No more macho-emotional drive styles and no more stopping at traffic lights, and hence no more traffic jams either. Automated cars are driven by an artificial intelligence-managed form of social distancing. Similar to automated vehicles, people who will be informed by a health distancing app will walk around more smoothly than today. Corona informed people to keep a proper distance, avoid large crowds, to avoid skin-to-skin contact with strangers. The scalable distancing app may at first sight force robotized movements onto the people, but, more likely, it will become a new cultural expression, a new cultural rule on how to behave in public, even when the pandemic is no longer raging among us. As people always have been either attractors or repellers, the Corona era increases the repelling forces. The health app knows which Corona state you are in and communicates that to you and your immediate neighbors. The neighbor's app receives the signals, processes the data, and advises the app user on what distance to keep. In China, a simple COVID-19 app has been applied successfully to reduce the transmission of viruses. The Chinese app notifies you when you have been in close contact with someone infected, and the app bears the user's immunity status, which can be checked by reading the QR code. China uses the app for a top-down determination of whether your QR code shows green, meaning free to go places, yellow, meaning seven days of self-quarantine, or red, meaning 14 days of self-isolation. One will be denied access to hotels and other places where people gather, and one is summoned to stay home in self-quarantine.

The positive side of a ubiquitous distancing app is an orchestrated smoothness of movements; potentially, it makes endless queuing superfluous, and the idle waiting times before cash desks redundant. We have become accustomed to long waiting times and long queues, which we have somehow accepted as normal in today's society. A social distancing app will reduce waiting time to a minimum and streamline the free flow of people at airports, stations, post offices, and banks. While no one will mourn the absence of long queues and waiting times, many will remain reluctant to accept top-down interference in one's

447

social behavior. Every type of swarm is top-down informed by some crucial data on the external conditions, but they are, at the same time, in balance with the strong internal drive of the members of the swarm. One question is: which data are fine to be received top-down, and what instructions are out of the question? Only non-political and non-discriminatory data are acceptable. The potential of AI-orchestrated crowd control eventually liberates the nomad in ourselves. Societies like Qatar, where every inhabitant has grandparents who lived as nomads, will most likely be among the first countries to adapt to such new AI-augmented nomadic citizen lifestyles. The Qataris have experienced, within the timespan of two generations, the transition from tribal nomadic to globally nomadic. Thanks to apparently unlimited financial resources, at least according to our European standards, Qatari citizens travel the world and yet maintain their private cultural habits. The rare combination of their global perspectives with restricted traditional family values makes the Qataris the ideal tribes for becoming the first species of international nomadic citizens, labeled as such in the MANIC research. Qatari citizens might more easily accept that public life is occasionally revisited according to new distancing rules than people from other cultures. Most Qatari families have their winter camps in the desert. They rent a piece of limestone desert and set up their traditional tents. Qatari families spend their cool winter weekends in their traditional winter camps, while in the heat of the summer, they fly out to the modern world. Typically, the women do much of the work in setting up the tents. Today's tents are fully air-conditioned and televised – so much for tradition. They are not disconnected from the world as in earlier nomadic days. Modern Qataris are nomads and global citizens in one body. It is my educated guess that Qataris will be the first truly new international nomads. The same will eventually apply to Emirati, Saudi, and Kuwaiti citizens, who have a similar tribal desert background. The Middle East has the potential to become the prototype society of Another Normal. The most successful Islamic countries like the UAE, KSA, Qatar, Oman, Kuwait, and the Southern coast of Iran are reshaping the Gulf lifestyle into a new Mediterranée, investing huge amounts of money in education, health, sports, culture, resilience, self-sufficiency, and renewable energy sources. The Gulf countries deeply invest in becoming a global tourist destination. To become independent in agriculture, they are establishing modern farms in the desert, based on permaculture, and urban farming. Soon, they will no longer be solely dependent on oil and natural gas. The Middle East realizes that the fossil fuel industry is at its peak and will be phased out in favor of renewable energy sources. The Middle East redirects its focus toward investments in science, technology, economic diversity, culture, sports, tourism, congress facilities, and hospitality to effectuate the transition. When the transition toward resilient renewable energy, food, and water security can be made in the Middle East, where the climate is harsh and the CO_2 emissions are higher than in other parts of the world, the change can be made anywhere.

Home is Where You Are

What kind of life are the new international nomadic citizens living? How do they earn a living? On the one hand, we see trends to return to local nomadic romanticism, while on the other hand, we see global travelers, living for shorter periods in different countries. The renewed romantic interest in a cottage, these days typically labeled as tiny houses, amid a natural environment is telling. Tiny houses are far from a solution to the acute housing shortages since they are typically built by the happy few who can afford to own a large piece of land. Tiny houses are almost without exception presented in pastoral environments, without any neighbors in near sight. Tiny houses presented like this refer to a possible off-grid autonomy, basically an extreme version of individualism. As if the world will become a better place, when a few privileged people can live their happy green lives, far away from their neighbors. In the Netherlands, there is a remarkable development called the Minitopia project for temporary houses on an empty industrial plot in Den Bosch, a project by the architect-developers Rezone. Rezone claimed a deserted tarmac site in the periphery of the city, on which to build 20 tiny houses, and invented a smart solution for the financing of the house. The Minitopia tiny house builders pay for their house directly; they do not request a mortgage, and they are independent of the business models of banks. They enjoy their freedom and accept the fact that the use of the land is temporary, for 5 years, 10 years, or maybe 15 years. These new homeowners are new nomads, and they do not fancy living in one place forever. After five years, they are already financially better off than if they would have bought the land plus the house under the exploitative regime of mortgages. Such a business model is well suited to the new nomads, who welcome temporary solutions to their housing needs. A simple national rule could make the change; one simple rule of law makes all the difference. Fortunately, there are low-cost versions as well, developments that appeal to a more social form of living together, closer to their neighbors. The tiny houses are by and large self-designed and self-assembled by the inhabitants; their costs vary between 5K and 50K EUR. Some houses are parametrically designed and produced according to the principles of the Wikihouse. A typical Wikihouse is locally produced at a workshop with a three-axis router machine and is assembled by a non-professional, typically the inhabitant together with their friends. They only pay for the house, while the land remains the property of the city. The lands belong to the community, not to an individual and not for investment or profit. As in the socialist paradigm, the land must be used for the good of society, not owned by a happy few. It is the cost of land and the artificial scarcity of land that has made housing in the Netherlands, as in any profit-driven capitalist country, extremely expensive. While agricultural land is still relatively cheap, the mechanisms of allocation of land for building are to blame for the enormous cost increase. Typically, the agricultural land is bought in the early stages of upcoming zoning plans by the project developers, who have no other primary goal than to make a maximum profit within the legal margins. A social agenda is merely a byproduct of the bigger project developers. The biggest hurdle for affordable housing is the artificially created scarcity, caused by the combination of restrictive laws for the use of the land and selling out designated areas to project developers. Land that is available for a fair housing policy has become so scarce that a new approach to land ownership is much needed.

If we only would cut in half the areas of climatically unsustainable pastures for cows, as proposed in my thesis project the Strook door Nederland back in 1979, and dedicate these lands for truly low-density social housing, the problem could be almost effortlessly solved. Can we project such a business model of home ownership in combination with almost free use of the land at a larger scale? To underpin the concept for the Strook door Nederland, I looked up in the Dutch statistical yearbook how many cows we have on how many pastures. It shows that there are only three cows for every hectare of land, meaning that these ruminants have a lot of space to move about in summer, at least 30 times more than we humans allow ourselves to use. Agricultural pasture is very cheap, perhaps today *ca.* 60K EUR per hectare, and used to be much cheaper in the seventies when I finished my studies. What if we just switch roles with the cows? We, humans, live in rowhouses or multistory buildings in a density of at least a 100 persons per hectare; in other words, in a 30 times higher density than the cows. Why would we award the luxury of living in a low density to cows and not to ourselves? Cows, pigs, chickens, whatever, could be pampered in row houses with private gardens and in multistory apartments as well. The cows will love it; they would prefer that over the overcrowded winter stables and they are confined to for half of the year, and we the people would enjoy a 30-fold increase in available space for our housing. In combination with a financial model, one would only pay for the house and not for the land; it would be a bright and green future for the superfluous pastures in the Netherlands. Think of the tremendous positive side-effects the strategic swap may cause, and replacing the ecologically almost dead pastures, one would see a rich flora coming up instead. The ecological balance would be quickly restored, so much more O_2 would be produced, and CO_2 absorbed. Reducing the areas of monocultural pastures is a hot topic today, the general opinion in society is gradually shifting toward the reduction of the polluting, cruel practice of meat production, especially the reduction of meat production that is meant for export. Currently, the small densely populated country of the Netherlands is Europe's largest exporter of meat products, which is completely disproportionate.

That unsustainable level of meat production should be cut in half to start with.

Although it might seem contradictory at first sight, the cows-to-people land swap would, in a similar fashion, stimulate the inhabitants to live a more nomadic life. The main prerequisite is that they will not own the land but accept that they are temporary users of the land instead. The nomadic international citizens will reconsider their home base every couple of years and move to other places in the world, where similar business models are introduced. While in the Netherlands, the pastures are the ideal areas for such a land use swap, while in countries like Qatar, the desert is the obvious environment to develop the same. Many Qataris already have their temporary winter camps, but the desert has a lot more potential for structurally temporary settlements, for winter and for summer. In countries with milder climates, the forests may offer the best potential for vast temporary housing schemes. This form of settlement is ultra-light urbanism since it does not need heavy time, energy, and money-consuming additional infrastructure. The arguments typically used against extensive land use for low-density housing are that it will create a new form of suburbia, that it will increase car movements, that new infrastructure is needed, and that it ultimately would cause more pollution and destroy nature. That would surely happen when in the hands of project developers under current capitalist conditions, but there is a potential for the better, on the condition of the implementation of some simple rules. In the first place, abundantly present trees, bushes, and plants absorb much more CO_2 and produce more O_2 than a cow-grazed pasture, which performs badly concerning the O_2–CO_2 balance. Second, when we combine several currently emerging strategies like the local production of energy, food, and water, electric automated vehicles including scooters and bicycles, and affordable housing for the many, in combination with internet shopping and ubiquitous home delivery, the picture looks very different. Solar panels, water extraction from the air, wind energy, and household farming of food, connected in a robust distributed network, reduce the need for the transport of energy, water, and food to a minimum. Third, ultra-low density housing should go along with a rule that restricts the size of their homes, like a maximum of 50 m² per person, like one big classroom, potentially including programmable furniture. Restriction in the size of the personal footprint in combination with extremely low density makes housing attractive and affordable for everyone, while not favoring the rich over the poor. And fourth, the new relatively small homes must be designed and built as an assembly of performative components to facilitate disassembly, rebuilding at other locations, or consciously recycling respectively upcycling its subcomponents into new components for future structures. Not only will the people live in a more nomadic fashion, but also the building materials and building components will be assembled in a more loose and more nomadic way. The circular economy is nomadic. As a net result of simple nomadic rules for connecting people and things, there will be a reduced need for top-down infrastructure and a reduced need for road structures. The existing roads only need to give access to the plot of pasture that was previously occupied by the cows; no new roads need to be made.

Living on Mars

The new nomads think beyond traditional schemes of living. While some are dreaming of a tiny house in a remote forest, others are ready to imagine a 3D-printed house on Mars. An early opponent to house ownership was the American inventor-architect Buckminster Fuller, who argued for replacing home ownership by renting high-performative structures. Buckminster Fuller lived much of his life in rented (hotel) apartments. According to Buckminster Fuller, the high performance of any product is only possible when the components are replaced as soon as technology has advanced. One no longer buys a commodity for life but rents the best available performance for as long as it takes. The rent of a house must include maintenance and timely replacement of underperforming parts by improved products, much like the maintenance and updates of software products. Current business models of software companies no longer support selling their products, but leasing on annual fees, guaranteeing the users are up to date for performance and security. Signify, a spin-off from the Dutch multinational Philips, has introduced a system of leasing performance. Philips guarantees a certain light performance; the user does not buy the light fixtures but pays a rental fee for the guaranteed performance, freeing themselves from the nuisance of broken parts and their replacement. Light as a service. Permanent maintenance is the core part of the contract, which ensures that the user will always enjoy the latest updates and the best available performance. Buckminster Fuller was very clear in his statement, by renting a house that is a product the inhabitants would always have access to the latest technologies, instead of being stuck in old houses with obsolete energy performance and poorly functioning installations, and recurrent expensive maintenance. It would give a boost to the economy as well since the incentives to renew will be evident; there will be a constant urge for better solutions at all imaginable levels in society. Better energy performance and better comfort performance, adaptive to modern lifestyles.

Ubiquitous lease instead of ownership can be an antidote for the current unbridled capitalist overproduction and overconsumption. The paradigm of sharing instead of having will inevitably lead toward the "when needed, as needed" economy. Performance-based production will be more capital intensive, and economic growth by producing more of the same would no longer be the leading aim. Leasing instead of having led to growth in quality, as opposed to growth that is measured in terms of quantity as is the actual troublesome situation. However, I do not appreciate how the big software companies use a cloud-based business model of leasing to bind the customers to their proprietary products, while their main priority is to increase their profits, rather than for sharing. The principle of renting performance instead of owning commodities, be it software or hardware like a house, should be established under the rules of creative commons, thereby bypassing the grip on the market by the capitalist greed of multinationals, simply by bringing the principle of shared ownership back to their users.

Once accustomed to a more nomadic lifestyle, while enjoying the improved performance levels, the people are ready to turn their backs on overpriced apartments and row houses that are typically too close to each other, not allowing for enough privacy. When too closely packed, there is no room to move, and there is no way that people and things, the acting components of the swarm, can behave freely as in the airborne swarm. Nor would there be space for the local production of food, energy, and water. Integrating food, energy, and water production to the basics of dwelling, the space needed will sort of double. In the current suburbs, citizens are forced to sit shoulder to shoulder, neatly in a row as the birds on the powerline. When set free from market-driven consumption lifestyles, people will reinvent their nomadic lifestyles; they will turn toward the desert, the forests, the plains, the swamps, cold regions, and eventually even beyond the atmosphere of the earth. Another part of the population would still prefer the cultural density of the metropolis, especially at the beginning of one's career, as establishing as many connections as possible to invest in future prosperity. Super high-density living in compact programmable homes and super low-density living in autonomous detached homes on a generous plot of land are two sides of the same coin. I am in strong favor of small city centers with high density, in combination with large areas of low-density populated areas. Everything in between, which includes all major city extensions of the last century, I consider suboptimal.

Looking beyond the capacities of Earth, space colonies might become a reality sooner rather than later, and living on Mars is a tempting prospect. NASA has held a competition for 3D printing on Mars, and the interest from designers is overwhelming. I especially like the entry by AI Space Factory for the 3D-printed inhabitable structures called Marsha, Mars Habitat. The shapes that AI Space Factory proposes are elegant and well-suited for standard horizontally deposited 3D printing. The design has a small footprint, leaving lots of space for the surrounding natural environment, whether on Mars or somewhere on Earth. It reveals a general tendency of world-makers, who are typically at the forefront of future developments in society at large, to imagine spatially distributed forms of living. Yet, the international nomadic citizens are at the same time deeply networked and intensely connected in cyberspace. Our favorite movie is The Martian, where Matt Damon survives after having been left alone on Mars after the other crew members evacuated due to a Martian sandstorm. The movie shows the importance of becoming connected, in whatever way possible, as to find a way back home. Damon managed to survive in the harsh environment of Mars by finding out how to grow food in a controlled closed environment, applying similar closed-loop principles as in hydroponic urban farming on Spaceship Earth. But not surprisingly, the most compelling reason why I like this movie is the fact that the NASA headquarters is located in our Bálna building in Budapest.

Expat Life

Not only will the new nomadic lifestyle affect our housing needs but equally on our workplaces shortly. As homes will offer temporary shelter for as long as it takes, also the use of the workplaces is varying over time. The number of self-employed professionals is increasing rapidly, all over the world. Young professionals are no longer looking for a lifelong bond with a company. They are seeking agile cooperation with peers rather than having a steady job as subordinate to a boss. The freestyle professionals are mobile; they can work at any place, and from anywhere, over the internet. The pandemic has shown that partially working from home has become a valid alternative to the daily commute to the office. At the same time, there is a growing number of firms that work internationally and send out their staff into the world, to the many distributed offices around the world. As in our profession, successful architects and designers have branches in many cities around the world, which was hardly the case some decades ago. ONL established the 100% Hungarian office ONL Hungary kft in Budapest, independent from ONL bv in Rotterdam, to realize the Bálna. By having a presence locally, the cooperation and mutual understanding with local experts and their way of working are more effective and allow for taking on more tasks and responsibilities. We could never have managed the preparation for the building permit from our office in Rotterdam. Only the fact that we were responsible for around 20 subcontracted experts simply means that a local presence is needed. The building permit procedure in Hungary is more bureaucratic than in the Netherlands; we had to submit 40 copies of a box stacked with hundreds of signed building permit drawings to get the approval. We hired promising young Hungarian architects, notably <u>Attila Bujdosó</u>, <u>Judit Márku</u>, and <u>Béla Kali</u>, to connect us to the Hungarian way of thinking and operating. The formation of international working groups is becoming the new norm, a sign of the demise of design colonialism.

Our two years in Qatar between 2017 and 2019 taught us how it is to live like an expat. Ilona and I experienced what the benefits are and what the downsides are. Instead of accepting an unfurnished compound villa that was offered by Qatar University, we rented a furnished luxury apartment in the <u>Falcon Tower</u> at a prominent site along the corniche, between the Hilton and the Four Seasons, with a splendid view of the yacht harbor, the famous <u>Nobu Restaurant</u> co-owned by Robert de Niro, and the iconic <u>Sheraton Hotel</u> designed by the American architect William Perreira, completed in 1982. The Sheraton is the first building to mark the new city of West Bay, a business district for the new Doha, built with revenues from the booming oil and gas industry. Befriended

engineers in Doha, who have experienced and contributed to the building boom, have witnessed the construction of the Sheraton. One positive side of renting is that one is entitled to a high service level, and one enjoys regular cleaning, immediate maintenance, omnipresent security, the coffee shop downstairs, including fitness with a fitness trainer, a spacious pool, and a barbecue lawn at hand; typically, the rent is all-inclusive. On the weekends, the new country is explored, new venues, and new restaurants, together with fellow expat friends; no time is spent on time-consuming fixing things in the house, and expats are exempt from doing odd jobs to maintain the property. Of course, the all-inclusive rent comes with a cost, but that is usually generously compensated by housing and transport allowances. One of the few downsides of expat life I have noticed is that, generally, expats do not take an active part in culture, at best as a consumer. In general, expats live a poor cultural life, busy with careers, and entertaining themselves, either with or without kids. Being an employee for a limited period in a foreign city does not give one the peace of mind to be creative.

For most expats, life is limited to working, eating, and relaxing. Rented homes do not inspire expat people to buy art, except perhaps for artisanal products, tourist trophies that look indigenous but probably are made in China as most products in <u>Souq Waqif</u>, the reimagined old souk. Expats are employees, not business owners. At best, expats are predecessors of the new nomads, while the new international nomads are their own bosses. They share similar benefits as the expats, but without the downsides, their work and family life is one indivisible whole. The success of the new nomadic lifestyle depends on whether they can make a living as independent self-employed professionals, perhaps working in an international team of like-minded self-employed peers.

We were something in between employee and independent architect and artist. I had some commissions and established some business relations; together with Ilona, we held workshops, delivered lectures, and exhibited interactive art installations, while Ilona produced quite a number of large acrylic paintings and exhibited her Q series and the Pattern series in the best places in town, the <u>Sheik Qassim al Thani Museum</u> and the <u>Doha Fire Station</u>. We touched upon being a new nomad. The new nomads are not on a permanent holiday; they know how to combine work, home, family, and leisure. The new nomads are not necessarily rich; they earn a living by working from home, by collaborative working in open collective well-equipped working spaces, where a fast broadband Wi-Fi is available, a perfect espresso and a choice of café-restaurants nearby to enjoy a healthy lunch. In the bigger cities worldwide, whether in democracies, dictatorial states, or socialist states, many such open networked work hubs have been established in recent years. The new open workplaces are inclusive of the new nomadic lifestyle that allows workers to work at any time at any place for any client, by and large in the service industry, which represents an ever-growing part of the world economy. The new nomad works when needed and as needed, as in a globally distributed swarm. Not looking at a physically nearest neighbor, but virtually connected to their nearest peers.

Connected Autonomy

A possible byproduct of the COVID-19 virus is the further development of smart homes located in natural environments. As people will more and more want to have a retreat in natural environments, and by that, I mean millions of common-class people, not the happy few, it seems more than likely that new forms of autonomous and networked affordable housing projects will be developed. Many prototypes for autonomous homes have been proposed and realized in the sixties of the last century; back then, it was, by and large, part of a luxury leisure movement. A perfect example from the sixties is the small 50-m² Futuro home designed by Matti Suuronen and decades before the Dymaxion houses by Buckminster Fuller. Buckminster Fuller designed the Dymaxion House to autonomously produce electricity for its power needs and to dispose of its waste, while providing a high level of comfort, regardless of the geographic location. Many autonomous houses of the past share specific features, of which one of the most prominent common features is that small feet houses are graciously lifted from the ground, as to minimize interference with nature. Flora and fauna are not to be obstructed. Somewhere between the sixties and now, I proposed some small-feet autonomous dwellings for the groundbreaking City Fruitful project for a productive neighborhood in Dordrecht, the Netherlands, designed together with Ashok Bhalotra. The most radical prototype was what we named Heron's Nest. The idea of the Heron's Nest was to offer a tiny 50-m² house on conical positioned high stilts, expandable to triple its initial size one level up. The outer walls of Heron's Nest are fully glazed and sloping forward to reduce overheating of the interior. Small feet homes like the Futuro, Dymaxion House, and Heron's Nest guarantee the continuity of the flora and fauna on the ground. The Heron's Nest concept has the extra quality of using a flat rooftop for growing food, whereas the expanded house would have a bigger roof for household farming. One of my other detached home concepts for City Fruitful is the Reflex Home aka the paprika house, small feet, with a glasshouse on top to grow food.

What is usually not acknowledged, but not less relevant, is that autonomous houses should not be off-grid but connected to their immediate neighbors, to general lite infrastructure, and to the internet, to work together, to learn from each other, to form a swarm of connected homes that respond to changing environmental conditions. Today's tiny house movement may look escapist but, in fact, is more focused on connectivity in cooperative grids. The urgency of being connected in autonomy is greater than it was back then because of the rapidly increasing climate instability, no longer to be considered a sign of global warming, but an alarming fact. Accelerated by the COVID-19 crisis, the climate crisis is now a bit higher on the political agenda, yet still without the necessary concrete steps to be taken by the decision-makers. Now is the perfect time to go full circle, update the prototypes, and roll the autonomous houses out into the countryside and programmable pop-up apartments into the metropolis. Rightfully, a set of urban questions arise when looking at these small feet homes in the outdoors, like, how do we reach these homes? By private car, automated vehicles, or ubiquitous taxi service? For City Fruitful, we planned a large parking garage at the entrance, whereas the CO_2 produced by the cars is directly tunneled into the glasshouses to feed the fast-growing plants like peppers and tomatoes. City Fruitful is a carless city; it is a walkable city designed for pedestrians, bicycles, and small electric vehicles. Almost everything we are talking about to fix our climatic problems was known back then in 1991, as it already was when the book Limits To Growth was published in the early seventies. We cannot afford to propose ideas for ubiquitously distributed houses without looking into the implications for the urban layout. I will propose some simple social and planning rules that, in an intimate interlaced connection to each other, could respond to current concerns about building in low density on agricultural land.

Distributed City

We do not need Koolhaas to make us aware of the potential of the countryside. His latest show at Guggenheim New York titled "The Countryside, the Future" is a curious mix of ignorance and self-proclaimed clairvoyance. It feels like Koolhaas finally discovers something that everyone has known for decades, while he exclusively was focused on the metropolis and had no interest in the other 95% of the country. Would he ever have taken a bike and crossed the countryside? If he would, he would have noticed the endless areas of closed-system glasshouses, the extra-large distribution centers consuming the open land along the highways. He would have noticed the many large data centers swallowing disproportionally big amounts of local energy for the data- and thus energy-consuming internet. A recent example is a scheduled huge data center for Facebook near Zeewolde in the Netherlands that supposedly consumes as much energy as the fleet of all Dutch trains together. And he would have noticed, the many distribution centers and other businesses have left the cities and populated the countryside for decades in industrial areas outside of medium-sized cities. My strategy for the countryside has been, since my thesis project in 1979, to combine extremely high-density metropolitan living and production with extremely low-density countryside living and production, with as few settlements as possible in between. I opted for hard edges between the density of the city and the openness and greenness of the countryside. When we swap roles with millions of cows, a nuclear family could live as spacious as 1 ha. When abolishing a mere 50% of the heavily polluting cow industry, the freed land of the Netherlands can be redistributed among approximately one million households, each living on a spacious plot of 1 ha. For distributed low-density housing, no extra investment will be needed while using existing infrastructure and networks. They may build their own zero energy small footprint home, taking advantage of the latest insights concerning the food, energy, and water nexus.

The city of the future is a distributed city, a city that is everywhere and nowhere. The future city is a city that includes both its dense city center and the extensively developed countryside, inclusive of two complementary extreme urban concepts. I am condemning metropolitan living like Frank Lloyd Wright was. I want to see both sides of the coin. The super low density may appear like a hyper Broadacre City but with even broader acres for the inhabitants. The difference with nowadays vision on transportation is that we will share automated electric vehicles rather than owning 1–5 cars. Would such a grassroots lifestyle chain us to the land? Would we keep traveling around the globe? Would we give up the newly acquired nomadic lifestyle, and would we isolate ourselves and alienate ourselves from the world? I know from personal experience that having a comfortable green home base in the countryside forms an ideal basis for traveling the world. It takes even less time to reach the main airports. Living life in the density of the metropolis reduces one's desire to travel to other metropolises, while you have the metropolitan culture already at hand. Many city inhabitants hardly ever leave the city and know little about the countryside. People like me appreciate contrasts, like the contrast between fluid "broad-acred" homing and hard-edged upward-moving cities. Indeed, I believe that we will keep on moving around the globe more and more, notwithstanding the climate crisis and rising travel costs. The new nomadic international citizens tend to stay much longer in the other worlds and work from there too. Airbnb has made us share and exploit our private homes; no longer is the home sweet home exclusively reserved for ourselves. I foresee a future where we as the new international nomadic citizens will book homes in remote places for longer periods, for months or perhaps for years, while continuing working from the new temporary home, as self-employed consultants or in multinational teams, smoothly communicating via videoconferencing and video-lecturing. COVID-19 has shown us how to work from home, wherever that home is. In my professional niche of design, it is exactly the digital parametric design to robotic production paradigm that allows working and producing anything from anywhere. Design teams are formed by multinational groups of individuals, and production will be activated when needed and as needed, using local materials and locally available routers and robots.

Post-COVID-19 Society

COVID-19 urges us to focus on possible scenarios for a reimagined society. When science and art rule the spatial configurations of cities and buildings, when people adopt digital distancing, when we are actors in top-down and bottom-up informed societal swarms, when we live at the same time locally and globally as the new nomads, when measuring-is-knowing is driving our actions, when we follow science, and when we interact with resilient environments, how will a post-COVID-19 new normal society look like? How will the urban fabric change, how will the streetscape change, how will buildings adapt, how will production and consumption patterns change, and how will our habits of shopping, sports, and visits to museums look like? What direction society might want to take? Many of the discussed societal undercurrents were already slumbering before COVID-19. Only when interlacing and combining early harbingers of change simultaneously, not one by one, may cause a radical evolutionary jump. Whether the hope for a radical societal change remains wishful thinking or an adequate evaluation of current developments, time will tell.

To bring on the radical change, only a few ubiquitously implemented top-down simple rules of law would need to be introduced, creating an open framework to unleash bottom-up developments, by and large, effectuated by local communities and individual households. Radical change to one level up in evolution typically emerges from a critical mass of 20% of change makers. It is never the majority that brings about a change; typically, majorities are the buffer to withhold changes as long as possible, until a rebellious and proactive minority breaks the ceiling. Such proactive minorities sometimes work for the better or the worse. We can only hope that the currently noisy rightwing radicals will not reach that critical mass of 20% but that the science-based, art-driven paradigm of transparent programming and automation, of which parametric design to robotic production is just one of the

463

proactive components, will take over in due time and render the anti-vaxxers, fact-free fascists, and extreme neo-liberal self-declared freedom-fighters powerless. Another Normal must be governed by a simple set of strong rules for health, climate, social justice, food, energy, and water, to reach the desired condition of robust resilience. A rule of law is an active component of the legislation. Only those laws that create bigger freedom while empathetically respecting others will hold. As is widely acknowledged, an overly complex set of rules with a lot of exceptions, in other words, bureaucracy, creates a relatively static society. We know from computer programming, design concepts, and urban planning that only a simple set of transparent and well-defined rules will result in a well-appreciated complexity. In general, a minimum number of imposed constraints will lead to a more beautiful, more diverse, sustainable, and resilient society, open to proactive participation by individuals. I certainly do not allude to the neo-liberalist capitalist view of having fewer rules to aim for maximum corporate profit, but on the contrary, for some strong egalitarian rules to give everyone an equal opportunity to develop as a collaborative individual. The economy of society 2.0 should no longer be governed by neo-liberal greedy exploitation but rather be based on real-time scientific data to define the framework for the well-being of all beings and things. Another Normal thrives in a stable climate, reinforcing the natural ecology of Earth by interlacing Nature 2.0 with Nature 1.0. Since scientific data monitor dynamic systems, the parameters driving the new rules of the new normal must adjust in real time, and our social, financial, and cultural systems must respond immediately to every minor change in the data patterns. I am done with greenwashing by empty non-committal political promises; what we need are concrete plans and immediate execution of the plans. No more promises to perform better in 10 or more years, but actions in the here and now. No more static diagrams but living diagrams that visualize the performance in real time, like in digital twins. Living diagrams that visualize improvements in real time and living systems that run the performance meters, in your house, in your neighborhood, and in your city. A non-disruptive balance between production and consumption must be established, both in the metropolis and in the countryside, whereas the production of almost anything must happen as close as possible to where it is consumed. Parametric design will involve the clients and users in the participatory design process. Robotized local production on command will replace mass production for discriminatory overconsumption. As with virus outbreaks like the COVID-19 pandemic, the only way to control a crisis is to continuously measure, adapt, and legislate immediately to maintain a livable balance. We need to develop ubiquitously implemented systems to monitor real-time health risks, climate change, and societal inequality and implement worldwide immediate solutions based on the measured data and activated by some simple adaptive rules of law, immediately coping with environmental threats. The simple rules that create diversity-in-equality are the constituent components of Another Normal. The simple rules are not a playground for the elite but a tool to liberate the majority from the elite.

To reach a thriving resilient global economy, an important component of a reimagined society will be to stimulate a collective production of large works of art. Works of art possess a unique capacity to energize the public. When looking back in history, it is not the reeling and wheeling of the big companies that represent value but the achievements of science and the works of art. It is the monuments, the global arrangements of mosques and cathedrals, the museums, the concert halls, the paintings, the literature, the scientific discoveries, the cultural specifics, the technological artifacts, the well-designed vehicles, and gadgets that characterize an era. To advance society by a law that reserves a high percentage of the national budget for art, science, and education, the sheer size and ambition of large works of art would entail a serendipitous marriage of science, entrepreneurship, and art, thus shaping the future artifacts of the new society. It is encouraging to see how much money, effort, and belief is invested in healthcare, culture, renewable energy, and in resilience in Qatar, to a smaller extent also in the UAE and Saudi Arabia. They have the oil and gas money and there is a growing tendency to spend the money wisely. When referring to art, I am notably referring to abstract or better concrete art, while artists that work after nature today are not the ones to put a dent in their culture, and they are not the ones who would use science to realize their ideas. Artists who work after nature and other storytellers do not produce a new universe but are self-imprisoned in versions of past universes. The fusion of science and art requires constructive minds, creating the unknown and exploring the alien. Artificial intelligence, popularized by engines like MidJourney, DALL-E, and Stable Diffusion, assists predominantly non-artists or would-be artists to explore the alien. But, since they use large databases of existing images, I consider the often-surprising outcomes a sort of modern alchemy. Whatever AI does, it surely reveals possible universes that had not been seen before, yet potentially existed. Abstract constructive thinking that surprises oneself and stimulates others is a prerequisite for a successful merge with scientific thinking. Ambitious art proposals may seem unreal at first sight, but when scientists and engineers join in from scratch, things are made to work. Science provides the data on which the intuition of the artists feeds, in a collaborative design process where artists, scientists, technicians, and civil participants respect each other's expertise. Practicing abstract artists, not storytellers, should enter the boards of directors of companies to develop intuition-driven outlooks for the company, to co-design the shape of future products and environments, and to imagine how the products and the environments could interact with the public.

We are approaching a more and more information-dense society. Information technology will invade the finest fibers of society and its inhabitants. Each one of us will monitor the self and be monitored in real time by our immediate neighbors in minute detail. Ubiquitous monitoring, knowing, and communicating will form the basis for social interaction and hence for social justice. As long as there is a bidirectional exchange of data, there is not a big brother problem with ubiquitous monitoring. If Big Brother is looking at you, then look back at the Big Brother. The bidirectional exchange of data will become as natural and self-evident as the exchange of words in our daily communication. Language in itself is a perfect example of how the post-Corona rules of law might look and feel. Language is per definition bidirectional; listening, thinking, and speaking are driven by language. Data is the new language. Citizens must have free access to all data that are derived from being monitored. The government does not need to know your data but your neighbor needs to know, and your fellow citizens need to know. Citizens must be free to use the open data to develop their bottom-up local strategies concerning climate change, health, and prosperity, while governmental bodies impose simple top-down rules to facilitate equal social opportunities for all.

Capital Intensive

Future developments must not imply producing less of the same while producing less polluting or less dysfunctional things at best somewhat reduces the dire actual condition of the global climate and the omnipresent inequality. The climate crisis will not be solved by a simple reduction of what we currently produce and consume. We will need to implement structural changes in the way things and goods are produced and consumed. The world does not need to use less energy; using energy is perfectly okay as long as it is clean energy. We do not need to eat less, as long as the food is fairly distributed among all world citizens. We do not need to accept less comfort if we adopt the lifestyle of the nomadic international citizen. We do not need to refrain from producing new vehicles and industrial products, as long as they are parametrically designed and robotically and locally produced. We do not need to travel less, as long as we choose the proper vehicle that is running on the cleanest energy for the distance traveled. A simple rule could be to introduce a progressive distance tax; the longer the distance traveled from product to consumer, the higher the tax. That would stimulate producing things as locally as possible. It certainly does not mean that we should not consume products from the other side of the world; it just will come at a cost. Another simple rule could be that the ones who produce waste have to progressively pay for it, i.e. not the ones who are the victims of the polluted areas. That would certainly stimulate waste recycling as close to where it is produced.

When looking at the global scale, only a minority of the world population has access to the best that technology has to offer. The deprived majority will rightfully request to have access too. That is

467

why migrants are coming to Europe and North America; they come to claim their share, after having been colonized for centuries. China is already catching up, India is as well, and Africa is on the rise too, struggling to overcome the oppressive colonization. As of yet, Africa, home to 15% of the world's population, produces only 3% of the global CO_2 emissions, which is five times less than average. The USA produces disproportionate amounts of CO_2, far above average. Logically, the inhabitants of Africa, India, and Asia want to level up to at least average; they want to have their fair share. Asian, Indian, and African people will want to reach a similar level of luxury and comfort as we enjoy in the West. How can the world cope with that inevitable future reality? How will wealth and quality of life for all 7.7 billion inhabitants of Spaceship Earth be distributed in a fair and just way? When we continue the way we produce and consume, even without Asia, India, and Africa catching up, it will mean nothing less than a disaster, of which we see the early symptoms already. The prediction was already made by the Club of Rome in 1972. They were foreseeing the collapse of the global economy and the stable climate in the mid-21st century, starting around 2030. Why did not the politicians act back then? Why did governments decide to continue on the opportunist path of business as usual, thus reaching the inevitable collapse sooner than later? Why was neo-liberalism so successful when that is knowingly dancing on the volcano? Or is, as Paola Antonelli the curator of MOMA stated at the Broken Nature exhibition (2019), that we designers do not have the power to stop our extinction, but that the power we have left is the power of design, to design a "beautiful ending" for ourselves aka humankind. It is a comforting thought that life on Earth will not cease to exist after humans have designed their demise.

As I am active in the building industry, could I raise similar questions for a building? Why is business as usual still dominant in the building industry, and why do we continue to exhaust resources at a rapid pace and produce unbearable amounts of CO_2 into the atmosphere, knowing that we contribute to at least 25% of the global climate disruption? Why did not the building industry adopt parametric design to robotic production, producing only when needed and as needed, knowing that the strategy would greatly reduce waste production and energy consumption? Why is the building industry still relying on building with concrete, which is the most polluting way of building, for almost 99% of the realized construction projects? The Liwa Tower is the first high-rise building in Abu Dhabi that effectively uses steel for the load-bearing structure. During the same period, the Capital Gate tower was developed in the Capital Centre district, designed by the design firm RMJM, also the master planner of the Capital Centre district. The Capital Gate design almost killed our project because it is coquettish, inefficient, and, hence, far more expensive per m² than the Liwa Tower. The execution of the Liwa Tower was demanded to be put on hold for some years, pending the complicated construction of the Capital Gate to finish first and to claim to be the first diagrid steel structure in the UAE. Efficient or not efficient, in general steel and wood offer much better climatic perspectives than concrete, while cleaner in production and assembly, and much more lightweight, reusable, and recyclable. Why then do we not see more lean designs using parametric design methods, and mean production using CNC machines and robots? This is partly because designers often leave it to the construction industry whether to build in steel or concrete and partly because the main contractors typically stick to their old habits. Since the early nineties, steel and wood are the materials that I work with – with minimal foundations in concrete – robotic production is the main technology for the execution of components and dry assembly for putting the prefabricated components together. Without labeling my work as sustainable and/or resilient, I have intuitively chosen the path of lean and mean parametric design to robotic production of components, the shortest roadmap toward sustainability and circularity.

The main question remains: how can we feed everyone, how can we provide housing and offer high-quality lives for all earthlings? We will have to produce much fewer unhealthy products, much fewer mass-produced clothes, and burn much less fossil fuel. I will unfold my vision of how that situation may be reached, within the decades to come. The design task is huge, while we currently have an Ecological Footprint of 1.7, and we will need three

Earths in 2050 to regenerate what we use in one year, three times more than Earth can handle. And it will get much direr when Asia, Africa, and India are catching up to the same level of luxury and comfort as we in the West are enjoying. We do need to do things at least three times better. To reach such a goal, we might want to implement a sort of half-life strategy for polluting less, which works in combination with Moore's Law for improving the building technology. Implementing a Half-Life strategy could mean that, despite the rapid growth of Earth's population, every political term pollution must be cut in half, until the excessive pollution of the atmosphere and the oceans has effectively disappeared. The building industry's version of Moore's law means that in every political term, the percentage of clean, lean, and mean production of products, food, energy, and water must be doubled until all production is effectively clean after four to five terms. This does not mean that we will produce less in total, but that we will do it better, cleaner, more efficiently, and above all, only when and as needed.

On the one hand, people behave more and more like global citizens, while on the other hand, one wants more and more to contribute to the well-being of local communities, wherever, whenever, and across the continents. Can these two seemingly contradictory tendencies be combined into one resilient strategy that brings equality and quality of life to everyone? And how does a possible way out of the climate crisis relate to the effects of the current pandemic? Can we imagine a simultaneously local and global economy that feeds the needs and ambitions of the billions without the devastating polluting side-effects? Especially since the weekly school strike actions and public appearances of Greta Thunberg, we are well aware of the urgency. We can no longer allow ourselves to exploit Earth and compromise the lives of future generations. The "when needed as needed" paradigm, as is practiced in parametric design to robotic production in the building industry, is fit to be applied as a general strategy to society as a whole. As needed means producing what is needed to guarantee the quality of life for all, just that, while when needed means delivering those products just in time to support the quality of life for all, just there, and just then. We need to develop the components for Another Normal society, which is based upon ubiquitous equality and accountability, guarantees a stable global climate, and provides accessibility for all to the best that technology has to offer. During the Corona crisis, we are witnessing again how much work has to be done, while the selfish West secures the COVID-19 jabs for their population in the first place, like a viral version of America First, and deprives the poor of being jabbed in time. Does the West not realize that maintaining their stubborn colonialist mentality will have a boomerang effect on our well-being? Fairness and equality for all are in long term in the best interest of the West as well.

One of the predictable side effects and aftereffects of the COVID-19 pandemic is intensified virtual connections, partly in addition to the usual physical connections, partly replacing them. In the long run, intensifying virtual connections on Skype, Zoom, Webex, Meet, and Teams will eventually stimulate intensified cross-continent physical contact as well. Intensified physical contact will apply to nomadic international workers, who seek cultural exchange in a collaborative international working environment, and to global travelers, who seek participatory forms of cultural exchange in leisure programs. In the end, all of us, whether working or relaxing, will become prosumers rather than consumers, feeding on a higher level of being informed, a higher level of personal activity, and thereby a higher level of invested capital per individual. The logical consequence of an information-rich society is that it will be substantially more capital-intensive. The challenge for society 2.0, Another Normal, is to be more capital-intensive and more resilient at the same time. Capital needs to be invested in human resources, advanced distributed technology, robotic production of almost anything, renewable energies, urban farming, and water management. It will lead to more investments in locally produced clean energy, more investments in resilient neighborhoods, more investments in healthy local food production, more investments in smart customized traffic systems, more investments in personalized cross-cultural leisure activities, and more investments in online work applications. Work and leisure will merge into a reimagined New Babylon-style game of life, partly lived in physical space and partly in cyberspace. It is the task of scientists, artists, designers, manufacturers, lawmakers, politicians, and world-makers in general, to design the components and the simple rules of law that make the Another Now work.

Cyclic, Not Circular

The circular economy is not an economy that is less capital-intensive. Circular design and building are well-informed forms of design, whereas all aspects of the base materials, the design, the production, and the assemblage are known. Being well-informed means being more capital-intensive, while relying on a massive number of connected databases. An information-rich capital-intensive society is a prerequisite for the circular economy. A full circularity implies that everything from mining to assembly is

known and executed in a controlled transparent process, to be supported by peer-to-peer (P2P) technology. Each node in a P2P network is both a client and a server. Each node has a copy of the file. Each change in the file is administered and negotiated between the peers. A digital society, where datasets are accessible and verifiable, can fulfill that full transparency of the origin of materials, performance, design, production, assembly, reuse, and accountability. A digital society, given shape by parametric design to robotic production, is inevitably a capital-intensive society. Circularity does not mean a return to a simple autonomous natural life but rather a leap forward to a networked society with full accountability of all players involved.

Yet, the term circular economy is not very well chosen. Going circular sounds like going in circles, meaning returning at the same point from where one has departed. The arrow of time does not allow it to be circular, while evolution proceeds in discrete steps, nature is cyclic, not circular. Circumstances will have changed when the beginning of the cyclic process meets the end. In the literal sense of the word, circularity is like a snake that eats itself and eventually suffocates, like a dog chasing his own tail and biting itself.

Economies are, by definition, cycling forwards like a spiral galaxy, but, unfortunately, currently, most economies are spiraling downwards with respect to the climate, like water swirling down through the sink. Currently, economic growth means more loss than gain; it works at the expense of the environment. Not only at the expense of local environments by polluting the air and/or the soil locally but also at the expense of the global climate. The global climate displays all the signs of going wild and of going out of control. Not out of control in the positive sense of an emergent property as described in Kevin Kelly's brilliant books <u>Out of Control</u> and <u>The Inevitable</u>, and not going positively wild into complexity and interactivity, based on real-time programming of parametric design to robotic production methods, as suggested in my book <u>Architecture Goes Wild</u>, but seriously out of control in a negative spiral. Out of control in the sense that we, the human population, actually are spoiling the earth that is feeding the Internet of People and Things, without a clue of how to stop the destructive process. We are polluting our own nest. The key question remains as to how can we spiral the economy up into an upward movement, respecting natural and human resources? Where do we go to construct an alternative information-rich and capital-intensive Another Normal from where we are now? Where do we go to divert the current downward spiral into a quantitative and qualitative upward cycle?

Since perpetuum mobiles do not exist in the observable Universe, we cannot go in circles without adding value, value in the form of energy, information, and capital. Upward cycles do not produce more of the same, nor do they produce fewer polluting versions of the same. Upward cycles synthesize until yet unseen simplexity-driven design concepts and innovative hybrids. An upcycling economy does not rely on exhausting Earth but rather on the synthesis of informed and performative organic and inorganic materials, a collaborative fusion of Nature 1.0 and Nature 2.0. More and more, the global society performs as one single distributed global brain. My optimistic assumption is that billions of highly connected human brains and AI-driven exo-brains, exchanging tons of data in real-time, will eventually constitute a highly informed global layer of consciousness that will proactively prevent the global climate from becoming toxic for us humans. The problem is not that Earth will not survive; the point is that we humans might have a hard time. We can only go forward, not back to where we once belonged. The way forward means adjusting the course of things through cultural, social, and technical innovations, through the careful implementation of proactive components of science, art, design, and rules of law for Another Normal.

10

The Chinese Patient

The Chinese Patient

In a documentary about a psychiatric hospital in China on Dutch television in the eighties, a Chinese psychiatric patient revealed his vision for a better society in four essential points:

1) clothes of wool!
2) food must be better!
3) transportation by limousine! and
4) China, one big villa park!

This Chinese patient is a genius, an enlightened visionary rather than a psychiatric patient, while society is the real patient. First, our clothes are made by child labor and mountains of waste of cheap clothes fill the landfills. Second, food is all too often mere junk-food causing obesity and other health issues. Third, the transportation sector is one of the most polluting and lethal sectors. Fourth, affordable housing for everyone is what even the richest countries badly missed. Our oceans are polluted and saturated with microplastics, our atmosphere is heating up, containing an overdose of CO_2, the earth's crust gets exhausted, natural resources are getting scarce, and viruses are thriving, infecting people and animals. In short, society is sick and needs serious treatment. We, the people who are responsible for exhausting the earth, are confronted by the younger generation with the need to take immediate action to adjust the course of things. We should not leave it to the next generations to clean up the mess; we must act now and create the possible Another Normal.

I present here some proposals for ubiquitous action, all of them firmly related to the component-based design to production strategies that we have developed for art, design, and architecture. It is not a retroactive justification of what I have done, but it offers a possible road to reverse the down-spiraling course of our societies, which as of now shares all the characteristics of a derailed cargo train. The ideas are not utopian or futuristic of some sort; they are more of an extrapolation of current societal developments, in sync with the swarm logic of component-based parametric design to robotic production. The four points of our Chinese friend are a pure form of art by its simplexity, just four simple rules to transform China into a paradise, with justice and quality of life for all. The Chinese patient opted for components that bring luxury and comfort. Today, we would add another social component to it, that of resilience, which is an empathic form of luxury and comfort that does not rely on the exploitation of underpaid workers and would not be lived at the expense of natural resources.

I discussed the notion of the component and the connections between the components in language, science, the universe, art, nature, life, synthetic products, and the dynamic relations between the components in the built environment. Here the possible role of proactive components in societies at large is discussed, adding a political dimension to the notion of the acting component. Which are the relational building blocks for a fairer and more just world? Which simple rules will create diversity? Which simple rules will guarantee justice for all, and which measures will create a resilient society? Is it possible to introduce new rules that do not work at the expense of oppressed groups of people? Can we imagine strategies that do not rely on colonist exploitation of remote land? The Chinese patient had some ideas that I will propel into the here and now, taking advantage of current technologies that he could not have foreseen. Many of the components for a better society arise from similar paradigms that shape the rules for Hyperbody's interactive installations and ONL's realized architecture and art projects. Rules that are drivers for change are rooted in a society that embraces the digital, which in my view will inevitably lead to enhanced transparency and verifiability of the performing component while leading to proper accountability of their authors. I have identified only those world-making rules for active societal components that have a strong link to my designs, to avoid becoming speculative. I do not suggest that the below ubiquitous rules of law are the only ones or the most relevant ones for an efficient transition toward an alternative present, toward Another Normal, but they sure will advance society in a more fair and just non-polluting direction.

I will focus on those performative components that are linked to the basic concept of the interaction components with their nearest neighbors in the corresponding swarm. I will focus on the interaction between people and things, the emergence of collaborative hive-minds, and the interplay between humans and robots. I will look at how components perform, how they connect, and how they interact. I do not present the usual measures that are typically propagated by climate activists, which are often arguments to do less of the same, not forward-looking, and not understanding the democratic power of ubiquitous digitalization. Hence, I do not argue for green roofs or balconies with bushes and trees, which I consider a form of greenwashing of otherwise traditional methods of operation. People will naturally do these things anyway when the circumstances are framed well. Typically, environmentalists focus more on a change of personal attitude than changing the rules of the game of life. And by doing so, they voluntarily position themselves as subject to the current capitalist rules of play. Typically, environmentalists are addressing the individual citizen and the politicians to follow the 7 Rs of sustainability: rethink, refuse, reduce, repurpose, reuse, recycle, and rot. Politicians typically try to transfer the responsibility to the citizens alone to free themselves from taking action, mindful of the famous quote from the inaugural speech of J.F. Kennedy, in which he urges us not to ask what the country can do for you, but what you can do for your country, for the freedom of men. While I am not contradicting the validity of the 7 Rs, the strategies that we are forwarding here are focused on the implementation of simple top-down drivers for change that are intended to dissolve the currently poorly performing capitalist paradigms. With due respect to the green movement of climate activists and environmentalists, I use a different language and apply different techniques to reach a similar goal, which is a better quality of life and better-performing products in a healthy environment. None of these drivers for change alone would be a game-changer. Only when intertwined and executed in parallel, the future could start to look brighter for the younger generations. The order of components for social change that are proposed here are not meant to be implemented in a specific order, since they are all equally relevant and should be developed ubiquitously and in parallel. Nor should societal action be limited to the rules of law proposed here.

Ubiquitous basic income (UBI)

Politicians in Finland, France, Scotland, the USA, and many posts on Twitter leave no doubt: only a ubiquitous basic income (UBI), commonly known as a Universal Basic Income system, can avoid an abrupt economic collapse during a virus outbreak, or any other major disruption. There

have been successful experiments in providing basic income in Canada in the seventies and more recently in Finland and other parts of the world. The idea is simple; why should we have complicated laws and programs to set up bureaucratic monitoring systems to support the unemployed, and the homeless, to check whether they comply with the restrictive conditions? Why would we have a system where people have to work for money to survive, instead of a system that guarantees a minimum living standard? Such a system could replace the complicated law and rules that are necessary to decide who is entitled to get how much support when losing a job or otherwise not being able to work. Many administrators are at work to interrogate the unfortunate, categorize them, control their doings, and guide them through their misery. The system makes jobless people feel inferior, as if they have done something wrong, and sometimes they are outright treated as criminals, tunneled into forced labor at the lowest end of society. By forcing them into mind-numbing work situations, the jobless are deprived from developing themselves and are doomed to stay at the bottom. Why such a complicated system that divides people into haves and have-nots? Success is rewarded and failure is punished, and in the current system, success sets the norm, and failure is for losers. If we only had one simple rule to provide everyone with a basic income, none would be considered a loser, and everyone will be equally respected, whether one works for a boss or does other things. Everyone, who lives, buys food and household products, and anyone who goes places makes things and contributes to the economy. You do not need to have an employer for that. Universal Basic Income should, by the (international) rule of law, become a universal right. It is already stated in article 25.1 in the Universal Declaration of Human Rights: "Everyone has the right to a standard of living adequate for the health and well-being of himself and his family, including food, clothing, housing and medical care and necessary social services, and the right to security in the event of unemployment, sickness, disability, widowhood, old age or other lack of livelihood in circumstances beyond his control." Why we do not have Universal Basic Income is not because there is not enough money in society; it is simply because there is an equally capitalist as well as a communist view that one as a grown-up has to earn a living, in other words, not as a basic right. In a capitalist society, this view creates for the majority of people a dependency on the entrepreneur, while the highest good is to become an entrepreneur oneself, from bellboy to general manager. In communist systems, it creates a dependency on the state, where the highest status is to become a member of the ruling class. In our current political system, we live somewhere between the capitalist and the communist, upheld by many complicated rules with many exceptions that are meant to compensate for the inherent unfairness of the system. Why not have a fair system to start with, based on one simple rule that creates diversity and activity?

Our dear Chinese patient was well aware of the Declaration of Human Rights and shared with us his poignant interpretation. The component question is how to obtain that right to a good standard of living, by simple rules that create exactly this. When we have the simple rule that each earthling will be entitled to a basic income, which guarantees the intended quality of life, technically, we will need to qualify and quantify how much is needed. Following the Chinese Patient, quality of life means open access to a daily portion of good food, access to quality clothing, open access to generous housing, and access to comfortable means of mobility. Ubiquitous means that it must be introduced all over the world, in parallel. To avoid mass migration, the conditions should be equally good in all nations. Such qualities need to be qualified and quantified for each situation in each of the 195 countries in this world. The economic capital that is generated needs to be fairly redistributed before it gets accumulated in the pockets of the multinationals and oligarchs in any of these countries. Common sense, validated by real-life experiments, has shown that people prefer work over staying idle. UBI could work as follows. Everyone must be entitled to get their UBI. Making exceptions to the rule would substantially complicate the rule, and we are looking for simple rules that create diversity. As in nonstandard architecture where every component needs a unique ID and is addressed by a simple algorithm, the UBI rule needs to address the citizen's ID. Everyone will be addressed in the same way, without exceptions. Each person in the world gets a unique UBI identity and a UBI account. This one simple global UBI rule addresses all unique identities. The specific levels of income in hundreds of different countries are dynamic parameters of the equation that includes the net income of those who have a job. Everyone will be, in principle, entitled to have the UBI, but those who prefer a job above the UBI will have their net income progressively subtracted from their UBI account, based on a simple formula. For example, when one earns twice the UBI, the UBI support will be set to zero, and everything linearly in between. It is my educated guess that in the Netherlands, it might cost around 18 billion EUR yearly, which equals 1000 EUR per inhabitant but saves at least half of that amount by zero administration costs. The youth and the elderly should also be entitled to get the UBI. A newborn child should get its bank account at birth but would first be allowed to use the money when officially grown up, for example, to finance their studies or start a business. Retired people would get the same UBI, replacing the state pension, which would mean another substantial reduction in state administration costs. Naturally, the rich should be taxed progressively to finance the UBI, not the lower average incomes.

How such a more fair and just society could work we can read in Yanis Varoufakis' *Another Now*. To start with, it will take courage from activists to address the problem and provide practical solutions; then the UBI movement needs to reach the 20% tipping point, before the people in charge will give in and implement the new simple rule. Any simple rule of law must consider the local costs of living index. A UBI in the USA would thus need to be 27% higher than in the Netherlands, while in Hungary, the UBI would amount to almost 60% of that in the Netherlands and 45% of the UBI in the USA. The current reality shows bigger differences between the minimum wages in these countries. As Hungary has a relatively low minimum wage of *ca*. 500 EUR per month, the Netherlands has some 1650 EUR per month. In the USA, the richest country in the world, the minimum wage currently is a mere 7.50 USD per hour, which amounts to 1200 EUR per month. Recent developments show that although in Hungary, there will be a 20% rise in the minimum wage, they still would be stuck at the lower end of the European market, especially since inflation is substantially higher than in neighboring countries. When we combine the GDP, the cost of living, and minimum wage into a UBI, and when we take the Netherlands as the default value, then in Hungary, one would expect around 1000 EUR, and in the USA, 2100 EUR. The Democrats' aim to raise the minimum wage to 15 USD per hour represents just the minimum it should be, in sync with the rest of the developed world. UBIs in Central Africa would be considerably lower; yet, still, they must be balanced for GDP and costs of living.

There are many reasons why a UBI is an obvious way forward. We simply cannot continue the current trend toward extreme differences between the rich and the poor, an alternative political idea is badly needed, and it is certainly not radical neoliberalism that will reverse the global and local unfairness. Extrapolating the current trend, the rich gets exponentially richer and the poor substantially poorer. No one would be able to defend that as a strategy; even the most hardcore neoliberalist would need to acknowledge that this would lead to a situation as most accurately described in the dystopian Hunger Games by the author Suzanne Collins and transcribed to film by Gary Ross. As in the Hunger Games, a very small super-rich elite will lavishly entertain themselves by violent exploitation of the unprivileged who live in slums, while compromising a selective number of representatives by training them in combat and villainously giving them some privileges in preparation for a 1:12 chance to survive. Technology has something better to offer than working for the rich and famous only. Ubiquitous digitalization and robotization, augmented with AI, will drastically reduce the number of low-level workplaces to begin with in the Western regions, then China, then the Middle East, India, and Africa, and eventually infectiously spread over the globe. Ubiquitous digitalization and robotics promise that products can be easily copied, and as Kevin Kelly has convincingly demonstrated in his book The Inevitable, everything that is digital and can be copied will eventually become almost free and abundantly available. The same will be true for things that can be copied by robotic devices. The current trends that are leading to a global form of ubiquitousness that Kelly notices are, among others, interacting, cognifying, flowing, screening, accessing, sharing, filtering, remixing, tracking, and questioning. Supported by a ubiquitous basic income, the ubiquitous digital connects everyone to everyone, and everything to everything, regardless of one's wealth. The inevitable future as Kelly sees it leaves time and space for improved ways to the lives of the seven, eight, and eventually ten billion crew of Spaceship Earth.

Ubiquitous land right (ULR)

Another challenge that needs to be addressed in Another Normal is the scarcity of land created by feudalism and capitalism. Only a happy few people are entitled to use for their selfish good a sufficiently large piece of land. In the Netherlands, only 17% of the total number of the 7.9 million homes is built as detached homes on a private plot of land. Special for the situation in the Netherlands is the ineradicable tradition of row housing; 58% of the Dutch live in row houses. Surprisingly, in the former socialist republic of Hungary, there are a whopping 65% freestanding homes out of a total of 4.5 million homes, five times as much as in the Netherlands, and twice the average in Europe. It was one of the first things I noticed on my first trip to Hungary at the end of the seventies. Every village in Hungary has exclusively detached freestanding homes; no multi-story apartment blocks at all, all of them on a generously big plot of land of 700–800 m², big enough to grow one's food. In Holland, a typical row house has 120–150 m² of land, leaving only space for a small backyard. Hungary has a five times lower population density compared to the Netherlands; many more people live in the villages in between the agricultural land, and they typically have five times more outdoor space to live. This fact could explain the relatively conservative political position in Hungary through the ages, whether socialist, monarchist, or nationalist. Besides their detached primary homes, many Hungarians, and certainly not only the rich, own a second home, a weekend house, usually in beautiful settings against gently sloping hills, close to one of the many remaining patches of forest. Under the rule of socialism, many larger plots were divided and distributed among the many. The abundance of private homes still is surprising while Hungary was under a socialist regime for 50 years. The high percentage of private homes is not a heritage from earlier Austrian–Hungarian regimes but realized under socialist rule. Notably, in the sixties, under the goulash-communist Kádár regime, more than 800,000 typical pyramid-roofed bungalow family homes, the "kocskaházak" or "Kádár kockák" (pyramid-roofed cubic homes), were built in the rural villages and the suburban peripheries of the bigger cities. Let us assume that each kocka house had, on average, three or four inhabitants, meaning that one-quarter of the total Hungarian population has lived in such a charming 60–80 m² house. One could buy the blueprints for almost nothing and many have built their houses themselves, with the help of friends, neighbors, and local craftsmen, and one could get cheap loans from the government. In the villages and the suburbs, the plots of land are big enough to support a family with vegetables and fruit, and small livestock. To Dutch standards, the suburbs of Budapest and other cities are generous garden cities, worthy of Broadacre City. On the housing market in Hungary, today, one can still find a Kádár cube for 20,000–30,000 EUR and renovate it for double that amount. Because of the affordable homes and plots of land, because

483

of the low general costs of living, and because of the advantageous tax regime – there is no tax on a state or private pension income, and property tax in Hungary – already tens of thousands Dutch pensionados have emigrated to Hungary to adopt a semi-autonomous lifestyle, especially in the Pécs region in the South of Hungary.

The UBI strategy should go hand in hand with another rule of law that gives the birthright to a plot of land, a piece of land that can be used to feed one's family, especially useful in times of crisis. Advocacy for the planning of lust garden cities typically comes about after a period of misery, in response to being treated unfairly in times of crisis. I call the rule of law of birthright to own or use a piece of land the ubiquitous land right (ULR). One simple piece of legislation states that each newborn child has the right to own a piece of land and gets a plot of land assigned for future use. These plots of land are subdivisions of abandoned pastures as a byproduct of the half-life strategy of the unsustainable Dutch livestock population. The total area of farmland covers a bit more than 50% of the country or 2.2 M ha. When the ULR gives you 400 m^2, potentially the freed land has a capacity of 55 M plots, three times more than the population of the country. Let us assume that a maximum of half of the Dutch population would opt to live in the countryside, and the other half in the metropolis. This would mean that only 16% of the freed pastures are needed for the ubiquitous land right program. That other 84% might be used for further non-polluting societal developments, sustainable food production and fine-meshed distribution, local robotic production, and distribution of consumer goods, for scientific research, educational facilities, public spaces, sports facilities, leisure facilities, public forests, and much for nature reserves, stimulating the return of wildlife in the Dutch river delta.

In the Middle East, a form of ULR law is installed for native inhabitants, in Qatar and the United Arab Emirates. These Islamic governments typically offer their citizens tax-free income, free high-quality healthcare, subsidized fuel, generous government-funded retirement plans, access to land to build homes with interest-free loans, and free higher education, even when pursued abroad. The socialist-communist Hungary of Kádár meets Islamic countries. Both systems top-down care for their citizens and distribute land among them. Ubiquitous land ownership should not be inheritable, while that could easily lead to another form of accumulation of capital, to another form of elitist privilege. Inheriting land is not necessary either when newborn individual citizens already have the right to own and use a plot of land as long as they live, from the day they are born until the day they pass away. I suggest counteracting the principle of the heritage of common goods with a system that I will call the ubiquitous booking app. In short, it means that any property can be booked for any period by anyone, the maximum amount of time being one's lifetime, the minimum being perhaps 1 hour. A consequence of the here-proposed simple ULR law is that a family of, say, four people would be entitled to have four units of land during their stay together, while each person is treated as an individual by law. No project developers are allowed; one person can only own one piece of land. When we assume that a basic unit of land for an individual is in the order of magnitude of 400 m², the average household of 2.5 persons would thus be entitled to 1000 m², inclusive of the infrastructure that is minimally necessary to reach the private domain from the general existing light infrastructure.

Realizing that cows in the Netherlands live on a generous density of three cows per ha or 10,000 m², the simple rule of law of the ULR proposes to swap roles with the cows. First of all, we will need much fewer cows, pigs, and chickens once stock breeding has been adjusted to sustainable business operations and support healthy patterns of eating habits, meaning that we should eat considerably less meat, but of higher quality and with much better quality of life for the livestock. Meat has an enormous water footprint, for a bigger part caused by the production of food for cows, pigs, and chickens. We will need to drastically reduce our meat consumption to reduce toxic antibiotics, fertilizers in the wastewater, and methane production. The production of 1 kg of meat requires dozens of times more water than 1 kg of protein-rich vegetables. The green pastures that are liberated from livestock may be partly transformed into a permaculture forest, balancing between edible trees, bushes, and plants. Other areas must be occupied by extremely low-rise affordable housing, following the ULR rule of law, which guarantees a green environment, planted by the inhabitants themselves, as in low-density garden cities. A rich flora and fauna will emerge from what once was a monocultural pasture. Driving through the Netherlands from East to West and from South to North, one notices the endless acres of pastures. These pastures are covered with perennial ryegrass, a protein-rich type of grass that increases the production of milk and this is excellent for the farmer's profit. But it comes at a cost; in ecological terms, this type of grass is almost dead – no flowers will grow, and no bees around, just grass. These pastures are virtually fenced green factories; the single positive quality is that they offer the open views the Dutch have become accustomed to and are keen to preserve for nostalgic reasons. Despite the usual narrative, the Netherlands is not very densely populated at all. Japan has a density of 332 people per km², while the Netherlands has a comparable population density of 423 people per km². The Dutch use 150% of their land when including the polders that are reclaimed over time from the Rijn Delta. Japan, on the other hand, uses practically only 20% of its land for agriculture, infrastructure, and cities, due to uninhabitable forests and mountains, meaning that the Netherlands is *de facto* six times less populated than Japan. Holland is home to a lot of people, but they do not live as densely packed together as in Japan. There is so much space available in Holland, the form of occupancy of the land just needs to be redefined. New ecological qualities will replace the visual qualities of the open land. The ubiquitous land rights habitation-swap strategy feels like "Broadacre City Revisited," applied to the country as a whole. The ULR is a political mix between a liberal view on individual ownership of a piece of land, and a socialist view of an equal distribution of the quality of life.

 The land-swap may also become a home-swap, as the cows may populate our row houses and our multi-story residential buildings and thrive there in relative luxury. We, humans, may take control of the pastures and settle there in a very low density of one household per 1000 m², as compared to the 120 m² a household that is offered in newly built satellite towns, in the Netherlands referred to as VINEX (Vierde Nota Ruimtelijke Ordening Extra) towns. I have built some housing projects in VINEX neighborhoods too, notably TGV housing in Leidsche Rijn, 8-Bit housing in Lelystad, and Daken and Dijken in Groningen. The perfect byproduct of such a massive land-swap would be that city centers will no longer suffer from the suffocating rings of socially dense, architecturally anonymous, and functionally monocultural suburban areas. There will no longer be something like a "sub-urb"; it will be either a metropolis or countryside. The inner cities will reinforce their proper high-density inner-city quality, while the land will be parceled out for the extremely low-density flora- and fauna-rich cultivation by individual households. The areas between the inner city and the low-rise garden settlements may be repurposed into food, energy, and water production facilities, catering to both lifestyles.

A simple law like the here-proposed ubiquitous land right gives ample space to low-density housing on former pastures and will create a ubiquitous lush green environment, full of variation and diversity, in types of homes, types of home-bound enterprises, and many species of animals and plants. Simple rules, complex outcome, simplexity. The expected diversity and richness illustrate the importance of installing simple rules that create an avalanche of positive effects. There will be no necessity to nail down exceptions to the rule through specific laws, and no exceptions to the rule will be needed. Simplexity not only works in the realms of science, art, and architecture but also works well when applied to structuring the land. In the end, every law is an act of design, following the rules of design processes. Only when each citizen is addressed equivalently as any other citizen, every citizen must be considered to be an equivalent member of the same swarm. The members of the low-density affordable housing swarm, once the ubiquitous land right rule has established the distribution of the plots, will look at the neighboring plots in real time and collectively decide in which direction to evolve, to make things work, to produce energy, food, and water, with minimal interference of a central authority.

Ubiquitous food production (UFP)

A post-World War I study in Germany calculated that one family in a garden city layout would need 500 m² to grow their vegetables and fruits, to be able to survive during another long-lasting catastrophic event. In the Broadacre City project published in his book The Disappearing City (1932), Frank Lloyd Wright generously grants each family a 1-acre plot of land, which equals 0.4 ha or 4000 m^2, to support itself. Today, trusting smarter by and large semi-closed automated hydroponic, aquaponic, or aeroponic components-based systems in combination with programmable LED lighting, we would need substantially less than that, perhaps only 50 m^2 per household, meaning that home production of basic food potentially is just the size of one classroom. The smallest food production units are the hydroponic indoor gardens like the Urban Cultivator, growing small plants from seeds. Kitchen-based hydroponics systems typically have the size of a refrigerator and will eventually become a ubiquitous component of the average household. In larger quantities and sizes, lettuce and spice cultivators will populate the restaurants. I saw an early example of the Urban Cultivator in the Banff Conference Center in 2010 when I was invited to lecture at the Building Dynamics Conference in the Banff Centre in 2013, organized by Branko Kolarovic and Vera Parlac of Calgary University. A perfect example of such a self-supporting restaurant, in terms of food production and waste management, is the

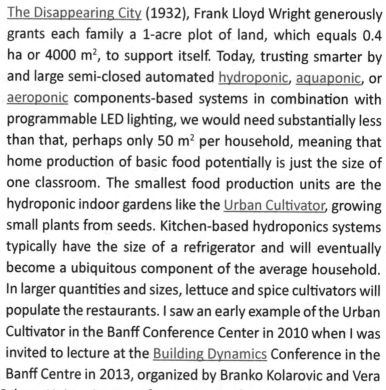

circular restaurant The Green House in Utrecht, established in 2018. The basic idea is that food must be produced as closely as possible to where it is consumed. And the waste must be used as closely as possible to where the food is prepared and consumed. On the plots of land distributed by the ULR law, waste must be controlled locally; basically, no generic waste should be allowed to leave the food production units of the plot of land the household is sustaining, while technological hardware waste must be taken back by the original producer for reuse or recycling. A healthy climate-friendly balance between producing and consuming food, water, and energy is not difficult to reach, provided one has enough space in and around the house, as foreseen by the ULR rule.

The basic idea behind ubiquitous food production (UFP) is to produce food as close as possible to where it is consumed. Locally what can be produced locally, while not refraining from consuming products that need to be imported, but that will come with a price that increases with the distance traveled. Trade remains the most important factor in keeping societies alive – trading individually and locally with the nearest neighbors, trading between neighborhoods, and trading between cities and countries. Trading is essential for cultural exchange, but there must be a cap on it to avoid overproduction and overconsumption by the privileged, which relies heavily on the current exploitation of developing countries. We need a new balance between what is produced locally and what is produced for the global market and traded internationally, whereas rules must be put in place to use renewable energy sources when, where, and as needed for the production, storage, and transport of the goods, and for the costs of polluting transportation. When something can be produced locally, it must be produced locally. My UFP rule of law intends to share the responsibility for growing food. One-third of the food production should be home-grown, one-third grown by medium-sized automated facilities integrated into facilities inside the neighborhoods, and one-third by large-sized automated facilities in the cities, respectively, countries. Each home, each neighborhood, and each city will specify the level of their food production according to their needs and interests. Specification of what to produce forms the natural basis for trade and cultural exchange with peers. Homes trade with other homes, neighborhoods with other neighborhoods, and cities with neighboring cities. Currently, new methods of growing food in medium-sized facilities are being developed, as in the recent initiative of the Smartkas, a 6000-m^2 facility with 10 levels of elevation up to 12 m, a fully automated farm with no personnel for growing and harvesting. In the pleasantly innocent yet knowledgeable early days of my early post-graduate life, I proposed in my entry for a national urban design competition to build one XXL food production center in the Haarlemmermeer polder south of Amsterdam, to provide food for the whole country. It was public knowledge back then in the eighties that an island like Texel, measuring 463 km^2, could feed all citizens of the Netherlands. Extrapolated to automated hydroponic production methods that are at least ten times more efficient, one could reduce this to 50 km^2. Back in 1980, I proposed a 2 × 2 km and

150-m-high automated food production center that would grow all the food for the country. I delivered my entry in the format of a U-Matic videotape, made at the video lab at TU Delft. As I could have expected, my competition entry was neglected by the jury led by a well-known Dutch architect, most likely because they had no equipment to play the video or because I did not follow the submission rules. But the concept and the technology are real; it could even be done back then, and it certainly can be done today. Half of the area of agricultural land could become available to become a giant nature reserve, given back to the rivers and the tides, and the other half for extreme low-density habitation, a super-Broadacre City. Needless to say, such an ultimate idea should only be considered a possible reality in sync with the other rules of ubiquitous basic income and ubiquitous land ownership; otherwise, people would become unemployed and homeless. The philosophy and technology of Another Now are intrinsically linked to social fairness and justice. The more robotically automated using low-energy systems the food is produced, the less physically bound it is to a certain location, and the more likely it is to become integrated into the households, in the fabric of the neighborhoods, and the spatial organization of the land. The more local the production, the less the transport necessary to bring the food to where it is consumed.

Due to the UFP rule of law, cities must be reprogrammed to be the home for an increasingly high number of inhabitants while the countryside must be dedicated to ubiquitous very low maximum density habitation, following the ULR rule. Cities, neighborhoods, and individual households must absorb urban agriculture and local food production to maintain a connected form of autonomy, organized in robust bidirectional networks. A dip in production at one place will be compensated by an increase elsewhere to stabilize the production in the networks as a whole. Now that in the post-COVID-19 society, working from home has become accepted as part of the new normal; even more, office towers will remain largely unused. At least half of the square meters of all office towers would not be needed anymore; they become available to be repurposed into facilities that grow food in support of the lifelines of the neighborhoods. Some unused offices may be transformed into housing units, and others into urban agricultural units, providing food for the neighborhood. Other abandoned office towers could be transformed into clean energy production plants, producing energy and condensing water from the air. Bringing food production back into the cities, into or on top of the existing buildings, frees most of the agricultural land to become repurposed. Besides the 16% repurposing of agricultural land into low-density multifunctional residential areas, another first and foremost choice is to turn another high percentage of abandoned agricultural land into national nature reserves to enhance the country-specific natural wildlife. Reforestation would establish new nature reserves within 25 years. We have seen periods of reforestation in earlier centuries. After much of the woods were burnt to build ships and used for heating, in the 19th century, the feudal rulers wanted to create domains for hunting. Along with the summer palaces, many big forests were planned, for their elitist pleasure. In our days, we will plan the new forests for the common good, not for the elite. Intuitively, a logical intervention to combat climate change is to opt for massive reforestation of cleared lands, monocultural pastures, and even the desert, as happens in China's reforestation projects and Africa's Great Green Wall project. Deserts used to be forests in earlier days, as were the pastures and fields that are now dedicated to large-scale agriculture and livestock. Sea-level deltas like the Netherlands used to be species-rich swamps. Europe used to be one massive forest, but much of it has been cut down to heat up homes, for shipbuilding, and to facilitate mass production of food, growing grains, dairy, and meat to feed the millions. Brazil is blamed for cutting the Amazon forests, but Europe has done the same; now, Europe must show a resilient way forward. Using modern multi-level closed, automated farming systems, we can grow the same amount of food in a fraction of the space needed. Ubiquitous food production, companies by a ubiquitous low-density habitation of deserts and pastures, facilitated by the birthright to a piece of land, would almost, without any top-down forestation effort from the government, transform the monocultural lands into ecologically rich areas that have abundant flora and fauna, thereby vastly contributing to a firm basis to stabilize climatic change.

The above three simple yet strong and fair rules of the UBI, the ULR, and the UFP would radically change the Dutch landscape and the cityscapes completely, for the better. These rules would create much freedom to determine one's road map to life, in combination with a shared responsibility to produce food, energy, and water. I do not foresee a return to traditional ways of farming and harvesting; traditional methods will be taken over by modern component-based distributed and downsized methods of producing food in controlled environments. The energy needed will be generated directly from the sun, the wind, and the tides, and the need for water to feed the plants by computerized pipetting may be generated by continuous harvesting of water from the air, using controlled condensation methods. Waste will not leave the food production facilities but will be recycled into the systems.

In general, for every kind of production, whether products, food, energy, or water, the new paradigm is production on demand. We know the principle of printing on command for books, and 3D printing on command for customized small products. But production on demand might apply to larger-scale production as well, household products, food production, energy production, and water production. A production-on-command strategy could work like this:

1) each individual, which is an acting component in the Internet of Things and people, has a demand for food, energy, and water,
2) each person and facility communicates the demand when needed, and as needed, with their immediate neighbors, and
3) the principle of the informed swarm, based on acting components informing neighboring components in real time, takes care of the immediate propagation of the individual demands, painfully precise in the type of food, amount of food, amount of energy, and amount of water that is demanded.

Taking the small individual demands together accounts for the total demand of a household, a neighborhood, a city, a country, and the world. The demands are measured in real time, and the food, energy, and water products are produced according to the demand, when needed, where needed, and as needed, to the most minute detail. No longer will products be mass-produced without an immediate relation to the demand. Naturally, there will be a cap on the demand, based on simple principles of what is a balanced healthy diet. When production on demand functions well, there will be no more waste; only what is demanded will be produced. For some products, there will be a delay in production after the initial demand. Plants do not grow in one day; some take a few weeks, and others might take a whole season to grow. It means that for the growth of food, one needs foresight, and one needs to produce for a projected average demand, which is given by the individual users, not by a central authority. Production targets will no longer be set by politicians or commercial firms that work for profit in the first place, and certainly not by a capitalist international rat race that leads to the exhausting of the Earth and the exploitation of the unprivileged. What is demanded and ordered via a scale-free network of small, medium-sized, and large food production facilities is produced. Nothing more and nothing less – produced as close to where it is consumed. In the scale-free network, the production facilities inform each other, such that if one of the production units fails, the others can take over. Some nodes of the networks will perform better and attract more traffic, while others may slow down to balance out the demand. When one bird drops down from the swarm because of fatigue or general weakness, the swarm remains intact. The same applies to the production of energy. Instead of one huge central energy production plant, one needs a robust fine-mazed network of smaller energy production facilities, on the level of connected households, connected districts, connected cities, and connected national facilities, thus spanning the entire earth. When production units at all levels are in constant bidirectional communication with each other, it will be easy to produce following the demand precisely, thus avoiding any form of overproduction, and thereby any form of unnecessary waste. The rule of ubiquitous food production will optimally function in a scale-free network of connected small, medium-sized, and large robotic food production units.

Ubiquitous shared responsibility (USR)

The interdependencies of food, energy, and water production systems are obvious when there is a shortage of one of these essential life-support systems. A clarifying strategic direction to take is to bring the production and the distribution of food, energy, and water (FEW) as close as possible to the places where it is consumed, and as close as possible to the individual buildings, whether small households or high-rise towers, neighborhoods, and cities. A fair strategy would be to introduce a rule of law, as a social-juridical component, that distributes the accountability for not only food production but also for energy and water production. Then, the production tasks for clean food, energy, and water are shared proportionally between buildings, whether individually owned or by big companies, neighborhoods and villages, and cities. Such a ubiquitous shared responsibility (USR) rule of law could, in principle, be written and approved today. One-third of the task would be mandated to the individual buildings, providing one-third of their need for food, water, and energy, thereby using solar and wind energy, in-house hydroponics, and water condensation directly connected to their homes and/or their plots of land. Another one-third will be provided for by the community centers in one's neighborhood, exploiting modern multi-story urban farming methods, arrays of solar cells wrapped around public buildings erected as lightweight structures on top of apartment buildings, and water harvesting fabrics or machines, where and as needed. The remaining one-third would be produced by the cities in large-scale facilities, robotically controlled multi-level agriculture, vast solar farms, tidal turbines, and, for example, algae-based desalination plants to produce drinking water. The one-third, one-third, one-third distribution must be applied both to low-density areas with detached homes on a spacious plot of land and to high-rise buildings in the inner cities. I studied how to deal with the USR shared distribution law in a refurbishment project for the Corniche Tower in Doha, Qatar. The design strategy is to top up existing buildings with lightweight structures for the generation of food, energy, and water. Based on the knowledge that concrete structures can bear 20% more weight than in their initial years after completion, the addition of substantially big lightweight structures can easily be realized without reinforcing the existing structure. The Corniche Tower features three types of toppings, one solar tower wrapped in solar cells, which adds another 100 m to the existing height, one multi-story urban farm structure, and an ultralight towering structure for water condensation with sufficient areas of bamboo/steel mesh to condensate the humidity from the air. Although I did not perform exact calculations, it addresses the idea of how much will be needed for a proportional coverage of the energy consumption.

Following the USR rule of law, the design challenge for new buildings is to integrate the productive components into the design language. Imagine a modest freestanding house that produces its food, energy, and water. Three sectors of the house will be dedicated to growing food, producing energy, and making drinking water. In countries with moderate climates, rainwater storage and filtration could provide a substantial one-third of the usage of water. In the Netherlands, there is 800 mm of rainfall per year. A 50-m² roof would be able to catch 40 m³ per year. The average consumption of water is a bit less than 48 m³ per capita, meaning that the USR of water production would be one-third of 48 = 16 m³ per year. Any Dutch freestanding house would be able to pay its share in the USR. In Qatar though, one of the hottest and driest places on Earth, the annual rainfall is 75 mm. Catching that amount does not come would only count for one-fifth of the USR target value. Therefore, additional means of producing water must be considered in Qatar. There are two methods: one is the machinic method, which requires quite a bit of energy, and the other is natural condensation from humid air, which requires a substantial area of steel meshes to have the water condensate upon during night-time.

To fully provide a healthy variety of food in one's private garden, one needs per household some 10-m-high 50 m² of a technologically advanced closed system as a separate structure in the garden. The one-third USR rule shrinks that food production structure to a standard 3.3-m height. The positive thing about the USR is that your neighbor might grow other types of food and exchange them with each other. The USR creates a community of trading citizens and enhances the sense of belonging to a community. Suppose one is living as a single, living in a most essential 50-m^2 programmable classroom-type dwelling, who wishes to have the automated food production unit integrated into the architecture, one would need another 3.3-m high 20-m^2 unit on top of the house or directly connected to the kitchen.

The food production unit can be easily combined with solar cells on the facade to provide energy and water collection on the roof. The design challenge to integrate FEW production will not lead to a revolutionary change. The annual power consumption per capita in the Netherlands is 7 KWh, which equals 20 solar panels of 1.5 m^2 when working at full capacity. Power consumption in the USA is double that value. Following the USR, must be approximately 10 solar panels single solar panel would be enough to pay one's fair contribution to the electricity needs of the Netherlands, and two panels in the USA. The other two-thirds are produced by the walkable neighborhood, and by the city where one is registered. In each walkable neighborhood of 2 × 2 km^2 that is home to 4000 people who each have a ULR plot of 1000 m^2, there must be a cluster of buildings, most likely those containing collective functions like the local culture center, including the neighborhood's health practice, food trade center, and the local product distribution center, which provides for one-third of the food, energy, and water consumption of that neighborhood. The city will use the abandoned multi-story buildings in the areas between the inner city and the countryside to fit in the FEW production. The facades may be clad with solar panels, and the rooms inside made fit for hydroponic food production, and for livestock. All comes down to that simple rule to produce food, energy, and water on demand as close as possible to where it is consumed, with a shared responsibility of individual citizens, neighborhoods, and cities. The former cow's realms transform into lush green habitats for people, and the people's flats are turned into state-of-the-art agricultural facilities, considering the maximum well-being of plants and animals.

Ubiquitous waste treatment (UWT)

Where there is production, there is waste. Where there is consumption, there is a waste. The challenge for Another Normal is to treat the waste with respect, use it in whatever is possible, cycle the waste as close as possible to where it is produced, and avoid transporting waste over long distances. Nature 1.0 organisms are food-processing bodies. They eat, process, and deliver waste products. In Nature 1.0, organic waste is simply dropped and processed by smaller organisms. For Nature 2.0, we must use different methods. Natural bodies like us living inside Nature 2.0 collect waste and send it out for treatment. This must be done as close to where one lives. In each neighborhood, there must be a small-scale waste treatment facility, perhaps with alga-based systems, producing fertilizer and potable water. Household products that have become obsolete must be collected locally in the neighborhood in the first place to prepare for repair or future recycling.

Communal facilities inside the neighborhood require space. In the low-density countryside, I would plan the common facilities horizontally adjacent to the individual plots. The communal areas include food, energy, and water production, waste treatment, short-term food, energy, and water storage and trading with peers, delivery center for products that are ordered via the internet and produced on demand, shared electric vehicles, sort of pick-up golf carts, for transport inside the neighborhood, and shared electric limousines for AI-driven travels over longer distances. The total of individual plots and communal areas would double from 16% to 32% of the abandoned pastures, which still is a feasible percentage of reusing of the land. In high-density cities, the communal facilities must be planned vertically on top of the buildings, not adjacent to them. Thus, the simple UWT rule of law, like all other ubiquitous rules, would have a different spatial effect for neighborhoods in the countryside from neighborhoods in the cities.

For ubiquitous rules of law to be able to work properly, the constituting components that define the items must be identified beforehand, tagged, and stored in a global database. Taking household products apart into their constituent components, as I exercised with students at Qatar University, makes one understand how the components are assembled and give the clue how things are to be disassembled into their unique parts. In the parametric design to robotic production process, each component has an identity that can be read by the scanners. The data that is retrieved from the global database with all unique identities tells us what material it is made of, how much material it consists of, where it is made, and perhaps how it should be recycled or upcycled. In the neighborhood waste treatment centers, these disassembled components are either reused or transported to larger "cycling" facilities in the city.

Ubiquitous energy production (UEP)

Together with my former colleague, associate professor Djamel Ouhrani, with whom I shared office at Qatar University, we estimated the energy needs, mostly cooling needs, for the faculty buildings at Qatar University. With the current state of technology, each building would need to be wrapped completely in a solar panel jacket, roof, and facades alike. Or, put a lightweight volume that is as big as the building underneath on top of it. When we follow the one-third responsibility rule though, just an installation covering the whole roof would be sufficient. Besides producing energy, the lightweight additions would function as a heat shield, providing shade for the covered building, thus somewhat reducing the need for cooling. The second one-third would be provided for by Qatar University with a solar farm within their borders, and the third one-third by the State of Qatar through a large solar farm in the desert. The cityscape will completely change when food, energy, and water production will be integrated into each building. The visual effect will be comparable to that of the bigger Japanese cities as in the Shibuya district in Tokyo, where the upper part of the buildings is lightweight structures choked up with colorful advertisements. In the one-third distribution law, the advertisement makes way for or may be combined with the food, energy, and water production components.

How might the production of energy-on-demand relate to the existing technologies of solar cells and solar water heaters? How to produce exactly what is needed and when it is needed, with a small storage capacity for darker days and emergency usage. Every natural organism uses energy when and as needed. We eat food when hungry, and we drink water when thirsty. We, humans, have limited storage capacity; we can only live for a few days without water and only a few weeks without food. Synthetically produced nature 2.0 houses, community centers, and city facilities may work similarly to natural organisms. They generate and use energy only when needed and as needed and generate the energy as close as possible to where it is consumed. They store energy for shorter periods but cannot do without for long periods. Energy consumption is higher on hot summer days when in need of cooling and on cold dark winter days in need of lighting and heating. In the in-between periods, buildings typically use less energy. Switching to unproductive standby when no more energy is needed is not making optimal use of the facility. The excess energy that is produced may be used for other purposes, uploading into a decentralized peer-to-peer grid or for batteries that last a few days, for heating water and pumping the hot water deep into the ground to store for usage by the heat pump systems in winter. Every system has to be considered in dynamic relation to other systems to find an optimal balance in the networked complexity of the Nature 2.0 energy ecosystem. Every installation must communicate with its peers, and with other species of clean energy-producing systems, in a peer-to-peer scale-free network, without a central authority. An

example of such a local energy production concept based on a peer-to-peer energy network using blockchain technology is the Solar Ville project by Space10, in a study financed by IKEA. As of now, the concept involves a 1:50 model of a village, which simulates how communities could create their own circular clean energy system. Space10's Solar Village uses solar panels and blockchain technology to create a functioning energy system in a community-owned micro-grid and negotiates via a neighbor-to-neighbor trading scheme. Cutting out intermediaries and middle management is mandatory to have the ubiquitous energy production rule work to empower people with affordable and clean energy. One possible aspect of the USR is that

each home, neighborhood, and city should have an electricity storage capacity of one week. For storage of longer periods, one needs to store the energy in other forms of storage than batteries. An intriguing example is the Plan Lievense, a proposal by Dutch engineer

<u>Lievense</u> to use wind energy to pump up water – the Dutch are good at that – into a 12-m-deep elevated lake and use the energy when needed by hydro turbines in the surrounding dikes. Lievense proposed windmills to pump up the water, but the system can also be used to store excess energy produced on sunny days by solar panels and windmills, wherever they are located. All distributed energy-producing facilities, whether small or large, must be connected in a scale-free network, whereby the neighborhood's facilities function as a hub for the individual homes and the cities' facilities as the hub for the neighborhoods. Every facility is in real time informed by its connected neighboring facilities and acts correspondingly to manage the smoothest possible distribution of energy.

Ubiquitous water production (UWP)

Water is the oil of the 21st century; predictably, wars will be fought over access to scarce water supplies. Almost everywhere are the water levels in the ground dramatically dropping, whether in Qatar, the USA, or river delta countries like the Netherlands. The groundwater is usurped by agriculture, by production facilities that are producing building and household products, and by energy-producing plants. The earth is drying up; scarce water supplies are contributing to the current climate crisis. We cannot continue to live in denial of this inconvenient truth. On top of the water-hungry industries, wildfires caused by long periods of severe drought are threatening California's forests and Australians' bushlands. In Africa, millions of people and animals are on the edge of starvation, and the land becomes unexploitable. Irrigation projects for growing food for export make the situation much worse. What is a possible remedy? Those areas in Africa that are hit the hardest have the greatest difficulty in providing a remedy. How to reverse this downward spiral? As a politician, I would pass a ubiquitous water production law, to be implemented in due time, which states that each home, each neighborhood, each city, and each agricultural project that is connected to either scale must provide its own water production, using clean methods. One possible solution is to harvest water from fog and humid air. A warmer atmosphere contains a higher percentage of water. We certainly would need large-scale installations that are extracting the water from the air, drop by drop to eventually produce water by the gallons. Traditional tribes are familiar with the condensation of the air during cold desert nights. For the Warka Water project, Italian architect, Arturo Vittori has designed and executed affordable installations to harvest water by condensation from the night skies in the desert. Besides low-tech solutions, we can also look at capital-intensive solutions such as atmospheric water-generating machines, which require electricity to run. And that electricity must be produced locally by solar panels. As the low-tech solutions take up a lot of space, and the high-tech solutions are more compact, a combination of both seems the most likely in most situations.

Can enough water be produced by condensation for a household on its property? Whereas a typical UK-based family daily uses 500 liters per day, the average family of five in the USA uses at least triple that amount, while people in Africa often have to walk kilometers to find any water at all. How to guarantee access to clean water when and as needed for every global citizen in an economy that does justice for all? When all Asians, Indians, and Africans would use the same amount of water as the citizens of the USA – and why should not they? – it illustrates the real challenge. The real challenge is no longer to consume less water, as the responsibility evading authorities in the West typically are trying to convince their citizens, but to produce water where, when, and as needed, whereby every single world citizen must have the same opportunities to fulfill their needs as close as possible to where they live. Needs may differ from person to person, from culture to culture. Different cultures will have culture-specific data for the per capita consumption of water, and they will have to find solutions that respond to their immediate needs. Water production solutions need to be found in the immediate environment of the house, in the neighborhood, and the city. Water needs to be produced as locally as possible. To extract 100 liters per day from the air, quite a sumptuous installation is needed. For now, it is not easy to fathom exactly what such a construct would look like, and how big it must be. Using natural condensation installations, my educated guess comes down to a duplication of the current building volumes to provide for sufficient water, that is when providing for the full 100% of the need for water. The USR law reduces that to one-third for one individual home. The same is true for the super local production of food and energy. When we combine the rules of law for the UFP, UEP, and UWP respecting the USR, the doubling of the volume will be the most likely outcome of the local production of FEW. The three strategies must be intertwined into a FEW nexus.

The doubling of the volume does not mean doubling of weight and costs though. The FEW generating installations are super lightweight. To visualize my assumptions, I have proposed, in the context of the M-NEX research project, the concept of the HyperCUBE for the Qatar University living lab masterplan proposal. The HyperCUBE is based on a mathematical tesseract, which is a three-dimensional volume that projects and increases its faces in multiple directions to form a multi-dimensional hypervolume. The additional dimensions are physically translated into a filigree lightweight steel structure, holding the solar cells and the arrays of small wind turbines, the steel mesh to harvest water, and the stacked hydroponics to grow the food. The HyperCUBE concept gives an extra dimension to building in three dimensions. Prosaically, the outward projected dimensions can be labeled respectively as the food dimension, the energy dimension, and the water dimension. While no exact calculations have been made, I do not know yet to what size the outward projected areas would need to extend. Further research will be needed. Other geometric forms are possible to visualize the food, energy, and water (FEW) wraps. A FEW-wrap can be an open cage around the object; it could be a topping on top of the existing volume or an interwoven hybrid of open and

closed volumes to reach the desired FEW performances. Anticipating the FEW interlacing, I have introduced in the Participator urban design instrument a porosity factor that calculates the amount of open space mixed with the closed spaces that are needed for the integration of food, energy, and water production into the urban fabric.

Counteracting the climate crisis by production on demand for almost everything big and small using state-of-the-art digital technologies is bound to change the appearance of cities and the countryside completely. We should not and cannot go back to an idyllic image of a forlorn society; we will have to move forward to Another Normal, an updated Society 2.0 that is based on fair treatment of everyone and everything, a Society 2.0 where people and things are intrinsically entangled with each other to form a synthetic ecology of things and organisms. A Society 2.0 is where individuals swarm among individuals to form a neighborhood, where neighborhoods thrive among other neighborhoods to form a city, and where cities cooperate with other cities worldwide. Society 2.0 will establish a new digitally augmented layer of synthetic culture wrapped around the entire globe, offering equal opportunities for a fair quality of life for every one of the billions of global citizens, whether in Africa, India, China, The Middle East, Europa, or in the Americas. A good start would be to stay within the green doughnut as described by Kate Raworth of Oxford University in her book Doughnut Economics. Raworth argues for a safe and just space for humanity, where overshoot, waste, and shortfall are banned from society. Overshoot and wasting mean compromising the health of the earth, and shortfall indicates compromising the health and quality of life of its people. We need to quantify the overshoot and the shortages, turn waste into food, and find concrete solutions. A non-quantifiable diagram can easily be misinterpreted and used to argue to go back in time by doing less of the same. Instead of doing less of the same, or better versions of the same, I propose a dozen interlaced sets of simple rules for Another Normal, simple yet concrete and impactful rules to find a way toward ubiquitous fairness and justice, to see things from a different perspective, by a form of simplexity that would make today's many complicated rules redundant.

Ubiquitous robotic production (URP)

Production on demand is best facilitated by robotic production methods, using 4-5 axis CNC routers, 5-6 axis robot arms, 3D printers, and hot-wire cutting machines, basically any machine that cuts, mills, welds, rivets, screws, and paints at the instruction of digital data. As of now, 3D printing machines represent only a relatively small part of a much larger ecology of robotic production machines. The impact of 3D printing is overestimated; the technology is not suitable for large multi-story buildings, let alone skyscrapers. In the imagined world of Another Normal, which is governed by digital data in the design and the component-based realization process, production on demand means the ubiquitous robotic production (URP) of almost anything, at any scale, in any place, at any time. URP is scalable from the small scale of household products to the big scale of skyscrapers. Almost any building process can be post-optimized by robotization, but the real value lies in proto-optimized parametric design to robotic production.

Missing parts of household products can be modeled in popular 3D modeling software and printed using tabletop 3D printers. The home-based 3D printer is today's equivalent of the 2D printer annex fax machine. Customized designs for household products and furniture – whether designed in Europe or on the other side of the globe – may be printed in a small local workshop, close to the client. Collaborating swarms of many smaller geographically distributed machines may create larger structures built up from many unique pieces of 3D printed material, usually some form of plastic called polylactic acid, commonly known as PLA, like the Project EGG by Michiel van der Kley and the Maker Chairs series by Joris Laarman. At the larger scale of the neighborhood, a network of connected workshops with CNC machines, the fab labs, will produce anything between a piece of furniture and the components for a small pavilion. At the city scale, a network of bigger robotized factories will produce the components for larger building complexes, even skyscrapers, in any material that is reusable or recyclable, wood, steel, glass, and carbon. Production on demand is fully scalable and applicable to all sectors of the maker industries, from food to products, and from energy to water. What the facilities have in common is that the production is fully robotic and based on parametric design input. In the not-so-distant future, any piece of furniture, any car, any house, and any skyscraper will be the outcome of a collaborative and customized production-on-demand process, based on participatory parametric design to robotic production methods.

Artists, inventors, and innovators do not produce on demand. Artists produce art without explicit demands from others, expressing themselves through music, dance, poetry, journalism, fashion, painting, sculpture, prototypes, inventions, and innovations. Once their work is acknowledged, they might do commissioned work, whereas the NGOs, cities, and states by and large have taken over the role of the ruling classes of the feudal era. Scientists work out of a genuine curiosity to understand how things work and pursue research as an intellectual challenge, moderately paid for by an institution or university. Researchers working for the CERN, NASA, DARPA, and Big Pharma, technology-driven firms, in general, perform a form of research on demand; yet, they still are, to a large extent, driven by their intuitions. Individual innovators do not wait for someone to ask for innovation. Innovation comes from within; innovation comes from intuition. The intuitive work of the artists, scientists, and innovators will not be automated or replaced by robotic production. Still, artists, scientists, and innovators make ample use of robots as an instrument to explore their search for the unknown. The Portuguese artist Leonel Moura uses a swarm of small robots for the Robot Action Painter series, which he calls a new kind of art. In the Machining Emotion series of robotic paintings, Ilona and I, assisted by designer-programmers Serban Bodea and Ana Anton, linked intuitive 3D sketches done with the 3D digitizer to fractal algorithms driving the robotic output. In all cases of robotically produced art, the artist remains the initiator, the programmer, and the choreographer. AI-driven robotics replaces that kind of work that is mostly repetitive and predictable. The more emotional factors are involved in the design and production process, the less subject the work may be to robotization. As predicted by the Qatari architect Ibrahim Jaidah, the designer of the Al Thumama stadium, 90% of the current work done by low-wage workers in the building industry can and will eventually be replaced by robots.

Referring to those designs, those manufacturing and assembly tasks can be done by a swarm of robots, operated by an AI-driven hive mind. ONL in praxis and Hyperbody in academia have played a substantial role in the inevitable rise of URP; we have pioneered in the field of art, design, and architecture to explore the potential of the file-to-factory process, from parametric design to robotic production. We have

demonstrated that URP gives the team of designers and producers full control over the technical and esthetic values of the constituent components, within tight budgetary constraints. Many start-up companies are working within the new rules for the new URP economy, among others the Rotterdam-based company Studio RAP or TheNewMakers, both partly spin-offs from Hyperbody and ONL.

The essence of the ubiquitous robotic production (URP) rule of law is that every single constituting component is, in principle, unique, addressed as a unique identity in the design and fabrication process. Even if one component happens to be the same as one of the other components, it still retains its unique identity to fit in that particular place, at that particular moment in time, and for that particular reason. In the beginning phase of the URP revolution, where we find ourselves now, mainly prototypes are produced robotically, but in due time, in Another Normal, the production of the complete assemblage package of constituent components will be produced robotically. Traditional mass production of a series of the same elements will be made obsolete by massive, decentralized customization of a series of unique components to construct unique customized products. Raw basic products that feed the robotic factories are provided for by local, medium-scale, and large-scale production facilities, produced on demand, taking advantage of local materials wherever and whenever possible. Raw materials and products are produced on demand if needed, where needed, and when needed. Just that, just there, and just then. The three responsible societal levels of individual households, communities/neighborhoods, and cities are physically and virtually networked to stay resilient and to guarantee the continuous advancement of the quality of life. Just as in diagrid structures where the forces branch out and join in when one node of the components, food, energy, and water production network fails, information simply bypasses the defect node, and their immediate neighbors take over.

URP is a process that is open for participation in any phase of the design-to-production process, as long as the participation runs on a digital platform. ONL and Hyperbody typically use gaming software to build applications for participatory design. We have done radical participatory design experiments, like the Virtual Operation Room (2002) by ONL/Hyperbody, described in my book Hyperbodies, Towards an E-motive Architecture (2003). The recently developed participatory urban design instrument called Participator (2019) by ONL invites the user into the earliest phases of conceptual design. The user-friendliness of participatory design to production platforms is still a hurdle that must be overcome. In due time, anyone who can operate a design application on a smartphone will be able to participate in a URP design-to-production process, using sliders to create the input data. Successful future forms of participatory design will depend on the user-friendliness of the user interface (UI). The mouse, numerical input of data, coding, and scripting are not user-friendly. Hyperbody experimented with speech recognition and spatial gestures. Microsoft's Xbox 360 Kinect gesture controller will need to be further developed to create a level playing field for all experts and non-experts alike.

Future-proof design practices anticipate future ubiquitous robotization of components and complete products. The full URP revolution is achieved by the ubiquitous use of peer-to-peer connected digital design instruments and by a direct connection between design data output into networked local production machines. Hotlines are established between the design machines and the production machines. At any scale from a sole proprietorship to multinationals, with a shared responsibility of each participant, from conceptual sketch to the daily operation of the project, the validity of the data must be guaranteed, to avoid information getting lost in translation. We will need novel forms of contracts between the participating parties, probably backed by the distributed peer-to-peer type of transaction. In the URP contracts, it must be clear who has modified what, and thus who is responsible for that modification. Each step in the evolution of the design must be recorded, and even the smallest modification must be validated and capitalized. Each step in the collaborative evolution represents a value. Many small design decisions are made by many different stakeholders, conceptual designers, material designers, spatial designers, geometry designers, climate designers, structural designers, interaction designers, facility managers, maintenance designers, and lifecycle designers. By taking steps together in real time, the shared project evolves. We will need new types of contracts in the building industry that reflect the radical changes in how projects are conceptualized, how they are designed, how the materials and the ecological footprint are administered, how the components are produced and assembled, how the projects are inhabited, and how they are reimagined. Every single component is labeled, and the history and the actual performance of each component are known and communicated to their nearest neighbors in the first place, thus giving real-time insight into the overall performance of the facility. And every interaction between people and things is labeled, validated, and evaluated. AI algorithms keep track of all the minor modifications that eventually lead to its materialized outcome. Each of those modifications has a value, and that value determines the financial reward for the input of the participant.

In a participatory design game, all are in it together, experts and laymen alike, while some are more in it than others. The democratization of the production facilities, supported by a ubiquitous immediate banking rule, based on a peer-to-peer distributed banking system, may eventually shatter the monopoly of investors and contractors in the building industry, and bring back the responsibility to those who do the actual design work, the production, assembly, and maintenance work. The ubiquitousness of decentralized small-scale transactions is actors in the long tail of the economy, as described by Chris Anderson in his book of the same name.

While Anderson mainly analyzed the revolutionary change in the music industry caused by digitization, the URP of the building industry will lead to a similar long tail of numerous distributed small producers, producing customized components when and as needed, as close as possible to where the components are assembled.

Ubiquitous autonomous transportation (UAT)

Both living in extreme densities and living in extremely high densities is a challenge for the transport of people and goods. Both ways of living must be supported by zero-energy transportation of people and goods. The best way to optimize transport is to have automated electric vehicles at the different scales of individual plots, neighborhoods, and cities. The means of transportation need to be thoroughly reimagined. This will inevitably lead to another rule of law, to be implemented in parallel to the other rules, stating that the inhabitants will share the use of fully autonomous electric vehicles, the ubiquitous autonomous transportation rule. Sharing is the new having. No one needs to *own* a car when the car of one's choice will come on demand to one's front door. The means of travel and delivery come on demand. Sharing does not mean that everyone or every company uses a similar car. There can be and will be as many different kinds as possible of vehicles to share, ranging from small electric pick-ups aka golf carts to connect the individual plots to the neighborhood center, small nondescript functional boxes to more luxurious mobile lounge vehicles – up to the taste of the customer – to connect the neighborhoods with the cities, all of them AI-driven. Cities will be connected to other cities by fast driverless automated trains, up to 1000 km distances. Trains will be driverless, as in the Doha Metro system that was just completed before the FIFA 2022. From there, airplanes might take over to reach larger distances than 1000 km. Fleet operators will offer you a range of autonomous mobility possibilities, fully customized to one's preferences to invoke the feeling that it is *your* car for *your* trip that is particularly meaningful to *you*. The Chinese Patient would be very happy with customized ubiquitous automated transportation; he would enjoy the trips by limousine. Car sharing is potentially a dozen times more efficient than owning a car exclusively for oneself. The shared vehicle will only be temporarily "owned" when and where needed. In our daily reality, most cars stand idle for most of the time, parked in front of the house or parking garages. Imagine that cars will be exclusively used on demand; it would make the very existence of private parking garages superfluous. The customer activates the ubiquitous booking app and orders a car after choosing a specific period to go from A to B. Through the booking app, the user sets other preferences like the type of vehicle, level of luxury, arrangement of seats, depending on whether one travels for work or leisure, background music, and infotainment program. "I want it all, and I want it now", as sings Freddie Mercury of Queen. The customer pays per use, streaming in sync with the distance covered, as in the taxi metering system. In the ubiquitous banking app, the customer sees the bank balance decrease while traveling. Real-time banking is another implementation of the immediate. If one changes plans and wants to step out of the car, one just orders the car to do so and leaves the car free for someone else to use, as is common practice with the small electric urban e-scooters. Hop on, hop off, step in, and step out. The electrification, identification, quantification, and verification of UAT vehicles are a prerequisite for them to become aware of their peers in the urban mobility network of e-scooters, e-bikes, and e-cars. The UAT vehicles will be in continuous contact with their immediate neighboring vehicles in real time.

Traditional human-driven motorized traffic is inherently divisive; it works only by splitting the road into separate lanes for pedestrians, bicycles, cars, and buses. Crossing a street has become a challenge; from the pedestrian point of view, the busy roads are like deep dangerous abysses. Human-driven vehicles are a dangerous species, killing more than 40,000 people in the USA alone per year, percentage-wise three times higher than in the Netherlands. Automated cars will be substantially safer, while looking at people and peers alike, proactively avoiding crashes. Opposite to human-driven vehicles, AI-driven autonomous vehicles are connective by nature. They connect in real time; they keep a close eye on their nearest neighbors to synchronize their speed and direction. Real-time digital connection is many times faster than human connections from car to car. Human brains are relatively slow; they are good at processing multiple strategic considerations at the same time, but that comes at a cost – it takes time. Automated cars process information at the speed of light, which to us humans feels like proactive behavior. Autonomous cars are aware of people and any other form of obstacles; they will stop when a person crosses the street. People and cars are players in the same system, a bit like on the busy streets that we have seen in Mumbai and New Delhi, where literally every form of transport smoothly mixes with the other. People, tutus, mopeds, bicycles, cows, cars, buses, and trucks accept each other in a viscous mobility flow. Pedestrians cross at random without being run over. The dense traffic in India's big cities is the analog prototype of the behavior of autonomous vehicles.

Autonomous traffic will look much more ordered though, while the space reserved for pedestrians will be highly increased. Once the UAT has been established with more than the critical 20% of all traffic, the citizens will find many new ways to relate to each other and their immediate environments. Electric vehicles will quietly zoom around the extensive spacious green neighborhoods like busy bees, hopping from user to user. The electric vehicles know each other, they know the road, and they know their part-time users. The company leasing the vehicles knows which vehicle is where and constantly measures its performance. When there is a reported failure, they send the service team immediately. Measuring is knowing, and real-time measuring is real-time knowing, thus making the measured performance part of the adaptive dynamic system of the electrified mobility swarm. The company traces where the vehicle is at any moment in time, as Google knows where and when your mobile phone is. Naturally, it is up to the users how much personal data they want to share, and to what extent they want to make use of the e-facilities. Electric vehicles look at each other like the birds do in a swarm, thus preventing them from bumping into each other and adjusting their speed to their nearest neighbors. E-traffic will, at times, move slower than human-driven vehicles but in a steady flow, and thus grosso modo faster in the journey from A to B. When the traffic stream is electrified and by and large self-driving, traffic lights, and fixed pedestrian crossings are no longer necessary; pedestrians just cross the road anywhere, while the AI-driven cars would proactively adjust to avoid any casualties. It will feel like a controlled form of chaos, perfectly safe for all road users. Because of the absence of traffic lights and traffic jams, the overall itinerary will be faster and much more comfortable. The level of comfort of ubiquitous autonomous transport will be so high that the e-cars will be soon transformed into social meeting places, temporary offices, home cinemas, and always on the move. For the nomadic international citizen, home is a versatile place. Home is where you are, what you do, and with whom you are. Home is not a fixed frozen place. The e-car is one of the rooms of your home.

The ubiquitous use of electric self-driving vehicles will both facilitate the efficient operation of low-density environments and change the streetscape of high-density cities at the same time. Only a small surplus stack of vehicles would be needed as a backup at maximum capacity. Peak hours will be something of the past, as the demand for transportation can be fine-tuned to avoid peaks in traffic. UAT vehicles would keep slowly circling by themselves around the places where the most traffic is expected, like Uber and Careem cars do when they are waiting for new customers to sign up. Parking places in the cities will be repurposed into pedestrian spaces, space for bicycle tracks, green spaces with benches to sit and relax, space for cafés, restaurants, and small kiosks, into delivery stations to fetch your online shopping. A revolutionary change in the streetscape is already happening, inspired by the concept of the walkable city, or the 15-minute city as first described by the French-Colombian urbanist Carlos Moreno. In parts of Paris, the 15-minute city rule was successfully implemented under the leadership of mayor Anne Hidalgo.

Cities are preparing to have an electric cars network running in the foreseeable near future, proactively anticipating Another Normal. Streets become drastically smaller when automated vehicles have taken over. Especially in cities such as Qatar, there would be no more need for eight-lane city highways with parallel roads chopping the city into isolated parts that are only accessible by car. The speed limit in the city streets of Doha currently is between 60 and 100 km per hour, and many Qataris and expats in their SUVs and power cars give it a little extra. In the transition phase, AI-driven cars and human-driven cars will mix; when the critical threshold of 20% AI-driven cars has been reached, the driving behavior of the humans will adapt to that of the robots. And then the phase of new urbanism will hold and change the highways into safe pedestrian-friendly boulevards. The effect of the ubiquitous automated transportation rule on urbanism is that there will be a lot of activity on what once was a narrow pavement. Pavements become wider and populated with an abundance of trees and pavilions in support of an intensified city life. Trees will provide for shadow, produce oxygen, temper the winds generated by the high-rise buildings, and bring the outdoor temperature down by three degrees. Hundreds of thousands of trees need to be planted to transform the former eight-lane highways that were once dividing the city into an archipelago of isolated city blocks into an urban oasis.

Ubiquitous home delivery (UHD)

During the COVID-19 crisis, home delivery companies thrived, while people were asked to stay at home. Someone had to bring the food to their homes; restaurants switched to takeaway food. Streetscapes have changed since, e-scooters, food and product delivery bicycles, scooters, and small vans are populating the streets. The ubiquitous home delivery (UHD) rule implies that shopping for food, household products, cars, and, basically, anything will be purchased online and home delivered. The products will come to you, instead of you going to the products. You order when needed and as needed, it is produced when needed and as needed, and delivered when needed and as needed. Ordering is a digital process, whereas each product, each piece of food, energy, and water is tagged, producing is digital aka robotic, and delivery is digital aka automated. The UHD rule of law is as simple as it is efficient and convenient. Why take the effort to go to the takeaway restaurant yourself; order the food, wait there for a quarter of an hour or more, drive back, unpack, and heat the food that has been too long in your car, if you can have it fresh, warm, and right onto your dining table at home, without the hassle of going there and waiting. I do not see any social quality in waiting at shops, traffic lights, or airports. The social aspect of going out for fun shopping and going out for dinner with friends is a completely different story; the social factor will grow as well. Entire cityscapes have already changed into one big outdoor terrace. City centers, whether in Qatar, Hungary, or the Netherlands, whether in big cities or local hubs, are transformed into culinary paradises. Many of the simple rules that shape Another Normal go hand in hand with a complete change in the look and feel of cities; some of the rules are only briefly interrupted by the Corona lockdowns, and other latent rules are accelerated by the Corona measures. UBI means digital money in the pockets of people, and UAT means that streets are the domain of pedestrians and cyclists.

Why spend hours in traffic and spend gasoline for the car to go to that MediaMarkt or that Auchan shop if you can order it today and have it delivered to your home the next day or even the same day? In the long run, consumers will find out that home delivery is indeed convenient, while the food and the products almost immediately come to them, and it does not cost them their valuable time. UHD has many advantages over DIY shopping. You can organize your precious time as you wish, you will not lose so much of your time driving to the shops, and you may save hours per day on travel time. UHD will save you money spent on gasoline and maintenance for the car. Online, you can do shopping in parallel to other activities. Much of the new normal shopping time will be spent online in augmented reality environments, giving you a shopping experience plus precise specifications and performance of the products you are looking at and ordering. What exactly would you miss from strolling for hours through endless shopping malls? Would you miss passing by the shopping windows, or would you miss being diverted from your intended purchases? Would you miss getting "inspired" by watching the displayed items? Will you miss getting tired of hanging out? Or is that exactly what you were looking for, just hanging out? Will the augmented reality version of being seduced by customized advertisements – like being addressed personally as in the movie Minority Report when walking across a shopping mall – give you a better-informed shopping experience than online shopping from home or anywhere? For specific high-end products, a fusion between online shopping and onsite shopping will be the obvious future. Displayed products will send you additional information about their performance while looking at them. The product recognizes you by face recognition or by the proximity of your mobile phone and gives you the exact data on the performance, competitiveness, availability, price, and reviews as any commercial website gives you. In the integrated, physical, and virtual world of shopping, one can of course order it to be delivered to the home, instead of having to carry large and heavy bags. Today's consumers expect a multichannel-informed shopping experience. Physical stores, a website capable of handling purchases, a mobile site or app, and a presence on social channels such as Facebook, Twitter, YouTube, and Google+ are essential to establish a relationship with the consumer. Informed shopping builds social connections between people and things, in a cyber–physical environment, where people talk to things, things to people, things to things, and people to people. Beyond talking and exchanging predefined data, the logical next step is to have bidirectional discussions between things and people. The thing has a unique identity, as has the person. Only when all players, things, and people in the online real-time shopping game have an addressable identity, such discussion can begin to take shape. Finally, ordering the product means initiating the URP production on command, to be delivered according to the UHD rule. The proposed ubiquitous rules work in parallel, step by step establishing the inevitable Another Normal.

Economically speaking, it is much more efficient to have food, fashion, books, household appliances, and basically anything that a household or a small enterprise needs to be delivered to your home or the businesses than to go out shopping yourself. The total amount of kilometers when everything is home delivered to anyone, anywhere in the world is considerably less than when each household goes out shopping with their private means of transportation. Delivery companies combine many deliveries in one trip, following the traveling salesman algorithm, thus saving fuel and time wasted for the shopping kilometers. An implicit relation to production on demand must be established for the full efficiency of the UHD. Production on demand and home delivery are mutually reinforcing tendencies – two sides of the same coin. The product must go directly from where it is robotically produced via the automated delivery service to the place where it is consumed. The products must be produced as close as possible to where they are consumed. No more Amazon warehouses are needed, and no more mass production, trading, packing, and storing of food, household products, energy, or whatever product we are fancying to order. Mass production produces more than there is demand for, and much of it is dumped when not sold. Future profit will come from the long tail of the economy, by series of one, not by mass production of thousands of the same. The cyber–physical fusion of the producer and the consumer will reduce storage space inside shops and storage space in distribution centers to a minimum, minimize the need for packaging, and minimize the kilometers traveled by the product to reach the consumer. Trying on clothes is becoming an immersive customer journey, a cyber–physical experience; prototype versions of <u>virtual fitting rooms</u> are already in place, in the more exclusive shops, and at home. You upload your measurements first, similar to what the tailor does. Your data are used as parameters to model the proper shape, and you can look in the cyber mirror at how it fits you. No more sweating customers spoil the shirts and dresses, no more wrinkles, and no more waste of time by physical testing. Cyber–physical testing is clean, comfortable, precise, efficient, adventurous, socially interactive, and satisfactory for both client and salesperson. The dress will be produced on demand at the closest robotic manufacturing workshop and delivered to the home within days. Production of clothes on demand will solve for a great deal the contribution of the fashion industry to global warming. Currently, half of what is produced in the fashion industry ends up as trash. An extreme version of the virtual dressing room is what is called <u>virtual clothing</u>. The customer dresses up virtually in exclusive dresses designed by famous designers and pays for the image only. The virtual dress is only produced in cyberspace, only meant to be shared with friends and followers on digital social media.

Where is the invention to customize the production of the fabrics for fashion, to replace mass production of the fabrics? Why produce the fabrics by the meter; why not produce them directly in the desired shape? Customized robotic production of fabrics would make superfluous the cutting of the mass-produced fabrics and would allow for robotic assembly of the pre-shaped components of the dress. My customizable Body Chair is personalized following a similar procedure as the virtual dressing room. Upload your measurements, choose between the active, the active– passive, and the passive model, and make your choice for the pattern and the finishing. The Body Chair is produced on demand and delivered to your home. Personalized choices are made via the Body Chair (web) app. Despite well-received international presentations at the Salone di Mobile, the Dutch Design Week, at the Design Without Borders exhibition in Budapest, as of now, the Body Chair has not (yet) become a commercial success, nor has ONL's fully customizable Ecoustic parametric ceiling project that we developed for Ecophon in Sweden. The company Packhunt.io from the Netherlands, founded by Hyperbody connection Jeroen Coenders, offers cloud-native parametric and digital solutions for immersive customer journeys. Usually, only by persisting as a start-up for a decade or more, with dedicated focus and continuous efforts, a possible commercial success comes into the picture. ONL and Hyperbody have developed numerous concepts for unique parametric fully customizable product designs and have found the partners to build the prototypes but have not been successful yet to bring the products to market. Being innovative is one thing, and bringing the products to market is an aspect of entrepreneurship that requires commercial instinct, physical and mental endurance, a good portion of narcissism, and a focused obsession. These are aspects of the designer's job, in which I have not invested time and energy, a typical designer's dilemma. How much of one's time is spent on innovation, and how much of one's time bringing ideas to market?

Ubiquitous booking app (UBA)

More and more people are getting used to booking their working and living spaces online. We already book visits to general practitioners online and visits to dentists, we book holidays and hotels online, and we book rental cars online, smoothly without any hassle. Consumers are getting used to buying food products and household products online, while home delivery is getting more and more appreciated and immediate. Online ordering may bypass intermediaries and order directly from the farmer, called direct farming or crowd farming and order household products directly from the producers. Everyone has experienced personally how booking online and home delivery of products has helped them during the pandemic. We booked time slots for visiting shops during the pandemic, and that is where it becomes interesting. Time may be booked personally to discuss specific matters with general practitioners or book participation in Zoom-type webinars to directly interact with the speakers. Booking has become a ubiquitous digital tool to plan your life, almost everything, anywhere, anytime, with anyone. At the Department of Architecture and Urban Planning at Qatar University, I developed the concept of the ubiquitous booking app (UBA) in the context of the MANIC research, the concept of Multimodal Accommodations for the Nomadic International Citizen. The UBA facilitates the booking of every type of accommodation for any period, at any place, and for any price. The UBA is the personal interface with the world, as was in previous times the desk of the travel agent and the ticket office. Booking anything anywhere has come to us, instead of us coming to them. Through the ubiquitous booking app, we connect to almost anything and anyone in the world. Some of these connections may come at a cost, while many services are free.

The UBA facilitates a radical new use of buildings and public space. Starting with multipurpose buildings such as the MANIC tower facilitating the 24-hour economy, but soon expanding into many other types of buildings including museums, leisure facilities, homes, and apartments, the concept of time-based leasing of space will revolutionize the way the built environment is used. The basic idea behind the MANIC concept is that users can book the space for any period, everything between ultra-short and almost permanent. One may book for 1 hour, 2 hours, a day, a week, a year, or maybe even for a longer time, depending on the planned activity. The space itself is multifunctional as in the Pop-Up Apartment proposal, where different pieces of furniture pop up or sink into the double floor on demand. Suppose there is a business meeting; one may need a large table, chairs, maybe a couch and comfortable armchairs for an informal chat, a toilet, and a pantry with a refrigerator filled with a choice of drinks. The pre-ordered configurations are already in place when one enters the room, and they sink back into the floor after the scheduled activity. Afterward, the room is cleaned and automatically returns to the default state, the initial condition. Through the UBA, personal preferences are set, some of which the app already knows beforehand from the profile. When the same room is booked by a yoga teacher for yoga classes, the room may remain an empty Zen-like space, unspecified without visual disturbances. Later that day, that same room may be booked for a night stay in the hotel mode, or a one-week holiday stay, including a popped-up bed and a popped-up bathroom. That same room, perhaps in combination with some other units that are scattered around the MANIC building, is used as a distributed office for that day. At will, the functional items only pop up when used. The bathroom may be full 50 m² when other functions are not used. Or, during the night, only the bed appears in the room, thus providing for a very spacious luxury overnight experience. The MANIC tower is a lively mix of short stay and long stay, a programmable vehicle for the 24-hour economy; MANIC never has a dull moment.

In the cities, many buildings are heavily underused, and office buildings are not used for half of the day, as are apartment buildings where the inhabitant is out for work. Why not combine the office and the apartment into a new concept that allows for flexible time-based booking? The bookable unit in such a multifunctional accommodation must be flexible in use as the Pop-Up Apartment, which then transforms into a pop-up workplace. Imagine a building with thousands of programmable units, capable of transforming from home to office on demand. That building would be a very lively community of people coming and going, a lively mix of work and being at home, whereas none of the units are owned by anyone but just reserved for a certain period.

The question of intimacy quickly comes to mind when considering the personal aspects of a home-like or office-like working environment. Where do you put your personal things? Where are all the small stuff like pens, souvenirs, precious glassware, and favorite books? Many personal things are already digital, but the presence of physical things will remain important to feel connected to a place. As in the flex office, the workers may have a personal trolley that they bring with them to the otherwise neutral workspace. Only for that particular day do they customize their working place, as nomads do when they are moving from place to place. Your personal belongings fit into a backpack, and the environment remains open for anyone to use, at the desired length. No waste is left; functional things are only there, popping up when needed and as needed, and your personal items are either digitally embedded in the workplace or physically kept close to yourself, perhaps temporarily stored in a box, as travelers do when they are visiting a city and want to be free to move around, as bathers do when they are going for a swim. If one insists on having a place for storing glassware, paintings, closets, and other larger memorable things, one will need to lease a unit or a couple of units for a longer period, which *de facto* equals the feeling of home ownership. There should be a fine-grained diversity of time slots to use a particular space in a particular building, facilitating the range between ultra-short stay and ultra-long stay, covering the whole spectrum from being on the move and being emotionally bound to one particular place.

Ubiquitous immediate banking (UIB)

Today's citizens are more of a nomad than is commonly understood, and, now, in the aftermath of the pandemic, it is a good moment to evaluate our behavior and implement the findings of immediacy and ubiquitousness on the built environment at large. Ubiquitous basic income, use of land, food, energy, and water production, automated transport, robotic production, home delivery, booking, and immediate banking are some of the current tendencies that are radically transforming the post-pandemic society at an accelerated pace. The biggest challenge is how to implement the ubiquitous digital social and technical systems without adding another layer of consumption and accumulation of waste to existing dysfunctional systems, leading to even more global warming. How to replace one derailed system with another system? The road that I have mapped out for myself is to change the system of the building industry from within, by focusing on the constituent components and their mutual relationships, to consider the component a proactive member of a swarm of components, exchanging data with their nearest neighbors in real time. Parametric design to robotic production on demand is bound to step by step replace the old paradigm of mass production and mass consumption. Old divisive derailed systems will slowly die out, melt away, and eventually disappear to make way for the next level of the smarter, more just, and fairer.

The question remains though how to implement ubiquitous basic income and the other above-described rules of law from within in daily political practice? By addressing each person as a unique identity, by establishing immediate relationships between the millions of identities? How can the income, or rather the tokens as in a verifiable blockchain-type economy, be fairly distributed without depending on a top-down bureaucratic system, where the middle management unnecessarily complicates things? How to avoid obstructive middle management completely? In my practice, the straightforward answer is the efficient and democratized design to production paradigm, whereas no middle management plays a role. In the design-to-production process, there is a direct relationship and a feedback process between the design concept, the precise data, and the product, and between the producer and the consumer. Nothing and no one stands in between. When the ubiquitous basic income is a rule of law, and thus a basic right, the tokens will be derived directly from the economic activities of the community, of which the receiver of the tokens is an acting part. For that to happen, we need a radically different banking system, an immediate form of banking that translates basic income and additional economic activity into tokens that are distributed on the fly to all actors. Economic activity is per definition a dynamic adaptive system, and the new ubiquitous immediate banking (UIB) system must follow the continuous changes in the economic activities and its rewards immediately, in real time, without seconds of delay, without second thoughts, without discrimination, and without exceptions to the rule. Immediate banking comes down to a streaming continuous transfer from and to a personal unique bank account, perhaps administered in the form of micro-tokens in a blockchain system, effectuated in many micro-tokens per second. On a personal banking app, streaming income and streaming expenses are generated and visualized in real time. Like any other digital real-time monitoring system, one can immediately see in the token meter whether one earns more than one spends, calculated in real time. It is visualized over one day, one week, or in whatever time frame it is observed, like the value of an asset in the stock market. It comes down to a personal, virtual, real-time stock exchange monitor, whereby the stocks are the tokens earned and spent. In the current real world, anticipating spending in time comes at a price, in the form of a percentage of loans and mortgages. On the other hand, postponing spending your income comes with a bonus, in the form of savings and dividends from the company one is working with, which may level out the costs of spending before it is earned.

In the world of Another Normal, payments are smoothed out over longer periods, according to one's settings and preferences. Income is not transferred in one larger amount at a time but evenly spread throughout the revenue. All incoming and outgoing transactions are distributed in a continuous flow of micro-payments. The rule of law of ubiquitous immediate banking (UBI) is that financial algorithms facilitate the smooth handling of income and expenses in real time, without premature spending and without postponing the earned tokens. The advantage of such a balance seems to be that no tokens are accumulated at any time by anyone, ensuring that all tokens are immediately used and invested in new economic activities, making maximum use of available financial resources. No accumulation of capital in the hands of a self-privileging elite. All available capital will be active, and no capital is secluded from the active economy. A peer-to-peer distributed blockchain system does not concentrate the calculation power in large energy-consuming data centers; its computing power is distributed among the personal computers of the members of a family, company, or organization. A peer-to-peer blockchain network guarantees the transparency and authenticity of transactions. Data-intensive blockchains can be difficult to scale up while slowing down the process and requiring lots of computing power; however, there are technologies like the Zero Knowledge Proof (ZKF) protocols that minimize the size of a blockchain to scale up to millions of users of a distributed blockchain network simultaneously. I do have second thoughts about the word blockchain itself, both the "block" part and the "chain" part. The block part suggests something fixed, like an Autocad block that only allows repetitive use. The chain part suggests a linear chain, while all of my thinking and executed nonstandard architecture is based on a non-linear network of interactions. In my work, I strive to "unchain" the building industry. The traditional linear chain from concept to final design to tender to the contractor to subcontractors to operation makes it difficult to collaborate from scratch and to feed information from engineers, manufacturers, and users back into the design process. In the linear building chain, and the same is true of a blockchain, one cannot easily go back in the process, try another variation by changing variables, and remodel and recalculate the 3D model from design sketch to execution detail. AI and neural networks are great decision-making tools for dealing with thousands of alternatives, but how would that relate to a blockchain? I consider each step in a design process a transaction. Blockchain is designed to secure transactions in a virtual casino-capitalist world. In a collaborative design process, one has thousands of small decisions that eventually produce the outcome. The first decision sets the tone, then the cell division process starts, and parts of the design are step specified into components. In a dynamic adaptive design process, each step remains parametrically related to earlier steps, meaning that there is always a road back to an earlier phase and starts from there in another direction. In many different types of software, one has some form of versioning, meaning that earlier versions of the same file and the same filename can be traced back. Google Drive and

Google Earth have simple forms of versioning; the files contain history. Rhino has something called incremental save, but that still means a completely new file, and thus data consuming. At ONL, we use a concurrent versioning system (CVS) that keeps track of the difference between the versions only and thus is extremely economical with data. A major flaw in the blockchain is that there is no back button. One can redo a transaction only with the most elaborate efforts. In blockchain technology, each step backward would need a minimum of 50% consent from the hundreds of thousands of miners, which would only be possible in very rare cases. One could argue that, therefore, transactions in blockchain are a very natural process. In Nature 1.0, one cannot take steps back either; in other words, growth and evolution come in discrete steps, bound to the arrow of time. In that sense, blockchain technology is in sync with the discrete nature of Nature 1.0. But, perhaps, blockchain is just old-school nature.

While building Nature 2.0, from the digital and from bidirectional networks, a new form of logic emerges, which is distributed, peer-to-peer, and nonlinear. In a bidirectional network, one can always move in any direction; there is no such thing as forward or backward. I guess that one would need a cloud-based CVS-type system that gives access to all players in the design game, simultaneously. While a 100% simultaneous is impossible, the one who is a fraction of a second faster than his peers owns the modification, and that modification is the recorded transaction, transparent, verifiable, falsifiable, and reversible. In the thousands of virtual reality universes, currently referred to as the Metaverse, there is no up and down, and gravity is artificially simulated, which is not a natural feature. In the known universe, there is no left and right, there is no up and down, and there is no such thing as an *XYZ* coordinate system. So why would we copy the restrictions of the current modeling and building practice in Nature 2.0? Why would we have a simulation of gravity in the Metaverse? The Metaverse is the perfect opportunity to renounce mimicry of the natural world and start building a new synthetic form of life, which in due time will be perceived as natural again.

Another Normal

In my work that unfolds in the professional field of the built environment, I have chosen to build following the principles of a swarm intelligence of connected, at times real-time informed components, in search of Another Normal. It is up to other disciplines to find rule-based implementations of the swarm in their professional fields of knowledge. I have touched upon those other fields of knowledge, by and large as an outsider and therefore somewhat naive, in an attempt to impersonate the enormous potential for establishing a fair and just economy for all, whether in Africa, Asia, the Middle East, the Americas, or in the West, where I happen to be born and live. People act in the companionship of and dialogue with things; we live inside the co-evolution of people and things. The total of all physical and virtual transactions shapes the economy, and people and things are the actors in the game of life. Chaos theory teaches us that, unknowingly, one innocent actor may have a decisive influence on the course of things, even by the tiniest of acts. Without knowing it, the Chinese Patient has been instrumental in changing the path China has taken in the last decades. China has become a global player again. The Chinese Patient was not a fool; he turned out to be a genius. The emerging participation society does not think of people as patients but as self-aware well-informed prosumers who monitor and fix their health, and who are in control of their own lives, in a continuous dialogue with their immediate neighbors. Our Virtual Operation Room project offers an early provocative glimpse into Another Normal of digital self-healing.

This book is a forward-looking, at times self-critical, view of what a deep digitized society, based on the hive mind of the swarm of people and things, might have in store for us. I have pursued to live Another Normal in the here and now. Our executed works of art and architecture are living proof of the potential of Another Normal, with all its flaws and shortcomings. I educate myself by daily working with teams at ONL, with students and staff at Hyperbody (2000–2016), by assisting Ilona in the technical execution of her works of art and by working on a level playing field with engineers and manufacturers. And they learn from me. In due time, my work will be scrutinized and criticized, at times forgotten and ignored, but inevitably further developed by the next generations of world-makers, activists, scientists, artists, and designers, where needed and as needed. By "protospacing" Another Normal, I leave my traces in the here and now.

Bibliography

Aish, R, *Introduction to Generative Components*, Bentley Systems, 2005

Al Sabouni, M, *The Battle for Home, The Vision of a Young Architect in Syria*, London, Thames & Hudson, 2016

Alblas, J, Bhalotra, A, Huisman, A, Oosterhuis, K, Schuringa, W, *City Fruitful*, Rotterdam, 010 Publishers, 1992

AMO, Koolhaas, R [editors], *Countryside, a Report*, Köln, Taschen, 2020

Anderson, C, *The Long Tail, Rewriting the Rules of Culture and Commerce*, London Penguin Random house, 2006

Barabási, A-L, *The Formula, the Universal Laws of Success*, Brown, Little, 2018

Bier, H, Oosterhuis, K, [editors], *iA#5 Robotics in Architecture*, Prinsenbeek, Jap Sam Books, 2013

Bier, H, *System-embedded Intelligence in Architecture*, Delft University Press, 2008

Biloria, N, *Adaptive Corporate Environments: Creating Real-time Interactive Spatial Systems for Corporate Offices Incorporating Computation Techniques*, Delft University Press, 2007

Bojár, I, *Interjú Kas Oosterhuis-szal*, Octogon magazine, 2006/1, Budapest, 2006

Boxmeer, R van, Peters, T [editors], *Minitopia, ruimte voor je woonwens*, Rotterdam, NAi010 Publishers, 2020

Cachola Schmal. P [editor], *Digital Real - Blobmeister*, exhibition catalogue Deutches Architektur Museum DAM, Basel, Birkhäuser, 2001

Chang, G, *HyperCell, a Bio-inspired Design Framework for Real-time Interactive Architectures*, Delft University Press, 2018

Collins, S, *The Hunger Games*, New York, Scholastic Press, 2008

Dijk, M van, *Little Big Think*, Lulu.com, 2019

Feireiss, L, Oosterhuis, K [editors], *The Architecture Co-Laboratory, Game Set and Match II*, Heijningen, Episode Publishers, 2006

Feng, H, Oosterhuis, Feng, H, K, Xia, X [editors], *iA#4 Quantum Architecture*, Prinsenbeek, Jap Sam Books, 2011

Feng, H, *Quantum Architecture*, Delft, [in progress]

Feringa, J, *Architectural Robotics: Bridging the Divide between Academic Research and Industry*, Delft University Press, 2022

Fiore, Q, McLuhan, M, *The Medium is the Massage*, Berkeley, Gingko Press, 2001

Friedrich, C, *Immediate Systems in Architecture*, Delft University Press, 2021

Hakkak, A, *Enhancing [Spatial] Creativity, Enhancing creativity of architects by applying unconventional virtual environments (UVEs)*, Delft University Press, 2017

Hovestadt, L, Kretzer, M [editors], *Alive !: Advancements in Adaptive Architecture*, Basel, Birkhäuser, 2014

Hubers, J, *COLAB, Collaborative Design in Virtual Reality*, Delft, Delft University Press, 2008

Huizinga, J, *Homo Ludens*, Groningen, Wolters-Noordhoff, 1938

Icke, V, *Reisbureau Einstein, over buitenaardse buren*, Amsterdam, Prometheus, 2017

Jaskiewicz, T, *Towards a Methodology for complex Adaptive Interactive Architecture*, Delft University Press, 2013

Johnson, R, Holbrow, C, *Space Settlements: A Design Study*, Cape Canaveral, National Aeronautics and Space Administration, 1977

Kandinsky, V, Punkt und Linie zu Fläche, Weimar, Bauhaus Bücher, 1926

Kasteleijn, D [editor], *Het Van Doesburghuis*, Bussum, Uitgeverij Thoth, 2004

Kelly, K, *Out of Control, The New Biology of Machines, Social Systems, & the Economic World*, London, Addison Wesley, 1994

Kelly, K, *The Inevitable, Understanding the 12 Technological Forces That Will Shape Our Future*, London, Penguin Books, 2017

Kolarovic, B, Parlac, V [editors], *Building Dynamics, Exploring Architecture of Change*, New York, Routledge, 2015

Koolhaas, R, Fundamentals, catalog Architecture Biennale of Venice 2014, Venice, Marsilio Publishers, 2014

Lénárd, I, Oosterhuis, K, *Kas Oosterhuis, architect _ Ilona Lénárd visual artist*, Rotterdam, 010 Publishers, 1998

Lénárd, I, Oosterhuis, K, Rubbens, M, *Sculpture City*, Rotterdam, 010 Publishers, 1994

Lénárd, I, *Powerlines paintings*, San Francisco, Blurb, 2019

Lootsma, B, *SuperDutch, New Architecture in the Netherlands*, New York, Princeton Architectural Press, 2000

Lynn, G, *Animate Form*, New York, Princeton Architectural Press, 1999

Lynn, G, *Folds, Bodies & Blobs*, Bruxelles, La Lettre Volée, 1998

Meadows, D, *Limits to Growth: A Report for the Club of Rome's Project on the Predicament of Mankind*, Wadhurst, Earth Island, 1972

Migayrou, F, Simonot, B, Brayer, M-A, *Archilab: Radical Experiments in Global Architecture*, London, Thames & Hudson, 2001

Mostafavi, S, *Hybrid Intelligence in Architectural Robotic Materialization (HI-ARM)*, Delft University Press, 2021

Nieuwenhuys, C, *Constant: Nieuw Babylon*, Catalogus tentoonstelling Haags Gemeentemuseum, The Hague, 1985

Novak, M, *Liquid Architectures in Cyberspace*, Cyberspace First Steps [ed. Benedikt, M],Cambridge MA, MIT Press, 1992

Oosterhuis, K [et al. editors], *Hyperbody, First Decade of Interactive Architecture*, Prinsenbeek, Jap Sam Books, 2012

Oosterhuis, K, Ahmad, M.A., Fadli, F [editors], *Proceedings of the International Conference on Game Set and Match IV Qatar-2019 [GSM4Q]*, Doha, Qatar University Press, 2021

Oosterhuis, K, *Architecture Goes Wild*, Rotterdam, 010 Publishers, 2002

Oosterhuis, K, *Associative Information Modeling [AIM]*, www.oosterhuis.nl, Rotterdam, 2015

Oosterhuis, K, *Emotive Architecture* [inaugural speech TU Delft], Rotterdam, 010 Publishers, 2001

Oosterhuis, K, *Game Changers*, Journal Next Generation Building 1, Amsterdam, Baltzer Science Publishers, 2014

Oosterhuis, K, *Hyperbodies, towards an E-motive Architecture*, Basel, Birkhäuser, 2003

Oosterhuis, K, *Multimodal Accommodations for the Nomadic International Citizen*, paper, Doha, Qatar University Press, 2020

Oosterhuis, K, Slootweg, O, Xia, X [editors], *iA#3 Emotive Styling*, Prinsenbeek, Jap Sam Books, 2010

Oosterhuis, K, *Towards a new Kind of Building*, Rotterdam, NAi publishers, 2010

Oosterhuis, K, *Ubiquitous Symmetries in the Abstract Calligraphic Paintings of Ilona Lénárd*, Symmetry: Culture and Science Volume 32, Number 4, pages 525-541, Budapest, 2022

Oosterhuis, K, Xia, X [editors], *iA#1 Interactive Architecture*, Heijningen, Episode Publishers, 2007

Oosterhuis, K, Xia, X [editors], *iA#2 Interactive Architecture*, Heijningen, Episode Publishers, 2009

Oosterhuis, K, *Zaha's calligraphic sweeps*, blog www.oosterhuis.nl, Rotterdam, 2016

Oosterhuis, K, *Space Time Volume*, Wiederhall 12 [Meuwissen, J editor], Amsterdam, Stichting Wiederhall, 1990

Oosterhuis, K, What's up Bálna?, blog, www.oosterhuis.nl, Nagymaros, 2019

Oosterhuis, K, *XYZ*, Wiederhall 3 [Meuwissen, J editor], Amsterdam, Stichting Wiederhall, 1986

Parent, C, Virilio, P, *Architecture Principe*: 1966 and 1996, Paris, Éditions Verdier, 1996

Protetch, M, *A New World Trade Center*, New York, Harper Collins, 2002

Psyllidis, A, *Revisiting Urban Dynamics through Social Urban Data*, Delft University Press, 2017

Raworth, K, *Doughnut Economics, Seven Ways to Think Like a 21st-Century Economist*, London, Penguin Random House, 2018

Saggio, A, *The IT Revolution in Architecture, Thoughts on a Paradigm Shift*, Lulu.com, 2008

Schoeffer, N, *La Ville Cybernétique*, Paris, Tchou Galerie, 1969

Schumacher, P, *Parametricism - A New Global Style for Architecture and Urban Design*, AD Architectural Design - Digital Cities, Vol 79, No 4, London, 2009

Schumacher, P, *Parametricism as Style - Parametricist Manifesto*, London, www.patrikschumacher.com, 2008

Spuybroek, L, *Grace and Gravity, Architectures of the Figure*, London, Bloomsbury, 2020

Turing, A, *On Computable numbers*, Proceedings of the London Mathematical Society, 1936

United Nations General Assembly, *Universal Declaration of Human Rights*, Paris,1948

Varoufakis, Y, *Another Now, Dispatches from an Alternative Present*, London, Bodley Head, 2020

Vollers, K, *Twist & Build*, NAi Publishers, Rotterdam, 2001

Wolfram, S, *A New Kind of Science*, Champaign IL, Wolfram Media, 2002

Wright, F, *The Disappearing City*, New York, William Farquhar Payson, 1932

Index

About the Author

Kas Oosterhuis is a visionary, practicing architect, founding director of the innovation studio ONL, and founding professor of the Hyperbody research group at TU Delft from 2000 to 2016. Environments at all scales — from furniture to buildings to cities — are considered complex adaptive systems, in terms of their complex geometry and their behavior in time. The main focus of the current practice is on parametric design, robotic building, and AI in all phases of the design to production and the design to operation process. Featured projects the A2 COCKPIT building in Utrecht, the BÁLNA mixed-use cultural center in Budapest, the LIWA tower in Abu Dhabi, and the individually customizable BODY CHAIR are living proof of Oosterhuis' lean design-to-production approach, in terms of precision, assembly, sustainability, costs, and design signature. Oosterhuis' built projects are characterized by a strong component-based integration of structure, skin, and ornamentation, paving the way for the affordable iconic. In his previous book Towards a New Kind of Building, a Designer's Guide to Nonstandard Architecture, Oosterhuis revealed the fundamentals of his personal design universe, which embraces the paradigm shift from standard to nonstandard architecture and from static to dynamic environments as the initial condition. In this book, The Component, Oosterhuis dives deeper into the role of the components that interact to form bespoke designs.

Printed in the United States
by Baker & Taylor Publisher Services